Readings from
American Scientist

Earth's History, Structure and Materials

Edited by

Brian J. Skinner

Yale University

Member, Board of Editors
American Scientist

WILLIAM KAUFMANN, INC.   LOS ALTOS, CALIFORNIA

Library of Congress Cataloging in Publication Data

Main entry under title:

Earth's history, structure and materials.

(Earth and its inhabitants)
Bibliography: p.
1. Earth—Addresses, essays, lectures.
2. Geological time—Addresses, essays, lectures.
3. Plate tectonics—Addresses, essays, lectures.
4. Rocks, Igneous—Addresses, essays lectures.
I. Skinner, Brian J., 1928–   II. American
scientist.   III. Series.
QE505.E17   550   80-21698

ISBN 0-913232-89-0

Contents

SERIES INTRODUCTION *iv*

VOLUME INTRODUCTION *1*

PART 1. EARTH AS A PLANET *5*

James W. Head, Charles A. Wood and Thomas A. Mutch,
"Geologic Evolution of the Terrestrial Planets" **65**:21
(1977) *7*

PART 2. GEOLOGIC TIME AND HOW IT IS
DETERMINED *17*

Henry Faul, "A History of Geologic Time" **66**:159
(1978) *18*

A. E. J. Engel, "Time and the Earth" **57**:458 (1969) *25*

Robert L. Fleischer, "Where Do Nuclear Tracks Lead?"
67:194 (1979) *37*

Irving Friedman and Fred W. Trembour, "Obsidian: The
Dating Stone" **66**:44 (1978) *47*

Elizabeth K. Ralph and Henry N. Michael, "Twenty-five
Years of Radiocarbon Dating" **62**:553 (1974) *55*

PART 3. PLATE TECTONICS AND DRIFTING
CONTINENTS *63*

Dan P. McKenzie, "Plate Tectonics and Sea-Floor Spreading" **60**:425 (1972) *64*

William A. Nierenberg, "The Deep Sea Drilling Project
after Ten Years" **66**:20 (1978) *75*

*Richard K. Bambach, Christopher R. Scotese and Alfred M.
Ziegler,* "Before Pangea: The Geographies of the
Paleozoic World" **68**:26 (1980) *86*

Bruce D. Marsh, "Island-Arc Volcanism" **67**:161
(1979) *99*

G. Brent Dalrymple, Eli A. Silver and Everett D. Jackson,
"Origin of the Hawaiian Islands" **61**:294 (1973) *112*

PART 4. IGNEOUS ACTIVITY *127*

Charles L. Rosenfeld, "Observations on the Mount St.
Helens Eruption" **68**:494 (1980) *128*

James G. Moore, "Mechanism of Formation of Pillow
Lava" **63**:269 (1975) *144*

*J. R. Heirtzler, P. T. Taylor, R. D. Ballard and R. L.
Houghton,* "A Visit to the New England Seamounts" **65**:466 (1977) *153*

Grant Heiken, "Pyroclastic Flow Deposits" **67**:564
(1979) *160*

Donald Hunter, "The Bushveld Complex and Its Remarkable Rocks" **66**:551 (1978) *168*

INDEX *177*

Introducing

Earth and Its Inhabitants

A new series of books containing readings originally published in *American Scientist*.

The 20th century has been a period of extraordinary activity for all of the sciences. During the first third of the century the greatest advances tended to be in physics; the second third was a period during which biology, and particularly molecular biology, seized the limelight; the closing third of the century is increasingly focused on the earth sciences. A sense of challenge and a growing excitement is everywhere evident in the earth sciences—especially in the papers published in *American Scientist*. With dramatic discoveries in space and the chance to compare Earth to other rocky planets, with the excitement of plate tectonics, of drifting continents and new discoveries about the evolution of environments, with a growing human population and ever increasing pressures on resources and living space, the problems facing earth sciences are growing in complexity rather than declining. We can be sure the current surge of exciting discoveries and challenges will continue to swell.

Written as a means of communicating with colleagues and students in the scientific community at large, papers in *American Scientist* are authoritative statements by specialists. Because they are meant to be read beyond the bounds of the author's special discipline, the papers do not assume a detailed knowledge on the part of the reader, are relatively free from jargon, and are generously illustrated. The papers can be read and enjoyed by any educated person. For these reasons the editors of *American Scientist* have selected a number of especially interesting papers published in recent years and grouped them into this series of topical books for use by nonspecialists and beginning students.

Each book contains ten or more articles bearing on a general theme and, though each book stands alone, it is related to and can be read in conjunction with others in the series. Traditionally the physical world has been considered to be different and separate from the biological. Physical geology, climatology, and mineral resources seemed remote from anthropology and paleontology. But a growing world population is producing anthropogenic effects that are starting to rival nature's effects in their magnitude, and as we study these phenomena it becomes increasingly apparent that the environment we now influence has shaped us to be what we are. There is no clear boundary between the physical and the biological realms where the Earth is concerned and so the volumes in this series range from geology and geophysics to paleontology and environmental studies; taken together, they offer an authoritative modern introduction to the earth sciences.

Volumes in this Series

The Solar System and Its Strange Objects Papers discussing the origin of chemical elements, the development of planets, comets, and other objects in space; the Earth viewed from space and the place of man in the Universe.

Earth's History, Structure and Materials Readings about Earth's evolution and the way geological time is measured; also papers on plate tectonics, drifting continents and special features such as chains of volcanoes.

Climates Past and Present The record of climatic variations as read from the geological record and the factors that control the climate today, including human influence.

Earth's Energy and Mineral Resources The varieties, magnitudes, distributions, origins and exploitation of mineral and energy resources.

Paleontology and Paleoenvironments Vertebrate and invertebrate paleontology, including papers on evolutionary changes as deduced from paleontological evidence.

Evolution of Man and His Communities Hominid paleontology, paleoanthropology, and archaeology of resources in the Old World and New.

Use and Misuse of Earth's Surface Readings about the way we change and influence the environment in which we live.

Introduction: Earth's History, Structure and Materials

Space exploration, with its landings on the Moon, its remote sensing devices placed on Venus and Mars, and its explicit images and measurements of other planets, spawned the new discipline of planetology. One result of the new discoveries is that we can now compare Earth's history with the histories of other planets: each has its own special properties and surface features, but there is a clear family relationship that unites all the terrestrial planets. The planets were born together, evolved together, and are bathed by the warmth of the same Sun. It is appropriate, therefore, that the first paper in this volume, by Head, Wood, and Mutch, should be a comparative history in which they discuss Earth's evolution and contrast it with the evolutions of the other terrestrial planets.

Space discoveries have had a dramatic impact on our thinking, but an earlier geological discovery had an even greater impact, challenging not only geologists but all mankind to re-examine fundamental beliefs. The challenge was the immensity of geologic time. The realization that Earth is very old and that all the rocks we see contain fragments of still older rocks has been with us for nearly two centuries. But the discovery that Earth is 4600 million years old and has had abundant time for extraordinary and even unimaginable things to happen, is a 20th-century discovery. The magic carpet by which we traverse geologic time is natural radioactivity and the way we use it to date rocks. The way in which radiometric dating changed and influenced our old ideas about the Earth's development is now history—but what exciting history! Not only does the story reveal much about the way geological problems are approached and solved, but it also reveals how we respond to extraordinary new ideas. The papers by Faul and Engel discuss both the history and the consequences of the developing notions of geological time. The papers mention, but do not stress, the many different dating schemes, so two papers, discussing unusual new dating schemes have been included: R. L. Fleischer discusses how the counting of fission tracks produced in crystals during decay of entrapped radioactive atoms can be used as a dating scheme. Friedman and Trembour describe a different and entirely new technique—the hydration of obsidian. When obsidian, a natural volcanic glass, is worked for artifacts, the newly spalled surface starts to absorb water from the atmosphere and to hydrate. The width of the hydrated rind is a measure of the duration of absorption and hence of the age of the object. No period of Earth's history, save the earliest moments of birth 4600 million years ago, has proved so difficult to date accurately as the last 100,000 years. Portions of that period are even now a largely closed book. But obsidian dating and ^{14}C dating, as discussed by Ralph and Michael, are powerful tools for timing some of the events of the vital prehistoric human era during which modern man arose and developed the communities in which we live today.

One of the consequences of immensely long geological time is that ample periods were provided to allow for majestically slow events to change the face of the Earth. It may be hard to imagine that continents move slowly over the face of the globe, but move they do at rates of a few centimeters a year, steadily rearranging their positions as they are rafted on plates of lithosphere but leaving an interpretable magnetic record impressed on the rocks of the seafloor. Plate tectonics and seafloor spreading, discussed by D. P. McKenzie, are surely the most exciting and astonishing discoveries of the last 20 years. Through their agencies innumerable geological puzzles have been clarified and all of the earth sciences have been forever redirected and revitalized. But plate tectonics would only be a tenuous theory, subject to qualitative argument rather than quantitative analysis, were it not for evidence drawn from the deep-sea floor. W. A. Nierenberg's discussion of the decade-long program of deep-sea drilling that finally placed plate tectonics beyond the realm of theory is exciting reading.

Every geologist probably has his or her favorite selection for the place where plate tectonics has made the most dramatic impact; three examples are included. Bambach, Scotese, and Ziegler recreate the geography of the long-lost Paleozoic world, a period in which fragments of continental crust moved in identifiable but presently unex-

1

plained paths over the face of the globe, sometimes colliding, sometimes residing in polar regions, sometimes in equatorial, but never stationary for long. Marsh, by contrast, describes one of the consequences of ancient lithosphere being swept back into the asthenosphere, there to be heated, to melt in part, and to create the distinctive arc-like chains of volcanoes so common around the present Pacific Ocean basin. But if island-arc volcanism is strange, how much stranger are the long-lived magma sources deep in the mantle, origin unknown, that create chains of oceanic-island volcanoes, such as the Hawaiian chain, as a plate moves slowly overhead! Dalrymple, Silver, and Jackson discuss the phenomenon in a most unusual paper.

All igneous activities seem to be influenced by plate tectonics; but volcanos, volcanic eruptions and igneous rocks in general are so important, and have been the focus of study for so long, they require separate discussion. Certainly the most dramatic of recent igneous activities was the violent eruption by Washington's long-dormant Mount St. Helens. On May 18, 1980, the picturesque stratovolcano burst into explosive eruption and devastated a huge area to the north of the mountain. Charles L. Rosenfeld, a geologist from nearby Oregon State University and a pilot in the Oregon Army National Guard, had an unparalleled opportunity to observe the eruption, to see and study the fallout of volcanic ash and to observe the flow of deadly pyroclastic clouds known as nuées ardentes. Rosenfeld's eyewitness description and spectacular photographs of the first stratovolcano eruption in North America in nearly 60 years are extraordinary; the article should be read in conjunction with the article by Heiken concerning the many and much larger pyroclastic flows observed in the geologic record.

Another dramatic eruption occurred in the early 1970's when a basaltic lava that flowed from Hawaii's Kilauea Volcano entered the sea and provided a band of intrepid divers the opportunity to study, as described by James Moore, the mechanism by which pillow basalts form. Pillow basalts are one of the Earth's most common rock types; but because they tend to form along mid-ocean ridges, details of their origin have long remained in doubt, as indeed have details of all submarine igneous activities. Despite the hostility and inaccessibility of the environment, the seafloor is yielding its secrets; two articles discuss submarine volcanic features, the first by Moore and the second by Heirtzler, Taylor, Ballard and Houghton, who discuss seamounts viewed from a deep-diving submarine.

The final paper discusses one of the Earth's most massive and extraordinary igneous features, the Bushveld Complex. Far larger than any other layered igneous instrusion and a treasure-house of mineral wealth, the Bushveld is still the subject of unanswered questions and the continuing focus of intense research.

The papers in Earth's History, Structure and Materials are statements about topics of today's active, ongoing research. Tomorrow's topics may be quite different. But they will be built on and will incorporate the ideas and facts discussed in this volume.

Suggestions for Further Reading

Dietrich, Richard V. and Brian J. Skinner, *Rocks and Rock Minerals* (New York: John Wiley and Sons, 1979).

Drake, Charles L., John Imbrie, John A. Krauss and Karl K. Turekian, *Oceanography* (New York: Holt, Rinehart and Winston, 1978).

Eicher, Don L., *Geologic Time,* 2nd ed. (Englewood Cliffs, New Jersey: Prentice-Hall, 1976).

Flint, Richard F. and Brian J. Skinner, *Physical Geology*, 2nd ed. (New York: John Wiley and Sons, 1977).

Heezen, Bruce C. and Charles D. Hollister, *The Face of the Deep* (New York: Oxford University Press, 1971).

MacDonald, Gordon, A., *Volcanoes* (Englewood Cliffs, New Jersey: Prentice-Hall, 1972).

Shepard, Francis P., *Geological Oceanography* (New York: Crane, Russack and Co., 1977).

Short, Nicholas M., Paul D. Lowman Jr., Stanley C. Freden and William A. Finch Jr., *Mission to Earth: Landsat Views the World* (NASA SP-360: National Aeronautics and Space Administration, Science and Technology Information Office, 1976).

Sullivan, Walter, *Continents in Motion: The New Earth Debate* (New York: McGraw-Hill, 1974).

Ueda, Seiya, *The New View of the Earth: Moving Continents and Moving Oceans* (San Francisco: W. H. Freeman and Co., 1978).

Wyllie, Peter J., *The Way the Earth Works* (New York: John Wiley and Sons, 1976).

Williams, Howel and Alexander R. McBirney, *Volcanology* (San Francisco: Freeman, Cooper and Co., 1979).

Wood, John A., *The Solar System* (Englewood Cliffs, New Jersey: Prentice-Hall, 1979).

Authoritative and up-to-date reviews, summaries, and analyses of many of the topics discussed in this volume can be found in the volumes published by Annual Reviews, Inc., Palo Alto, California 94306.

Articles of special interest will be found in all volumes of *Annual Review of Earth and Planetary Sciences* starting with vol. 1, 1973.

PART 1 *Earth as a Planet*

James W. Head, Charles A. Wood and
Thomas A. Mutch, "Geologic Evolution of
the Terrestrial Planets," **65**:21 (1977),
page 7.

James W. Head
Charles A. Wood
Thomas A. Mutch

Geologic Evolution of the Terrestrial Planets

Observation and exploration have yielded fundamental knowledge of planetary evolution and have given rise to an exciting new view of Earth as a planet

Luna, Venera, Mariner, Ranger, Surveyor, Apollo, Lunokhod, Pioneer, Viking. These spacecraft with heroic names have taken us from the triumph of landing a pennant on another world to the reading of a weather report from Mars on the evening news. The accompanying avalanche of scientific information from 15 years of space exploration, combined with the simultaneous understanding of terrestrial plate tectonics, has led to an exciting new perspective on the nature and evolution of the planets of the inner solar system.

Physical, chemical, and dynamical studies of the planets have greatly contributed to the current renaissance of understanding, but simple observations of the types of terrain which form planetary surfaces and inferences of processes which pro-

James W. Head is an Associate Professor of Geology at Brown University, where his research centers on planetary geologic processes and history. He received his Sc.B. from Washington and Lee University and his Ph.D. from Brown and was involved with the Apollo Lunar Exploration Program from 1968 to 1972. Charles Wood is a Ph.D. student in the Department of Geological Sciences at Brown. His research and publications concern volcanology, lunar geology, and recent climatic change in Africa, where he taught and did research for four years. Thomas A. Mutch is a Professor of Geology at Brown and received his Ph.D. from Princeton. He is the author of books on the geology of the Moon and Mars and is presently the leader of the Mars Viking Lander Imaging Team. This work was performed under NASA Grants NGR-40-002-088 and NGR-40-002-116 from the Lunar and Planetary Programs Office. Thanks are extended to M. Cintala, R. Hawke, A. Gifford, R. Roth, S. Matarazza, and the National Space Science Data Center. Address: Department of Geological Sciences, Brown University, Providence, RI 02912.

duced the terrains have yielded fundamental knowledge of planetary evolution. In this article we review photogeologic and other evidence for each of the terrestrial planets—Mercury, Venus, Earth, the Moon, and Mars—but our discussion of each planet is grossly simplified, generally omitting descriptions of landforms such as craters and volcanoes (information readily available in the cited references); instead, we concentrate on the relationships between regional terrain types, ages, and planetary evolution.

The planets

Knowledge of the characteristics and processes operating on the Moon has been accumulating since the earliest visual observations centuries ago (summarized in Mutch 1972). Increasingly sophisticated Earth-based observational techniques have been complemented and extended by exploration by spacecraft and man. Soft-landing spacecraft and orbital vehicles provided abundant photographic evidence of the nature of the lunar surface and clues to the processes operating there. Manned exploration of the Moon during Apollo provided on-site investigations and sample return that opened an entirely new field—the petrologic and geochemical history of a planetary body other than Earth.

Galileo's early subdivision of the lunar surface into mare and terra is still significant in terms of lunar history and processes. The rough, cratered lunar highlands, or terrae, dominate the crust of the Moon, contrasting with the low-lying, dark mare units (Fig. 1). Geologic and petrologic evidence from Luna and Apollo samples (Taylor 1975) dem-

onstrates that the cratered highlands represent an early crust that resulted from global melting in the first few tens to hundreds of millions of years of lunar history. The global crust was continually modified by the impact of material from elsewhere in the solar system. The cratering record preserved in the early crustal units represents a distinct phase of intense cratering that began to decline rapidly about 3.8 billion years ago. Although other processes, such as volcanism, may have operated during this early period, the surface history of the Moon was written by craters of all sizes. Extremely large impacts excavated huge depressions, perhaps as large as 2,000 km in diameter, and spread ejecta over large areas of the planet, sometimes affecting a whole lunar hemisphere. The youngest of these large basins, such as Orientale and Imbrium, are surrounded by distinctive radially textured deposits that buried and gouged large regions of the lunar surface. Although seismometers placed on the Moon by Apollo astronauts prove that cratering continues to the present, relatively few large craters and no large basins have formed in the last 3.8 billion years.

The next stage of lunar history was dominated by the emplacement of the dark mare plains that cover approximately 17 percent of the lunar surface (Head 1976) and occur predominantly on the lunar Earth side. In general, maria are relatively thin layers of basaltic lava totaling less than one percent of the volume of the lunar crust. The time of initial mare emplacement is uncertain, but radiometric dates of returned Apollo mare rocks suggest that major outpouring of lava occurred from 3.9 to 3.2 billion years ago (Taylor 1975).

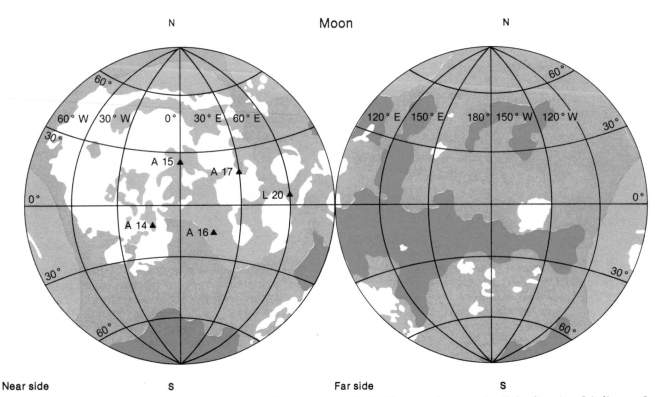

Figure 1. Geologic terrain units of the terrestrial planets have been synthesized, simplified, and modified from a variety of recent maps. The lunar surface is dominated by craters and relatively thin ponds of lava within frontside basins. There are no complex landforms such as the folded mountains on Earth or the giant canyons of Mars; the Moon is thus a primitive body whose surface was shaped by impact cratering and extrusion of lava flows. A indicates Apollo landing sites; L indicates a Luna site. (From Wilhelms and McCauley 1971, as simplified and extended by Howard et al. 1974.)

Although some mare deposits may be as young as two billion years (Boyce 1975), there has been no extensive igneous activity on the surface of the Moon for the last three billion years (Fig. 1).

Although major lineaments and fractures are observed in the lunar crust, there is no evidence of folded mountain belts or similar indications of compressional tectonic activity. Lunar mountain ranges were produced largely in conjunction with the formation of major impact basins.

The small size of Mercury (Fig. 2) and its proximity to the sun long hindered telescopic investigations, but in 1974 the Mariner 10 spacecraft photographed about 35 percent of the planet, revealing a cratered surface remarkably similar to the Moon's (Murray et al. 1975). Preliminary geologic mapping (Trask and Guest 1975) distinguished smooth and cratered plains and basin ejecta, as on the Moon, but there are a number of significant differences. Craters on the lunar highlands are densely packed, with the rims of the youngest superimposed on older craters. Lunar mare regions are sharply bounded and mostly contained in contiguous basins. On Mercury, by contrast, craters are often interspersed with relatively smooth plains, resulting in a speckled terrain map. At present there is a controversy concerning the relative ages and origins of the plains and the craters. Like most conflicts for which good evidence supports each side, probably each is partially correct: plains are probably older than crater units in some areas and younger in others, and have varied origins.

The most unusual terrain features on Mercury are lobate scarps that cut plains and craters alike. These scarps, which extend from tens to hundreds of kilometers in length, are interpreted as evidence of crustal shortening. Strom et al. (1975) calculated that the amount of shortening corresponds to a decrease in radius of 1–2 km. The same amount of contraction is estimated independently from theoretical modeling of lithospheric cooling (Solomon and Chaiken, in press).

Although Mercury has many impact basins (Wood and Head, in press), only a single large one, Caloris, is well shown in the area photographed. The interior of Caloris is surfaced by a lunar-mare-like plain wrinkled by concentric ridges. Much of the radial ejecta from Caloris is apparently covered by a continuation of the basin fill material. Whether this is basaltic lava similar to lunar maria or a vast deposit of impact-emplaced material is currently under debate.

The pull of gravity on Mercury is more than twice as strong as on the Moon, resulting in a restricted distribution of ballistically emplaced ejecta from impact craters (Gault et al. 1975). However, central peaks and wall terraces of craters are approximately as abundant on the Moon as on Mercury and Mars, which have much stronger fields. This suggests that, in addition to gravitational effects, other factors such as impact velocity or target strength influence the morphology of craters (Cintala et al. 1976).

Crater and basin formation has thus dominated the early history of Mercury, fracturing and mixing an ancient crust. Volcanism may have produced plains deposits concurrently with the early intense cratering and may also be responsible for the Caloris plain, which formed later. On the basis of crater densities and flux estimates, virtually all the major ter-

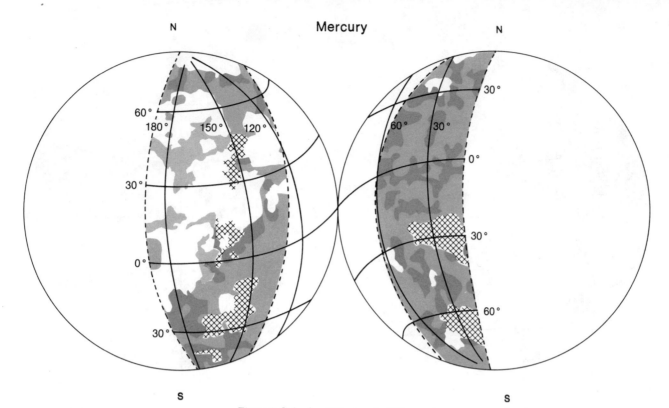

Mercury

Figure 2. Only about 35 percent of Mercury has been photographed, but the surface we have been able to get a look at is remarkably similar to that of the Moon. Cratered terrain is com-mon on the planet, and large tracts of smooth lava-like plains occur around the Caloris Basin (30°N, 190°). (From Trask and Guest 1975.)

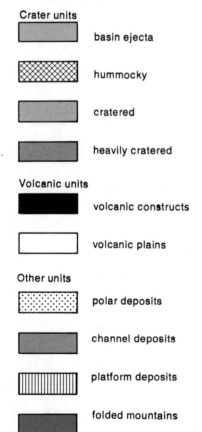

Key to terrain units

Crater units

basin ejecta

hummocky

cratered

heavily cratered

Volcanic units

volcanic constructs

volcanic plains

Other units

polar deposits

channel deposits

platform deposits

folded mountains

rain units (Fig. 2) formed in the first 1.5 billion years of Mercury's history.

Earth-based observers have long been intrigued by the red color, dusky markings, and the waxing and waning polar caps of Mars. Early Mariner

missions provided the first close-up views of the surface of Mars and por-trayed a cratered terrain not unlike that seen on the Moon. However, these missions had photographed only a small portion of the surface of the planet, and the global view pro-vided by Mariner 9 in 1971 quickly demonstrated that Mars was a geologically diverse and complex planet that had evolved to a stage considerably beyond the surface of the Moon (Masursky 1973).

The geologic diversity revealed by Mariner 9 is illustrated by the varia-tions within the cratered terrain, which comprises about 50 percent of the surface area of Mars (Fig. 3). Within this cratered terrain are plains units with variable crater densities perhaps reflecting early volcanic re-surfacing of ancient cratered crust. The cratered highlands contrast with much of the northern hemisphere of Mars, which consists of flat, low-lying, sparsely cratered plains that cover about 30 percent of the surface of the planet. The boundary between the cratered terrain and the plains to the north is marked by a chaotic and hummocky material indicative of collapse and erosion at the edge of the cratered unit.

Some geologists believe that melting of a buried permafrost layer initiated the formation of the chaotic terrain of Mars (Sharp 1973), and evidence for surface water at an earlier epoch in the history of the planet is abun-dantly displayed in Mariner 9 and Viking photography. Meandering channels, tributaries, braided streams: the full lexicon of terrestrial fluvial geology is represented on Mars. Additionally, permanent caps composed of both water ice and car-bon dioxide ice are found at each of the poles. Exposed layers of dust and ice in these caps are evidence for eo-lian erosion, transport, and deposi-tion—processes that appear to have dominated the planet Mars for a bil-lion years or more.

Martian dust storms and polar caps were known from Earth-based tele-scopic observations, but the nature of even the largest landforms, such as the 27 km high shield volcano Olym-pus Mons, was not appreciated. Such mountains represent the most recent volcanic activity on Mars; however, earlier shields exist, and flow fronts and wrinkle edges demonstrate that many plains units, in the northern lowlands and elsewhere, are volcanic in origin.

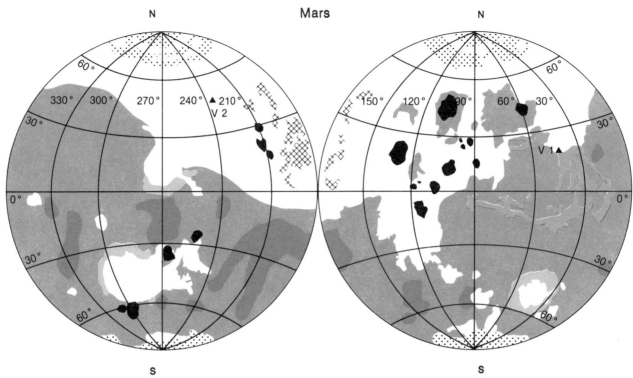

Mars

Figure 3. Although large areas are covered by cratered terrain, major updoming and shield volcanism distinguish Mars from the less complex Mercury and the Moon. The concentration of volcanic smooth plains in the northern hemisphere is a fundamental observation which must be explained in any theory of the evolution of Mars. *V* indicates the Viking landing sites. (From Pollack 1975 and Carr et al. 1973.)

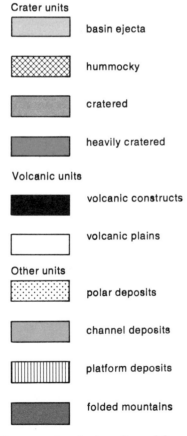

Key to terrain units

Crater units

basin ejecta

hummocky

cratered

heavily cratered

Volcanic units

volcanic constructs

volcanic plains

Other units

polar deposits

channel deposits

platform deposits

folded mountains

The ages of geologic units on Mars are somewhat uncertain due to a lack of detailed knowledge of the rate of impact crater formation. If martian cratering rates were similar to those of the Moon, the majority of the cratered terrain would have formed early in martian history, generally prior to about 3.5 billion years ago (Soderblom et al. 1974) and cratering has probably continued to the present, but at a much reduced rate. Volcanism has played a continuing role in the surface evolution of Mars, from the time of formation of the cratered terrain, through the surfacing of the northern lowlands, and, locally, by the building of large shield volcanoes in the last billion years (Carr 1973). The channels appear to be at least one to two billion years old. Tectonic features on Mars include radial structures associated with large impact basins; features similar to lunar mare ridges occurring in plains regions; and grabens and extensive fault valleys, such as Valles Marineris. The most prominent structural features on Mars are centered around the Tharsis region and formed approximately a billion years ago during the updoming of this region (Carr 1974).

The surface of Venus is obscured by a thick cloud cover, but the planet's similarities in size and density with Earth invite comparison of surface geology and processes. Photographs returned by Soviet unmanned landings in 1976 show the surface at the Venera 9 and 10 sites to be blocky, with indications of soil and bedrock. Recent analysis of radar data obtained by Earth-based observations (Rumsey et al. 1974) suggests that Venus may have a varied geologic terrain perhaps more similar to Mars than Earth.

Circular structures between 30 and 1,000 km in diameter have been detected and may represent impact craters and basins. Other large features include a 1,500 km linear trough similar in scale to the martian Valles Marineris and a large low circular dome with a central depression, similar in some aspects to shield volcanoes (Malin 1976). If the cratered terrain represents a surface modified by abundant impacts, then that portion of the Venus surface is relatively old and has not undergone the extensive tectonic recycling typical of Earth. Preliminary evidence of tectonic and volcanic processes suggests, however, that activity perhaps comparable to that of Mars has also taken place. More extensive high resolution radar coverage is required to determine if terrestrial-type tectonics have actually operated on the planet.

It is fortunate that mankind evolved and geology developed on our planet,

Earth

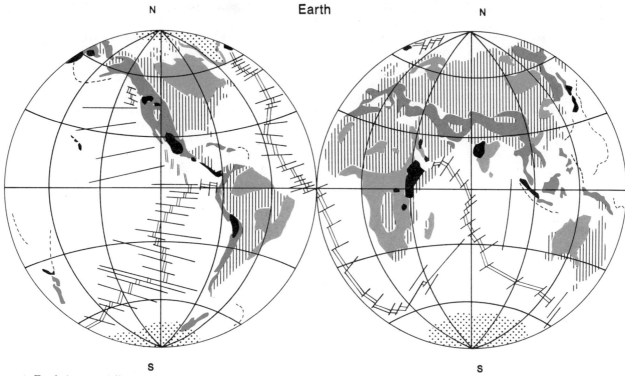

Figure 4. Earth is essentially a tectonically active and volcanic planet—young lavas erupt from spreading ridges (*double lines*), forming the ocean floors, which are sheared along transform faults (*single lines*) and are ultimately consumed at subduction zones (*dashed lines*). (From Wyllie 1971.)

since delineation of some unique terrestrial terrain units would have been difficult from space. Unlike other terrestrial planets, Earth has the majority of its crust concealed by liquid water and much of its exposed land mantled by a thin but dense veneer of vegetation. Atmospheric effects often add to surface obscuration. While the hydrosphere, atmosphere, and biosphere are not terrain units, they are dynamic geologic agents, which, with plate tectonics, make Earth a unique planet. Water, liquid and frozen, erodes and erases the land, recycling rocks as sediments that fill basins, form plains, and create new land. Such continental flatlands, perhaps, are the terrain equivalent of the smooth plains on other planets.

In the last ten years it has been recognized that plate tectonic movements of Earth's lithosphere are responsible for most of the large scale terrain features. Plate collisions produce folded mountains (the Alps, Atlas, and Appalachians), while continental rift valleys and vast plateau-forming deposits of basalts appear to be related to plate breakup. Where plates are consumed, lines of andesite volcanoes appear, and at

ocean ridges, lavas ooze out, creating new ocean floor.

Despite the dynamic crustal activity and resulting complexity of the Earth, some of the same terrain units occur as on the other planets (Fig. 4). The distribution of meteorite craters

Figure 5. The percent of the surface area of each terrestrial planet occupied by major terrain units is illustrated by the color code used in Figures 1–4.

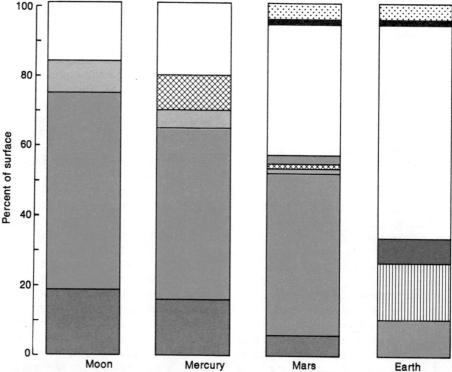

(especially large ones) suggests that the 10 percent of Earth surfaced by ancient rocks—the Precambrian shields—are the nearest terrestrial analog to cratered terrain, even though the shields' crater density is much less than that of cratered terrains on other planets. The most pervasive terrain unit on Earth is the basaltic plains of the ocean floor. These volcanic plains are among the youngest of Earth's rocks, ranging in age from 0 to 200 million years. Within the last 65 million years, separate phases of volcanism have built broad plateaus of basalts (Ethiopian and Indian Traps, etc.) as well as conical mountains and craters (Andean volcanoes, Hawaii, etc.).

Polar deposits similar to, but more massive than, those on Mars also occur on Earth. As on Mars, terrestrial polar terrain seasonally increases and decreases in area. The instability of Earth's polar terrain emphasizes that the map of terrain units is a snapshot of a particular geologic instant. Unlike the Moon, which has been largely unchanged during the last 3.5 billion years, Earth has a surface constantly in motion, with the positions and abundances of terrain units continually varying. Indeed, according to one hypothesis of the evolution of Earth's crust (Hargraves 1976), there was only a single terrain unit—water-covered plains—for the first billion years of Earth's history.

Comparative planetology

Our brief examination of the terrestrial planets reveals some important similarities and differences in the distributions of terrain types. Two terrain units occur on all planets. Cratered terrain and volcanic plains are ubiquitous, but their abundances vary from planet to planet. On Earth, volcanic units dominate 62 percent of the total crustal area (Fig. 5). Precambrian shields have an area of only about 10 percent. By contrast, the Moon has 70 percent cratered terrain and only 17 percent lava plains. Mars is somewhat intermediate, with abundant cratered terrain and significant volcanic plains. On the one-third of Mercury photographed, craters dominate, but smooth plains, largely associated with the Caloris Basin, occupy 25 percent of the known surface. Although there is disagreement on the interpretation of Mercury's smooth plains, their large

area, which is comparable to that of volcanic plains on Mars and the Moon, provides support for a volcanic origin.

Cratered terrain is generally believed to be the oldest unit on a planet, recording the impacts—and mass additions—of the terminal stages of planetary accretion and heavy bombardment. Preservation of cratered terrain implies a relatively stable crustal history. Volcanic material can only reach a planet's surface to form plains and build cones if the crust is fragmented. The ratio of volcanic to cratered terrain is a crude Planetary Evolution Index (PEI): high values represent highly evolved surfaces, and low values, primitive surfaces. The PEI for Earth is 6.2, an order of magnitude higher than that of any other terrestrial planet, and the PEI decreases from Mars (0.7) to Mercury (0.3) to the Moon (0.2). The limited extent of radar imagery of Venus apparently shows volcanoes and giant troughs, but craters appear to be the most prevalent landform (Malin 1976). Thus Venus appears more Mars-like than Earth-like, and Mars itself is more like the Moon and Mercury than like Earth.

Igneous processes

Evidence that both Earth and the Moon have undergone early global melting suggests that extreme igneous events accompanied the formation of all terrestrial planets (Lowman 1976). Subsequently, regional and local provinces of volcanic rocks were formed as a consequence of planetary melting associated with radioactive heating. The volumes and production rates of these postformation igneous processes varied from planet to planet and with time on individual planets, but their effects were widespread.

Igneous processes dominate Earth, considering that the thin veneer of water which occupies about 70 percent of the planet's surface is underlain largely by oceanic basalts and that perhaps one-third of the remaining continental rocks are granitic plutons, plateau basalts, and arc andesites. The northern lowlands of Mars (nearly one-third of the total surface area) are interpreted as volcanic flows (Carr et al. 1973), and various plains units on Mercury are probably of volcanic origin (Strom et al. 1975). Volcanoes have also been

interpreted from radar imagery of Venus (Malin 1976), 17 percent of the lunar surface is covered by mare basalts (Head 1976), and some meteorites have volcanic compositions (Bogard and Husain 1976). Thus all the terrestrial planets, at least one meteorite parent body, and perhaps some large satellites were host to volcanic processes.

The mode of occurrence of volcanic rocks on Earth varies according to the viscosity of the melt, which, in turn, is largely controlled by silica content. Silica-rich magmas tend to fragment explosively on eruption, producing plains and cones of ash, whereas silica-poor magmas usually extrude quietly, building shield volcanoes and basaltic plains. Geochemical evidence suggests that silica-poor rocks come directly from the mantle, whereas silica-rich rocks are petrologically complicated by interactions with continental crust (Carmichael et al. 1974). The distribution, chemistry, and form of terrestrial eruptive rocks are tectonically controlled, with lithospheric plate-margin volcanism more important than intraplate, hot-spot volcanism. Acidic volcanism is generally restricted to plate boundaries, but basaltic volcanism occurs within plates as well as at their margins.

The most widespread style of volcanism on the inner planets is basaltic plains. Terrestrial ocean floors, lunar mare fill, martian northern lowlands, and perhaps the smooth plains of Mercury are probably all basaltic plains. It seems safe to predict that basaltic plains exist on Venus. Mare ridges occur on all these units except the terrestrial examples, which instead are characterized by spreading ridges and transform faults. This emphasizes a major difference in the role of volcanism on Earth compared to other planets. Ocean-floor basalts are created as a major product of plate movement, whereas generation of basaltic magmas on the Moon and perhaps the other planets is unrelated to the tectonics of the depressions they fill. However, the common occurrence of flow fronts and sinuous rills/lava channels on various planets illustrates that the surface forms and processes of volcanic activity were similar. In fact, the variation in characteristic lengths of sinuous rills/lava channels—hundreds of kilometers on the Moon, tens of kilo-

Table 1. Properties of terrestrial planets. The symbol $\oplus$ indicates Earth; R, retrograde.

	Moon	Mercury	Mars	Venus	Earth
Diameter (km)	3,476	4,880	6,787	12,104	12,756
Diameter ($\oplus$ = 1)	.27	.38	.53	.95	1
Mass ($\oplus$ = 1)	.01	.05	.11	.81	1
Volume ($\oplus$ = 1)	.02	.06	.15	.88	1
Density (gm/cc)	3.3	5.4	3.9	5.2	5.5
Surface gravity ($\oplus$ = 1)	.16	.37	.38	.88	1
Rotation period (day)	27	59	1	243R	1
Distance from sun (10^6 km)	150	58	228	108	150
Main atmospheric constituent	—	—	CO_2	CO_2	N

SOURCE: Hartmann 1972 and Sagan 1975.

meters on Mars, and a few kilometers on Earth—demonstrates how a particular volcanic process is modified by differing environmental conditions.

A second volcanic landform that occurs on more than a single planet is the basaltic shield. Whitford-Stark (1975) has shown that heights of terrestrial shield volcanoes, lunar domes, and martian volcanoes are linearly proportional to their basal diameters. Hence, the shapes of the shields cannot be influenced strongly by gravity or atmospheric pressure, for these quantities vary widely from planet to planet whereas the proportions of the volcanoes do not. Since shield height is directly proportional to the volume of erupted magmas, the extreme heights of martian volcanoes may not be a consequence of a stable crust remaining over a hot spot, but rather may simply mean that a tremendous amount of lava was erupted from a single vent. In any case, Vogt (1974) has shown that the heights of terrestrial shield volcanoes are limited by plate thickness rather than by the movement of a plate away from a hot spot. Thus the giant martian volcanoes imply a lithospheric thickness on Mars much greater than on Earth. This observation, along with the lack of extraterrestrial stratovolcanoes or their calderas, such as occur above downgoing slabs, suggests that plate tectonic crustal motions occur only on Earth.

Tectonism

The Moon is a relatively small planetary body (Table 1) with a thick solid crust that formed early in its history. Tectonic features include grabens and arcuate mountains associated with major impact basins, lineaments located predominantly in the ancient cratered terrain, and compressional ridges associated with the lunar maria. The Moon exhibits no evidence of major lateral tectonic movement or major vertical displacements of crustal material. The Moon may represent one end of the spectrum of tectonic evolution of planets: a small, dry planetary body with a thick crust that precluded extensive vertical and horizontal tectonic movement.

Mercury is slightly larger and contains a large dense core. Although impact-basin tectonic features are seen on Mercury, extensive lobate scarps dominate the surface tectonic environment. These features appear to indicate a regional and perhaps global environment of compression. Little evidence exists for extensive tensional features. These observations strongly suggest that the surface tectonism of Mercury was dominated by volumetric contraction on a global scale, apparently related to systematic cooling of the planet (Solomon and Chaiken, in press).

Mars, on the other hand, is larger and, although it contains many of the tectonic features seen on the Moon, it is dominated by extensional features such as Valles Marineris and other lineaments related to the Tharsis uplift. This evidence suggests that, unlike the Moon and Mercury, Mars may have undergone planetary expansion during its history (Mutch et al. 1976; Solomon and Chaiken, in press). There is no compelling evidence for regional lateral crustal movement that might indicate plate tectonic activity.

In its present form, Earth is dominated by major lithospheric plates and compressional and tensional ac-

tivity associated with their lateral movement. There is no need to invoke, or strong evidence for, planetary expansion or shrinkage.

The presence of abundant impact craters on Venus implies that plate tectonic processes have not recycled the crust. The evidence of tension provided by the large trough suggests that tensional features occur on large planets but are lacking on small ones.

Planetary elevations

The three terrestrial planets whose surfaces have been adequately documented exhibit differences in their distributions of elevations as well as in terrain types. The bimodal distribution of elevations of ocean floors and continents on Earth is dramatically illustrated by the well-known hypsographic chart (Fig. 6). There are also profound differences of age, petrology, and morphology in the two major terrain types. Ocean floors are homogeneous plains of young basaltic rocks (neglecting the volumetrically insignificant sediments, ridges, and islands), whereas continents are diverse in age, composition, and form.

A naked-eye view of the Moon shows a similar dichotomy of light and dark areas, which corresponds to cratered highlands and mare lowlands. The lunar hypsographic chart does not, however, indicate a bimodal distribution of elevations; rather, the distribution is unimodal and skewed. Although the lunar chart is preliminary (being based only on topographic data within 45° of the equator) one possible interpretation is that lunar surface heights have a Gaussian distribution except for excess low areas resulting from basin formation. It follows from this interpretation that the Moon has only a single crustal type, highlands—an observation consistent with geophysical and petrologic data.

On Mars, a fundamental dichotomy exists between the cratered terrain common in the southern hemisphere and the smooth northern lowlands. A preliminary hypsographic curve for Mars (Mutch et al. 1976) is dominated by a peak at 2–3 km elevation due to the cratered terrain, but the northern lowlands, unlike the terrestrial ocean floor, are not concentrated at a uniform elevation. Plains occur at two different elevations (Saunders 1976). Thus the distribution of topography on Mars appears to be intermediate between the unimodal Moon and bimodal Earth.

Ages of planetary surfaces

Ages of various terrain units on Earth and the Moon have been determined by field and laboratory investigation, but ages for surface units of Mars and Mercury depend upon models of impact cratering rates, since rocks from these planets have not been dated radiometrically. Figure 7 compares the estimated distributions of surface ages as a function of surface area for the terrestrial planets. Evolution of the surfaces of the Moon and Mercury was essentially complete 2.5 billion years ago or earlier. In contrast, 98 percent of Earth's surface is less than 2.5 billion years old, and 90 percent is less than 600 million years old. Fifty percent of Mars is less than 2.5 billion years old, although only 15 percent is as young as 600 million years; thus age distributions on Mars appear to be intermediate between the Moon and Mercury and the highly evolved Earth. The extreme youthfulness of most of the surface of Earth is due to the destruction of old terrain and the creation of new by plate tectonic motion that may have begun about 2.5 billion years ago

(Siever 1975). We have previously noted that major tensional structures occur on Mars and perhaps on Venus; yet there is no evidence of plate tectonics on those planets, suggesting that the grabens may represent an aborted attempt to establish plate tectonics.

Planetary evolution

Several basic themes and questions emerge from a review of the geology of the terrestrial planets and provide a new perspective with which to view Earth and its history. What are the fundamental processes that formed the terrestrial planets? Because of the youthful nature of Earth's surface, we would never have listed impact cratering as a fundamental process. Examination of other planetary surfaces, however, shows its great importance, particularly in the early history of the solar system. Prior to exploration of the ocean floors, we also would have underestimated the significance of volcanic processes. In terms of areal coverage, volume, and time duration, impact cratering and volcanism are the two processes dominating the surface histories of the terrestrial planets. Atmospheric and hydrospheric processes often are important agents of terrain modification, as on Earth and Mars.

Do terrestrial planets share a com-

mon early history? Analyses of returned lunar samples strongly suggest that the outer several hundred kilometers of the Moon underwent extensive melting, the heating being provided by the terminal stages of planetary accretion (summarized in Taylor 1975). Were all the terrestrial planets characterized by similar "magma oceans," or has accretion been slow enough in some cases to preclude initial melting (see Weidenschilling 1974)? Does cratered terrain represent remnants of this solidified crust, as it appears to do on the Moon? Do volatile-rich planets undergo evolutionary paths different from those for volatile-poor ones (e.g. evolution of thick atmosphere/hydrosphere)?

What are the significant factors in planetary evolution? Lewis (1974) and others have argued that differences in planet density are largely related to position and temperatures within the cooling nebula, with the more refractory elements common toward the sun and the more volatile elements abundant toward the outer planets. Although these factors may be significant in terms of planetary bulk chemical composition, there is no obvious correlation between stage or style of planetary surface evolution and distance from the sun. Correlations do appear, however, between stages of planetary evolution (PEI)

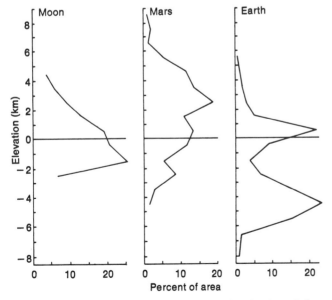

Figure 6. Hypsographic charts depict statistical distributions of planetary elevations. The terrestrial and martian charts are from Mutch et al. (1976). The hypsographic data were compiled from the lunar topography chart of Bills and Ferrari (1975). Data are available only for the zone between 45° N and 45° S.

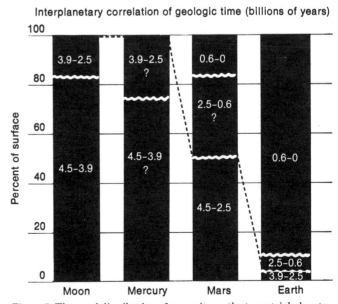

Figure 7. The areal distribution of age units on the terrestrial planets is based on radiometric dating of lunar and terrestrial rocks and on modeling of crater densities on Mars and Mercury. Here Mars appears to have age distributions intermediate between the Moon and Mercury on the one hand and Earth on the other.

and planetary mass and volume (see Table 1). The Moon and Mercury, with small volumes and masses, have a high proportion of primitive cratered terrain, while Mars shows a factor of two increase in both evolution index and mass and volume. Earth, on the other hand, shows close to an order of magnitude difference from the other planets in these values. Thus the abundance of cratered terrain decreases, being replaced by volcanic plains as planetary mass increases. Apparently, planetary size greatly influences thermal evolution (Kaula 1975). What factors relate these aspects to each other? Is there a minimum planetary mass below which mantle convection will not cause lithospheric rifting?

Do the present surfaces of planets represent stages in evolution experienced (or to be experienced) by each planet, or does each planet proceed along a separate path depending on its physical properties and position within the solar system? At present, the answer appears to be yes to both questions. Initial accretional heating seems to have produced an initial differentiation, and radioactive heating seems to have produced a second, but still early, differentiation in most terrestrial bodies. Cratering has dominated the early history throughout the inner solar system. However, on the basis of size and internal characteristics, it is unlikely that the Moon and Mercury will evolve past their present primitive state. Mars has a relatively high proportion of ancient cratered terrain, appearing closer to the Moon and Mercury in many respects than to Earth. But do the major uplifts and hemispheric asymmetry of Mars represent an incipient plate tectonic phase? Indeed, plate tectonics may represent a relatively recent stage in terrestrial evolution.

Exploration of the solar system over the past fifteen years has provided an exciting view of Earth as a planet. Man has traveled far enough into the solar system to marvel at the crescent Earth rising over the mountains of another planetary body. Planetary probes have unveiled surface features at scales far beyond our earthbound experience: craters the size of Europe, canyons for which our "Grand" Canyon would be but a puny tributary, and volcanoes nearly three times as high as Mt. Everest. The record of the early history of the solar system, which undergoes systematic destruction on the dynamic Earth, has been laid before our eyes. We will never again be able to view Earth history in isolation.

References

Bills, B. G., and A. J. Ferrari. 1975. *Proc. Lunar Sci. Conf., 6th Geochim. Cosmochim Acta.* Supp. 6, frontispiece.

Bogard, D. C., and L. Husain. 1976. A new 1.25 billion year old Nakhlite-Achondrite (abs.). *Div. Planet. Sci., 7th Ann. Meet. Am. Astr. Soc.*, p. 5.

Boyce, J. M. 1975. Chronology of the major flow units. *Conference on Origin of Mare Basalts and Their Implications for Lunar Evolution*, pp. 11–14. Houston: Lunar Science Institute.

Carmichael, I. S. E., F. J. Turner, and J. Verhoogen. 1974. *Igneous Petrology.* NY: McGraw-Hill.

Carr, M. H. 1973. Volcanism on Mars. *J. Geophys. Res.* 78:4049–62.

Carr, M. H. 1974. Tectonism and volcanism of the Tharsis Region of Mars. *J. Geophys. Res.* 79:3943–49.

Carr, M. H., H. Masursky, and R. S. Saunders. 1973. A generalized geologic map of Mars. *J. Geophys. Res.* 78:4031–36.

Cintala, M. J., J. W. Head, and T. A. Mutch. 1976. Characteristics of fresh Martian craters as a function of diameter: Comparison with the Moon and Mercury. *Geophys. Res. Let.* 3:117–20.

Gault, D. E., J. E. Guest, J. B. Murray, D. Dzurisin, and M. C. Malin. 1975. Some comparisons of impact craters on Mercury and the Moon. *J. Geophys. Res.* 80:2444–60.

Hargraves, R. B. 1976. Precambrian geologic history. *Science* 193:363–71.

Hartmann, W. K. 1972. *Moons and Planets.* NY: Bogen and Quigley.

Head, J. W. 1976. Lunar volcanism in space and time. *Rev. Geophys. Space Phys.* 14:265–300.

Howard, K. A., D. E. Wilhelms, and D. H. Scott. 1974. Lunar basin formation and highland stratigraphy. *Rev. Geophys. Space Phys.* 12:309–27.

Kaula, W. M. 1975. The seven ages of a planet. *Icarus* 26:1–15.

Lewis, J. S. 1974. The chemistry of the solar system. *Sci. Am.* 230:51–65.

Lowman, P. D., Jr. 1976. Crustal evolution in silicate planets: Implications for the origin of continents. *J. Geol.* 84:1–26.

Malin, M. C. 1976. Observations of the surface of Venus (abs.). *Div. Planet. Sci., 7th Ann. Meet. Am. Astr. Soc.*, p. 38.

Masursky, H. 1973. An overview of geological results from Mariner 9. *J. Geophys. Res.* 78:4009–30.

Murray, B. C., R. G. Strom, N. J. Trask, and D. E. Gault. 1975. Surface history of Mercury: Implications for terrestrial planets. *J. Geophys. Res.* 80:2508–14.

Mutch, T. A. 1972. *Geology of the Moon.* Princeton Univ. Press.

Mutch, T. A., R. E. Arvidson, J. W. Head, K. L. Jones, and R. S. Saunders. 1976. *The Geology of Mars.* Princeton Univ. Press.

Pollack, J. B. 1975. Mars. *Sci. Am.* 233:107–12.

Rumsey, H. C., G. Morris, R. Green, and R. Goldstein. 1974. A radar brightness and altitude image of a portion of Venus. *Icarus* 23:1–7.

Sagan, C. 1975. The solar system. *Sci. Am.* 233:22–31.

Saunders, R. S. 1976. Analysis of compensated regions of Mars. *Reports of Accomplishments of Planetology Programs.* NASA TM X3364, p. 53–56.

Sharp, R. P. 1973. Mars: Fretted and chaotic terrains. *J. Geophys. Res.* 78:4073–83.

Siever, R. S. 1975. The Earth. *Sci. Am.* 233:83–90.

Soderblom, L. A., C. D. Condit., R. A. West, B. M. Herman, and T. J. Kreidler. 1974. Martian planetwide crater distributions: Implications for geologic history and surface processes. *Icarus* 22:239–63.

Solomon, S. C., and J. Chaiken. In press. Thermal expansion and thermal stress in the Moon and terrestrial planets: Clues to early thermal history. *Proc. Lunar. Sci. Conf.* 7.

Strom, R. G., N. J. Trask, and J. E. Guest. 1975. Tectonism and volcanism on Mercury. *J. Geophys. Res.* 80:2379–507.

Taylor, S. R. 1975. *Lunar Science: A Post-Apollo View.* NY: Pergamon.

Trask, N. J., and J. E. Guest. 1975. Preliminary geologic terrain map of Mercury. *J. Geophys. Res.* 80:246–47.

Vogt, P. R. 1974. Volcano height and plate thickness. *Earth Planet. Sci. Lett.* 23:337–48.

Weidenshilling, S. J. 1974. A model for accretion of the terrestrial planets. *Icarus* 22:426–35.

Whitford-Stark, J. L. 1975. Shield volcanoes. In *Volcanoes of the Earth, Moon, and Mars*, ed. G. Fielder and L. Wilson. NY: St. Martin's.

Wilhelms, D. E., and J. F. McCauley. 1971. Geologic map of the near side of the Moon (Map I-703). Washington, DC: USGS.

Wood, C. A., and J. W. Head. In press. Comparison of impact basins on Mercury, Mars and the Moon. *Proc. Lunar Sci. Conf.* 7.

Wyllie, P. J. 1971. *The Dynamic Earth.* NY: Wiley.

"Actually, they all look alike to me."

PART 2 Geologic Time and How It Is Determined

Henry Faul, "A History of Geologic Time,"
66:159 (1978), *page 18.*

A. E. J. Engel, "Time and the Earth," **57**:458
(1969), *page 25.*

Robert L. Fleischer, "Where Do Nuclear Tracks
Lead?" **67**:194 (1979), *page 37.*

Irving Friedman and Fred W. Trembour,
"Obsidian: The Dating Stone," **66**:44 (1978),
page 47.

Elizabeth K. Ralph and Henry N. Michael,
"Twenty-five Years of Radiocarbon Dating,"
62:553 (1974), *page 55.*

Henry Faul

A History of Geologic Time

"It is perhaps a little indelicate to ask of our
Mother Earth her age . . ." (Arthur Holmes 1913)

Time was conceived in geology. I mean the absolute, continuous, endless time in which the history of planets and stars is reckoned. It happened in Edinburgh, in the 1770s, in a small·group of thinkers which called itself the Oyster Club and included Joseph Black (1728–99), the chemist; Adam Smith (1723–90), the economist; and James Hutton (1726–97), the physician-farmer turned geologist. In more conventional circles, the largest unit of time was still the human life span, and the age of the world was accepted as established by Bishop Ussher's artful arithmetic with biblical chronology, but the Oyster Club discussions were not bound by the limits of such preconceptions. Hutton had taken his friends to see the rocks along the Scottish coast, and they observed that every rock formation, no matter how old, appeared to be derived from other rocks, older still.

Uniformitarianism

Years went by before Hutton saw fit to put the idea in print. He finally put it this way in 1788: "The result, therefore, of our present enquiry is, that we find no vestige of a beginning,—no prospect of an end."

Henry Faul is professor of geophysics at the University of Pennsylvania in Philadelphia. He was trained at M.I.T. (S.B. 1941; Ph.D. 1949), served in the Manhattan Project (1942–46) and in the U.S. Geological Survey (1947–63), and was professor of geophysics at the Southwest Center for Advanced Studies, now the University of Texas in Dallas (1963–66). The author is indebted to A. O. C. Nier for candid reminiscences of his days at Harvard. Address: Department of Geology, University of Pennsylvania, Philadelphia, PA 19104.

Physical infinity is always relative, and Hutton was responding to the convention of the day, but compared to the lives of the Patriarchs, Hutton's time was infinite indeed.

Hutton's departure from biblical chronology was rooted in his uniformitarian approach to the history of the Earth. He was convinced that geologic processes in ages long past were no different from the processes now active and thus open to observation. Geology could be interpreted without recourse to postulated catastrophies such as the Noachian Flood. The Earth was forever changing, but its nature remained the same. The concept was radical for its time—and slow to catch on.

A rival school arose in Germany, at the new Mining Academy in Freiburg, where Abraham Gottlob Werner (1749–1817) developed an impressive following by teaching an orderly, simplistic kind of geology in which all rocks, including granite and basalt, were said to have precipitated from a primeval ocean in a regular, universal sequence. Without mentioning Noah or Moses, Werner presented a picture of the Earth's past that anyone could reconcile with religious opinion. That was the root of his success.

One of Werner's disciples, Robert Jameson (1774–1854), became Professor of Natural History at the University of Edinburgh in 1803 and held the post for fifty years. About 1800–20, the Freiburg Academy was a small provincial trade school, but the University of Edinburgh had more than fifteen hundred students and was one of the world leaders in medicine and science. Jameson's classes were large, usually fifty to a

hundred students, and so, less than a decade after Hutton's death, Edinburgh became the fountain of Wernerian ideology (J. Eyles 1973).

The Industrial Revolution generated new demand for coal, metals, and transport facilities. The sinking of mines and the building of roads and canals required geologic information where little or nothing had been known before. New studies produced new interest, geology became popular, and geologic knowledge of the country rapidly expanded. As geologic evidence accumulated, the Wernerian viewpoint lost ground and the biblical concept of time slowly faded. The discovery by William Smith (1769–1839) that strata, no matter how different they may appear, are of the same age when they contain the same assemblage of fossils provided the means of establishing the stratigraphic sequence and thus, indirectly, a newly expanded time base.

Largely through the writings of Charles Lyell (1797–1875), the uniformitarian view eventually displaced Wernerian ideas, and the conception of time, no longer tied to Mosaic chronology, expanded inordinately. Under the oldest lavas of Mt. Etna, Lyell found fossils that looked to him like shells now living in the Mediterranean, and he wrote of "an indefinite lapse of ages having been comprised within each of the more modern periods of the earth's history" (1833). Imbued with the same spirit, Charles Darwin (1809–82), who had studied with Jameson but came to follow Lyell's view of time, made an uncharacteristically hasty calculation of the time required for erosion to excavate the Wealden Valley in southeastern England (1859, p. 285). His

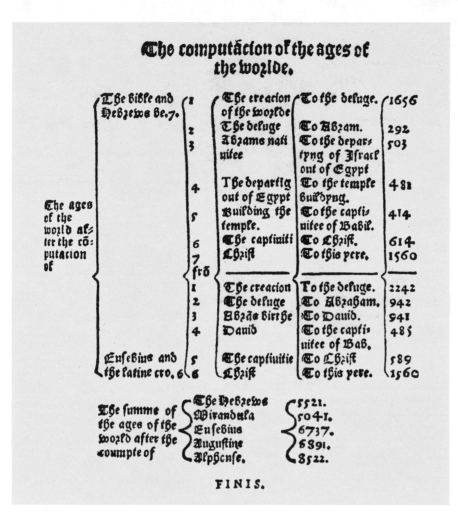

Figure 1. Calculations of the "age of the world" were made long before Bishop Ussher. (From *Cooper's Chronicle*, London, 1560.)

or ice—every kind of erosion—ultimately depend on the energy of the sun. What, then, was the meaning of uniformitarianism if the sun was cooling? All geologic surface processes must be slowing down.

Kelvin's geological ideas were influenced by his dour senior colleague at the University of Glasgow, Henry Darwin Rogers (1808–66). Rogers was from Philadelphia, had taught geology at the University of Pennsylvania, and, together with his brother William (1804–82), later the founder of M.I.T., had made a name for himself studying the Appalachian mountain chain. William had observed high temperatures in deep coal mines in Virginia and Henry very likely discussed these findings with his good friend Kelvin, who had long been interested in heat flow in the Earth. When Kelvin broke his leg, some months before he was to read his paper on the cooling of the sun at the meeting of the British Association in Manchester in 1861, he asked Rogers to read the paper for him (Thompson 1910, p. 421).

Knowing the temperatures at various depths and making assumptions about the thermal conductivity of the intervening rock masses, Kelvin was able to estimate outward heat flow. He concluded that the Earth also had to be cooling and that the time elapsed since a solid crust first formed on a presumably molten globe could be calculated (Thomson 1862b). As he saw it, the Earth and the sun were both about 100 million years old, but considering all the assumptions that had to be made in those calculations, he felt obliged to ascribe wide limits of error to his result: a neat factor of four. The Earth could be anywhere from 25 to 400 million years old.

Geologists paid no attention to Kelvin's attacks on uniformitarianism until he decided to take the trouble to advertise them in an address to the Glasgow Geological Society in 1868 (Thomson 1871). When he finally got his reaction, it was swift, elegant, and ineffective. Thomas Huxley rose to defend uniformitarianism against

result—about 300 million years—is many times larger than present estimates (Holmes 1965). Darwin needed lots of time for his natural selection process, but the uniformitarians had overreached themselves.

As the sequence of sedimentary rocks in Britain and western Europe was studied and gradually worked out, thinking about time became tied to the new stratigraphy. Lyell had repeatedly mentioned "indefinite periods of time" required for various geologic processes, but it was clear that strata and groups of strata were individually finite. There were gaps in the record as it was exposed in England, but some of the gaps could be filled by observing rocks of the same age in other countries. There was no evidence that any of the gaps could be of "indefinite" duration. The mid-nineteenth-century geologist had only a crude idea of the rates of deposition of sedimentary rocks, but it was enough to make clear that the whole magnificent layer cake of British strata represented an enormous period of time—but not infinity.

The cooling of the Earth

The Industrial Revolution also had its impact on physics—physicists became concerned with dynamics and energy. The origin of the sun's heat was an obvious problem, and by the middle of the nineteenth century it was clear that the sun could not remain hot forever. Its thermal energy was very large but clearly finite. William Thomson, later Baron Kelvin (1824–1907) was not the first to consider the problem, but he was certainly the most vigorous proponent of the new idea. He added up the energy emitted by the sun, assumed physical conditions and heat capacities that seemed plausible at the time, and calculated a rate of cooling that turned out to be surprisingly rapid (Thomson 1862a). All movement of air and water on the surface of the Earth is produced by energy received from the sun. Weathering of rocks, transport of sediment by wind, water,

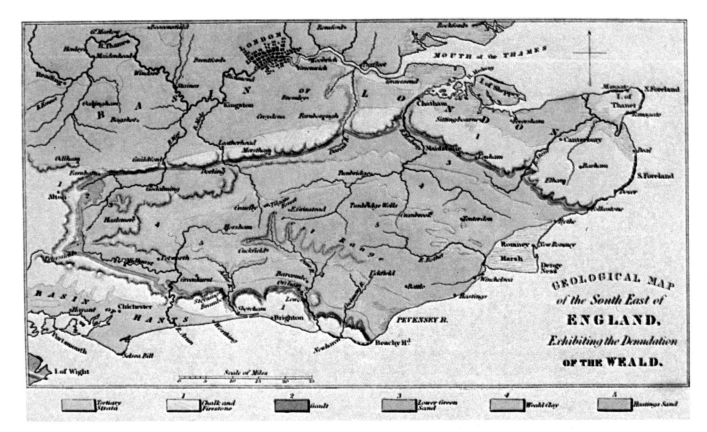

Figure 2. Surrounded by a chalk escarpment, the Wealden valley is a breached anticline with the soft rocks under the chalk removed by stream erosion. The uniformitarians greatly underestimated the erosive power of rivers and both Lyell and Darwin thought that the cliffs must have been cut by an invading sea. Darwin's home was south of Farnborough, just north of the edge of the chalk, and he chose the valley in a vain effort to illustrate the vastness of geologic time. (From Lyell 1833.)

Kelvin's attacks in 1869, but came up with little or nothing that Kelvin had not already considered. Then as now, it was difficult for geologists to argue with the results of physics.

Kelvin's approach was direct and his premises made sense. The uncertainty he quoted was large enough to allow most geologists to make an orderly retreat from their concept of very long time; but as the years went by, calculations tightened and Kelvin's time became shorter and shorter. In 1880, Clarence King (1842–1901) and Carl Barus (1856–1935) set up a laboratory in the U.S. Geological Survey for measuring physical properties of rocks, including thermal conductivity and heat capacity (Barus 1893). With the new results, King further reduced the uncertainty of Kelvin's cooling rate for the Earth. By the time most geologists had capitulated to the overwhelming power of Kelvin's physics, King had obtained an age of 24 million years (1893).

Not all geologists gave up, of course. C. D. Walcott on this side of the At-lantic and Archibald Geikie on the other both made their objections on geologic grounds, and a few others joined in, but it was a matter more of faith than of hard facts. (For a detailed and perceptive view of the controversy read Burchfield 1975.) Darwin had dropped the Weald calculation and softened his early uniformitarian stand in successive editions of the *Origin of Species,* but had been unable to bring himself to accept the short time scale. Kelvin's numbers were incompatible with the time required for natural selection. Something had to be wrong with them, and Darwin vainly sought what it could be. He remained convinced of the validity of his theoretical conclusions and simply refused to follow his contemporaries onto the Kelvin bandwagon.

Meanwhile, Kelvin's time became too short even for some of his supporters. Perhaps it was their picking that made him return to the subject of time in a grand speech to the Victoria Institute in 1897 (Kelvin 1898). By that time he had been Britain's leading physicist (in contemporary eyes) for more than a generation and also a great popular hero, thanks to the publicity that surrounded his work on the transatlantic telegraph. His lecture was a ceremonious revival of the old theoretical conclusions, now presumably vindicated by elegant experiments. It was a fancy "I told you so," hardly intended as a challenge for further debate.

But a challenge it turned out to be. T. C. Chamberlin (1843–1928), professor at the new University of Chicago, had been working on the hypothesis that the Earth and the other planets accumulated in a cold state from small pieces, which he called "planetesimals." In his view, the Earth had never been completely molten. The concept was not mature in his mind and was published only much later (Chamberlin and Moulton 1909), but it was clearly incompatible with Kelvin's model of an originally incandescent Earth. Not being bashful, Chamberlin attacked immediately (1899). "The fascinating impressiveness of rigorous mathematical analysis, with its atmosphere of precision and elegance, should not blind us to the defects of the premises that condition the whole process," he wrote, and he went on to take Kelvin's arguments apart, one by one. Much of what he said had been said before without great damage to the cooling concept, but then he reached beyond the limits of previous discussion:

What the internal constitution of the atoms may be is yet an open question. It is not improbable that they are complex organizations and the seats of enormous energies. Certainly, no careful chemist would affirm either that the atoms are really elementary or that there may not be locked up in them energies of the first order of magnitude. . . . Nor would he probably feel prepared to affirm or deny that the extraordinary conditions which reside in the center of the sun may not set free a portion of this energy. . . . A geologist begins to grow dizzy contemplating such thermal possibilities. Why should not atoms, atomecules, and whatever else lies below, one after another, have their energies squeezed out of them; and the outer regions [of the sun] be heated and lighted for an unknowable period at their expense?

That was written in the spring of 1899. Henri Becquerel's experiment with those fateful flakes of a common uranium salt (potassium uranyl sulfate) on a photographic plate wrapped in black paper had been reported on 24 February 1896. Kelvin had been aware of Becquerel's experiments when he spoke to the Victoria Institute (Kelvin, Carruthers Beattie, and Smoluchowski de Smolan 1896, 1897). Chamberlin, two years later, may or may not have been, but even if he had followed them closely, he would have found little in them to lead him toward postulating a nuclear energy source for the sun. Early nuclear physicists were concerned with the excitation of the puzzling "Becquerel rays" and had said practically nothing about the source of their energy by the time Chamberlin wrote his prophetic guess about the energy of the sun.

Also in 1899, John Joly of Dublin presented his first version of an elaborate new calculation of the age of the Earth based on the amount of salt in the sea, an idea originally proposed by Edmond Halley in 1693 (V. Eyles, 1973). Joly assumed that the oceans condensed from a primeval atmosphere as pure water and that the salt was derived from the weathering of rocks and brought down by

rivers over geologic time. From the present salt concentration in the sea and a weighted average of the flow rates and salt contents of major rivers, Joly obtained an age between 80 and 90 million years for the time when the surface temperature of the Earth dropped below the boiling point of water (Joly 1901).

Joly's papers bristled with numbers, in fair antidote to Kelvin's, and produced lively discussion for a few years. His approach was boldly quantitative but was beset with serious geological uncertainties. Radioactivity was in the wind and Joly himself soon abandoned the salt of the sea in favor of radium (Joly 1903).

Radioactivity

It was a long way from the "radiations emitted by phosphorescent bodies" to the discovery of radioactive decay, but the subsequent development of nuclear physics was explosive. The first glimmer of radioactive transmutation came from the emanation experiments of Elster and Geitel (1902), working at the Herzögliches Gymnasium in Wolfenbüttel, and Rutherford and Soddy (1902), at McGill University. Ernest Rutherford (1871–1937) wrote the basic equations of radioactivity in 1904, calculated the amount of heat released by radium present in ordinary rocks, and concluded that "the time during which the Earth has been at a temperature capable of supporting the presence of animal and vegetable life may be very much longer than the estimate made by Lord Kelvin from other data." R. J. Strutt (1875–1947) showed in 1905 that the amount of helium in uranium minerals was larger than what could have accumulated from alpha decay in Kelvin's geologic time. Before long, the evidence became overwhelming, but Kelvin remained unimpressed. He died convinced it was all wrong.

At Yale University, Bertram Boltwood (1870–1927) collected published analyses of pure uranium minerals (mostly from Hillebrand 1890) and calculated ages from the lead content, assuming a uranium half-life of 10^{10} years, more than twice the present value (Boltwood 1907). His ages ranged up to 2,000 million years, and some of them are not far from presently accepted values because his er-

rors tended to compensate. Whatever their inadequacy, they clearly reestablished the very long time scale. Boltwood himself was painfully aware of the wide uncertainty in the uranium decay-rate figure available at the time, the hazy understanding of the uranium decay scheme, and the severe limits of existing chemical techniques. Moreover, his health was failing and his interest in geology was marginal. He dropped the age work and preferred to follow his calling as Rutherford's chemist, studying radioactive-decay systematics as a primary problem.

The debacle of Kelvin's war on uniformitarianism had had little effect on his public image. His *Popular Lectures* (1894) had remained standard fare among the reading public. Thus it happened that, about 1905, a physics teacher in Gateshead High School in Newcastle-on-Tyne handed a copy to a very bright pupil named Arthur Holmes (1890–1965) (for a good biography see Dunham 1966). Thus the torch was passed, though one could hardly think of two characters more divergent than the overpowering, self-quoting, deadly-serious Kelvin and the mild, humorous, open-minded Holmes. The only thread between them was a lifelong passion for the study of geologic time.

Holmes began his long career in the laboratory of R. J. Strutt in Imperial College in 1910, and his first paper, "The association of lead with uranium in rock-minerals, and its application to the measurement of geological time" (1911) fairly presaged his life's work. There was, however, another side of him, and periodically he would go on long geological surveys in Africa and Asia, usually employed by commercial interests. In that way he developed, in addition to his exceptional mastery of the new physics, an impressive command of general geology (see Holmes 1965). In 1927 he proposed his first time scale in a popular booklet, *The Age of the Earth*, perhaps to show that he knew it was based more on his geologic intuition than on hard physical data. His subsequent revisions (1937, 1947) continued in the light vein, steadfastly undaunted by the shortage of reliable measurements.

Isotopic age determinations on rocks, securely tied to the fossil record, finally became available in the late fifties (including a few from this writer), and Holmes snapped them up with glee. "My 1947 attempt to construct a time-scale . . . has now outlived its usefulness. Of the two alternatives then proposed, the 'B scale' in particular enjoyed an unexpected success for ten years. . . . But now I come to bury the B scale, not to praise it," he began (1960), and he went on to develop his "1959 scale" with the new data. That time scale is still with us and is not likely to change very much in the future. It is amusing to contemplate that Holmes's old estimates, based on what now look like no data at all, were rarely off from currently accepted values by more than 20%.

The proper samples

On this side of the Atlantic, the study of geologic time took on the form of structured organization. In 1923, the Committee on Measurement of Geologic Time by Atomic Disintegration of the National Research Council was appointed, with A. C. Lane (1863–1948) of Tufts College as chairman. The Committee on the Age of the Earth, also of the National Research Council, was set up in 1926, with Yale's Adolf Knopf (1882–1966) in charge. Knopf's committee produced the much read *Bulletin 80* (National Research Council 1931), a hefty volume containing five short articles without much residual value and a masterfully organized 336-page review, ranging from Becquerel's experiments to the state of the isotopic art as it stood in 1931, by the ubiquitous Arthur Holmes. Lane's committee issued encouraging annual reports until 1955, but its main function, while the chipper chairman was still alive, was to serve as a base for his worldwide search for datable mineral specimens. Highly pure uranium minerals are rare and lack of suitable samples was a standard complaint among the early geochronologists.

In 1936, at age 73, Lane had refused to sign the Massachusetts teachers' oath and formally retired, but he continued to use his boundless energy to collect radioactive specimens and lead ores from every corner of the world and distribute them to every respectable analyst who showed interest. Analytical results accumulated, but their geological interpretation

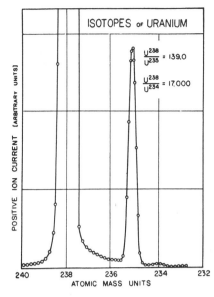

Figure 3. Nier's determination of the mass spectrum of uranium in 1939 made possible the calculation of the decay rate of ^{235}U and thus became the key to the Holmes-Houtermans (1946) computation of the age of the Earth. Each point in the spectrum represents a swing of the ballistic galvanometer and the height of each peak is proportional to the amount of the corresponding isotope. The value of the $^{238}U/^{235}U$ ratio still stands within his limits of uncertainty.

was something else again. Decades went by before the effort finally bore fruit.

One of Lane's good customers was Gregory P. Baxter (1876–1953), professor of chemistry at Harvard and a specialist in the gravimetric determination of atomic weights. He and

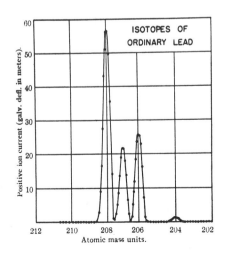

Figure 4. Nier's mass spectrometer cleanly resolved the lead spectrum (1938). ^{208}Pb is produced in the decay of ^{232}Th (thorium), ^{207}Pb from ^{235}U, ^{206}Pb from ^{238}U. ^{204}Pb, which is not radiogenic, is often used as the reference to express the variation in the radiogenic isotopes.

his students had studied dozens of common lead samples, but always obtained about the same value, unaware of the coincidental fact that isotopic differences in common leads tend to cancel out in the gravimetric determination (Nier, pers. com.).

The breakthrough

The systematics of the uranium and thorium decay chains were then reasonably well known, but analytical difficulties had not diminished much since Boltwood's time. Standard chemical techniques were still inadequate, and the decay constant of the rarer isotope of natural uranium, ^{235}U, was hazy because its content in normal uranium was known only roughly. The breakthrough came from a Harvard postdoc named Alfred O. Nier (1911–), who had come from Minnesota in 1936, bringing a new ingredient, the magic touch of a born instrumenter.

Mass spectrometers had been known for some time, but a new electrometer vacuum tube had just been perfected, with high gain and extremely low grid current, which made it possible to amplify and measure very small electric currents. Nier used it to measure the beam of uranium and lead ions in his mass spectrometer. The new instrument was accurate enough and stable enough to make the first isotopic analyses suitable for calculations of geologic time. Using uranium halides and lead iodide meticulously prepared by Baxter from Lane's judiciously selected mineral specimens (Nier, pers. com.), Nier determined the isotopic composition of natural uranium (Nier 1939) and thus was able to calculate the decay rate of the rarer isotope, ^{235}U, from available measurements of the combined alpha activity of natural uranium. He also measured the isotopic composition of the lead extracted from 29 samples of radioactive minerals and 25 lead ores (Nier 1938, 1939; Nier, Thompson, and Murphey 1941). Then the war came, Nier went on to other things, and everything pertaining to uranium became SECRET.

The age of the Earth

Nier's data remained available in libraries, of course, for anyone who would find them. Two independent thinkers went to work on them and both finally showed that such measurements permit calculating the age of the Earth as a planet. In Germany it was F. G. Houtermans (1903–66); and at the University of Edinburgh, where he had held a chair since 1943, who else but Arthur Holmes? The idea may seem obscure at first, but it is marvelously simple. We have in the Earth two different parent isotopes, ^{235}U and ^{238}U, each decaying into a different daughter isotope of lead, ^{207}Pb and ^{206}Pb respectively, at *a different rate*. Different isotopes of a heavy element are, of course, chemically identical and thus geochemically inseparable. Uranium and lead are chemically very different, however, and pure lead, cleanly separated from uranium by natural processes, is common in nature. We assume that each lead deposit was concentrated from a large rock system (containing some uranium) at some time. In principle, all that is needed is the isotopic composition of three natural lead deposits of different and independently known geologic ages, and we can extrapolate back in time to the origin of the Earth.

Houtermans (1946) wrote the equation this way:

$$\frac{\beta - \beta_0}{\alpha - \alpha_0} = \frac{Ac\,D}{Ra\,G} = \frac{1}{139} \cdot \frac{e^{\lambda'w} - e^{\lambda'p}}{e^{\lambda w} - e^{\lambda p}}$$

His symbols α and β stand for the ratios $^{206}Pb/^{204}Pb$ and $^{207}Pb/^{204}Pb$, respectively, in a lead deposit of age p. The subscript 0 refers to these same ratios in "primordial" lead ("natürliches Blei" of Houtermans), the lead at the time of the origin of the Earth, w, in other words, uncontaminated by the decay products of uranium. Ac D (actinium-D) and Ra G (radium-G) are now obsolete symbols for ^{207}Pb and ^{206}Pb, λ is the decay constant of ^{238}U, λ' the decay constant of ^{235}U, and 139 is the present ratio of these uranium isotopes (see Fig. 3).

Our present knowledge of the age of the Earth derives directly from these interpretations (Holmes 1946, 1950; Houtermans 1946, 1947). The values they originally reported were about 3,000 million years. Later refinements with many more lead analyses by other workers produced the presently accepted date for the accumulation of the planets: about 4,600 million years ago.

REVISED TIME-SCALE

Time-scale in millions of years

PERIODS	Since beginning of period	Duration of period
PLEISTOCENE		ca 1
	ca 1	
PLIOCENE		10
	11	
MIOCENE		14
	25	
OLIGOCENE		15
	40	
EOCENE		20
	60	
PALEOCENE		10
	70 ± 2	
CRETACEOUS Upper / Lower		65
	135 ± 5	
JURASSIC Upper / Mid. & Lower		45
	180 ± 5	
TRIASSIC		45
	225 ± 5	
PERMIAN		45
	270 ± 5	
CARBONIFEROUS Upper / Lower		80
	350 ± 10	
DEVONIAN Upper / Lower		50
	400 ± 10	
SILURIAN		40
	440 ± 10	
ORDOVICIAN		60
	500 ± 15	
CAMBRIAN		100
	600 ± 20	

Figure 5. In Holmes's final time scale (1960) covering the part of Earth history correlated by fossils, Holmes used known thicknesses of sediments to interpolate between isotopic age measurements made on rocks geologically tied to well-defined points in the fossil record.

In the words of Charles Lyell (1830, p. 71), "we have now arrived at the era of living authors, and shall bring to a conclusion our sketch." The supercharged endeavors of the Manhattan Project (1942–46) not only built the bombs, but also boosted nuclear analytical technology to a point where the minute quantities of daughter products in radioactive minerals could be handled. Accurate mass spectrometry, isotope dilution, flame photometry, and neutron activation developed into reliable analytical tools. Besides lead from uranium, it was now possible to measure the argon accumulated from the decay of potassium and the strontium from the decay of rubidium in geologic time. In the early fifties, laboratories for isotopic age-determination were established across the country, mainly by students of Harold C. Urey (1893–) and Nier. At last, Hutton's time was open to measurement.

The wealth of new data placed the history of the Earth and the moon on a quantitative basis for the first time. More than that, it provided the point of departure for a major conceptual breakthrough in geology, perhaps the greatest of all time: age determinations on volcanic rocks found on land, combined with measurements of the remanent magnetization of those same rocks, showed that the Earth's magnetic polarity has reversed itself frequently in the geologic past. The resulting paleomagnetic time scale was found to match the magnetic profiles measured by ships over mid-ocean ridges, thus proving conclusively that the sea floor is spreading and that the continents are really drifting, as had long been suspected by some. It was the beginning of a new era in man's understanding of the Earth.

References

Barus, Carl. 1893. The fusion-constants of igneous rock. Part III. The thermal capacity of igneous rock, considered in its bearing on the relation of melting-point to pressure. *Phil. Mag.* 35:296–307.

Becquerel, Henri. 1896. Sur les radiations émises par phosphorescence. *Comptes Rendus* 122:420–21; Sur les radiations invisibles émises par les corps phosphorescents, 501–03.

Boltwood, Bertram. 1907. On the ultimate disintegration products of the radio-active elements. Part II. The disintegration products of uranium. *Am. Jour. Sci.* 23:77–88.

Burchfield, J. D. 1975. *Lord Kelvin and the Age of the Earth.* Science History Publications.

Chamberlin, T. C. 1899. On Lord Kelvin's address on the age of the Earth as an abode fitted for life. *Science,* n.s. 9:889–901 and 10:11–18. Reprinted in *Smithsonian Ann. Rep. for 1899* (1900):223–46.

Chamberlin, T. C., and F. R. Moulton. 1909. *Science* 30:642–45.

Darwin, Charles. 1859. *On the Origin of Species*. Murray.

Dunham, K. C. 1966. Arthur Holmes. *Biographical Memoirs of Fellows of the Royal Society* 12:291–310.

Elster, J. P. L. J., and F. K. H. Geitel. 1902. Über eine fernere Analogie in dem elektrischen Verhalten der natürlichen und der durch Becquerelstrahlen abnorm leitend gemachten Luft. *Physikal. Zeit.* 2:590–93.

Eyles, J. M. 1973. Jameson, Robert. *Dic. Sci. Biogr.* 7:69–71.

Eyles, V. A. 1973. Joly, John. *Dic. Sci. Biogr.* 7:160–61.

Hillebrand, W. F. 1890. On the occurrence of nitrogen in uraninite and on the composition of uraninite in general: Condensed from a forthcoming bulletin of the U.S. Geological Survey. *Am. Jour. Sci.* 40:384–94.

Holmes, A. 1911. The association of lead with uranium in rock minerals, and its application to the measurement of geological time. *Proc. Roy. Soc.* A 85:248–56.

———. 1913. *The Age of the Earth*. Harper.

———. 1927. *The Age of the Earth: An Introduction to Geological Ideas*. Benn.

———. 1937. *The Age of the Earth*. New edition. Nelson.

———. 1946. Estimate of the age of the Earth. *Nature* 157:680–84.

———. 1947. The construction of a geological time-scale. *Trans. Geol. Soc. Glasgow* 21:117–52.

———. 1950. The age of the Earth. *Smithsonian Ann. Rep. for 1948*:227–39.

———. 1960. A revised geological time-scale. *Trans. Edin. Geol. Soc.* 17(part 3):183–216.

———. 1965. *Physical Geology*. Ronald.

Houtermans, F. G. 1946. Isotopenhäufigkeiten im natürlichen Blei und das Alter des Urans. *Naturwiss.* 33:185–86.

———. 1947. Das Alter des Urans. *Z. Naturforsch.* 2a:322–28.

Hutton, James. 1788. Theory of the Earth; or an investigation of the laws observable in the composition, dissolution, and restoration of land upon the globe. *Trans. Royal Soc. Edin.* 1:209–304.

Huxley, T. H. 1869. Geological reform. *Geol. Soc. London Quarterly Jour.* 25:xxxviii–liii.

Joly, John. 1901. An estimate of the geological age of the Earth. *Smithsonian Ann. Rep. for 1899*:247–88.

———. 1903. Radium and the geological age of the Earth. *Nature* 68:526.

Kelvin, Lord (see also Thomson, William). 1894. *Popular Lectures and Addresses*, vol 2. Nature Series.

———. 1898. *The Age of the Earth as an Abode Fitted for Life: The Annual Address*. The Victoria Institute. Reprinted in *Science* 9:665–74, 704–11.

Kelvin, Lord, J. Carruthers Beattie, and M. Smoluchowski de Smolan. 1896. On electric equilibrium between uranium and an insulated metal in its neighbourhood. *Nature* 55:447–48.

———. 1897. Continuation of experiments on electric properties of uranium. *Nature* 56:20.

King, Clarence. 1893. The age of the Earth. *Am. Jour. Sci.* 45:1–20.

Lyell, Charles. 1830. *Principles of Geology*, vol. 1. Murray.

———. 1832. *Principles of Geology*, vol. 2. Murray.

———. 1833. *Principles of Geology*, vol. 3. Murray.

National Research Council. 1931. *The Age of the Earth*. National Research Council Bull. 80.

Nier, A. O. 1938. Variations in the relative abundances of the isotopes of common lead from various sources. *Jour. Am. Chem. Soc.* 60:1571–676.

———. 1939a. The isotopic constitution of uranium and the half-lives of the uranium isotopes. *Phys. Rev.* 55:150–53.

———. 1939b. The isotopic constitution of radiogenic leads and the measurement of geological time. II. *Phys. Rev.* 55:153–63.

Nier, A. O., R. W. Thompson, and B. F. Murphey. 1941. The isotopic constitution of radiogenic leads and the measurement of geological time. III. *Phys. Rev.* 60:112–16.

Rutherford, Ernest. 1904. *Radio-Activity*. Cambridge Univ. Press.

Rutherford, Ernest, and Frederick Soddy. 1902. The cause and nature of radioactivity, part 2. *Phil. Mag.* 4:569–85.

Strutt, R. J. 1905. On the radio-active minerals. *Proc. Royal Soc.* A 76:88–101.

Thompson, S. P. 1910. *The Life of William Thomson, Baron Kelvin of Largs*. Macmillan.

Thomson, William. 1862a. On the age of the sun's heat. *Macmillan's Magazine* 5:288–93.

———. 1862b. On the secular cooling of the Earth. *Roy. Soc. Edin. Trans.* 23(1):157–69. Reprinted in *Phil. Mag.* 25:1–14.

———. 1871. On Geological Time. *Glasgow Geol. Soc. Trans.* 3:1–28. Reprinted in *Popular Lectures* 2:10–64.

"I'm beginning to understand eternity, but infinity is still beyond me."

A. E. J. Engel

Time and the Earth

"Due to a lack of interest, tomorrow has been cancelled."
—U.S. Government ad, Atlantic, July, 1969.

Earliest man was aware that the earth's history was catastrophically eventful. Some were forcefully laid to rest in a rain of volcanic ash. A much later generation of naturalists proved that the hills were ephemeral, that the seas have often covered the lands, and that the lands have commonly risen from or—as Herodotus noted 25 centuries ago—been built out into the seas. It is now clear that our unstable and probably drifting continental rafts represent a small, highly differentiated froth, derived with evolving oceanic crust, oceans, and air through vast periods of time from the interior of a turbulent earth.

Many of the geologically recent sequences of events of earth, including the evolution of man and other metazoan life, were deciphered in detail in the last half century by systematic techniques, especially an understanding of the geometric interrelations of minerals and rocks, the correlation of strata by mapping, and the use of index fossils.

In contrast, conclusions regarding the rates and dates of most earth events, their worldwide correlation, and the sources of the earth's enormous internal energy were largely intuitive, and erroneous, prior to the discovery and understanding of the radioactivity of the earth.

Systematic, radiometric studies of the earth have evolved in the last three decades, and proliferated in the last 15 years. They have revolutionized our concepts of time and the earth, as is briefly illustrated in the calendars listed in Table 1. Column I lists several important events in the history of the earth. Column II lists the "age" of this event that was widely accepted by scientists several decades ago. Column III lists the ages of these events as they are now dated by radiometric methods. The ages are given in years measured backward in time from the present. An aeon = one billion, that is, 10^9 years.

Growing up in Missouri, A.E.J. Engel obtained his B.S. and M.A. degrees from the University of Missouri, and his Ph.D. from Princeton. A geologist with the U.S. Geological Survey and Professor of Geology at the California Institute of Technology, he has been affiliated for some time with Scripps Institution of Oceanography as Professor of Geology. His studies have been largely related to the origin and evolution of the earth's crust and its fossil and contemporary environments. The present article is the text of Dr. Engel's Vanuxem Lecture at Princeton University, Spring 1967, a part of the Vanuxem series on "Time." Address: Scripps Institution of Oceanography, Geological Research Division, La Jolla, CA 92037.

Viewed in the perspective of the last three decades, it is obvious that the accepted ages of most important earth events have enormously increased. This also means that concepts of the rates of evolution of the earth, its oceans, air, life, and continents, have enormously decreased. Paradoxically, the evolving awareness of the great antiquity of the continents and of life has been accompanied by indications of the extreme youth of the present oceanic crust. Thus, in the three decades of 1935–1965, the ratio of the inferred ages of the "average" continental crust to oceanic crust has changed from about 1 to about 20. Yet features of the oldest rocks of the continents tell us that continents evolved from oceanic crust and mantle, and that migrant seas were invading the evolving continents over 3.4 aeons ago. The radioactive earth keeps changing the ocean's bed. Meanwhile, the continents appear to have wandered widely over the earth's surface as the sea floors have emerged and spread from the world-girdling oceanic ridges and rises [1]. The impelling forces may be large convective cells within the earth, activated and stalled at intervals, as the radioactive earth builds up and expels heat.

These cells, and related processes of terrestrial differentiation, have waxed and waned in a series of six to eight major pulses, or episodes, that are reflected on continents by major mountain-building, granite-forming events. During early or intermediate episodes, the earth may have had a much shorter (12-hour?) day. During more recent episodes the earth appears to have had a magnetic dipole that has repeatedly reversed itself. In the last 50 to 150 million years, the rate of spreading of ocean floors and migration of continents appears to have ranged from perhaps one-half to 10 centimeters per year.

Geologists have used their observations of contemporary earth processes and products as a key to those of the past. Yet any reconstruction of the relation of time and the earth from a contemporary vista encounters formidable barriers. This is in large part because the amount of time available to man for observation, measurement, and experiment is infinitely small relative to the rates of many major earth processes and to the length of geologic time. There is, of course, another imposing reason. Rock processes and the products we observe today reflect contemporary crustal environments. But these environments have changed with time, in space, in their proportions, and in kind. There was, after all, a time on earth when there was no "crust" at all—and very little earth.

Table 1. History of the Earth and Associated "Ages."

	I *Event*	*II* *Inferred pre-radiometric ages in the period 1900–1935* A. D.	*III* *Radiometric ages 1960–1967*
Origin of man		5000–50,000 yr	>2.5 million yr
Recent continental fragmentation and drift		Not widely recognized	0–200 million yr
Evolution of metazoan life		5000–400 million yr	600–700 million yr
Origin of life		5000 yr to one aeon	3.2–3.4 aeons
Oldest oceanic crust		200 million yr to 1.5 aeons	150–300 million yr
Oldest continental crust		200 million yr to 1.5 aeons	3.5–4 aeons
Origin of the earth		0.5–2.5 aeons	4.6–5 aeons

Hence, the past, as read from the oldest rocks, must be used as a critical key in understanding and evaluating the present; and the past and the present suggest the future of the earth and its life. Both certainly will change enormously and episodically with time. Man can and does accelerate some of these changes, none of which he understands. Curiously, man called himself *sapiens* (wise). But he has proved to be the dirtiest animal of the earth, presumptuous, even contemptuous of it in the midst of his ignorance. He is also, of course, the only one both capable and cussed enough to destroy it for all "higher" forms of life.

Compared to the age of the earth, man's tenure thereon is a momentary spark. Compared to most other animals, man lives a relatively long life. Yet many of us are not ready for death when it comes. This reluctance to shuffle off this mortal coil stands in curious contrast to our aggressiveness and busy fabrication of hot cars, guns, nuclear arsenals, and inflammable space hardware. There are, of course, nostrums for those who can swallow them. Some men hold out hope of life in other worlds, or of emigrating to them. Others seek death or drugs to gain Valhalla. Some of the current voyages in space and time are classified as progress and closely related to the economy, national prestige, and the Gross National Product. Clearly, they leave too little time or resources for studies of our own earth. Among the attitudes that evolve, as Dobzhansky and Iltis have noted, are those of the optimists and the pessimists [2]. The optimists believe, or claim to believe, that ours is the best of all possible times in the best of all possible worlds. We pessimists are those who fear that the optimists are right, and that man, if not the Earth, is running out of time. Here, at two thin protruding points from otherwise polarized positions, the attitudes of the pessimistic scientist and Jehovah's Witnesses almost touch.

The age of the earth

We have said that man's concepts of time and the manless earth, while consistently inelegant, have changed as abruptly and revealingly over the last half century as has the hemline. Studies of either or both provoke and excite. But one earthy topic at a time. The accepted age of the earth reflects some of the flux in our thinking about time and the earth. Figure 1 illustrates how rapidly the accepted age of the earth has increased since 1900 A.D. The large black points in Figure 1 are calculations of the age of the earth made in this century by some of the most knowledgeable earth scientists. The vertical coordinate indicates the year the calculation was published. The

horizontal coordinate shows the age of the earth in aeons or billion (10^9) years. Each of the earlier calculations, indicated by the cluster of points in the lower left of Figure 1, was remarkably precocious at the time it was made. Each was widely debated, and then accepted in the scientific community. All are almost surely wrong. The cluster of points of the upper right reflects the tumultuous but necessary wedding of classical geology and a chronology of the earth based on radioactivity. These are the so-called radiometric or "absolute" ages of the earth that have been determined by studies of the decay products, decay rates, and amounts of the radioactive nuclides, especially potassium, rubidium, uranium, and lead in crustal rocks and meteorites. Many additional points could have been inserted on this graph, especially at intermediate positions, to illustrate the halting search for old earthly rocks via less sophisticated radiometric and geologic mapping techniques. The open circle in the upper right corner of Figure 1 is an exciting but probably erroneous calculation by several of our Soviet colleagues as noted below.

The fact that the calculated age of the earth has increased by a factor of roughly 100 between the year 1900 and today—as the accepted "age" of the earth has increased from about 50 million years in 1900 to at least 4.6 aeons today—certainly suggests we clothe our current conclusions regarding time and the earth with humility. It also seems to say that if we just relax and wait another decade the earth may not be 4.5 to 5 aeons, as now suggested, but some 6 to 8 or even 10 aeons in age. In fact, there are presently cries of "Eureka" [3]. A group of Soviet geochronologists headed by Gerling argue that the potassium-argon age of the silicate mineral pyroxene in rocks from the Kola Peninsula is between 6 and 8 aeons. Their calculations are indicated very approximately in Figure 1 as the open circle on the upper right. Paradoxically, contemporary studies also show that pyroxenes tend to scavenge and hold excessive argon, the decay product of potassium 40. This extra argon would suggest, incorrectly, a much greater age for the host mineral and rock. Consequently, the calculated ages for the Kola rocks are probably in error and the age of the earth accepted by most earth scientists has reverted to some 4.5 to 5 aeons [4].

Some questions and conclusions

Figure 1 tells us that we have a lot of time on our hands and therefore a lot of explaining to do. It says very significantly that, as the earth has aged, our estimates of the rates of evolution of the earth's features, its oceans,

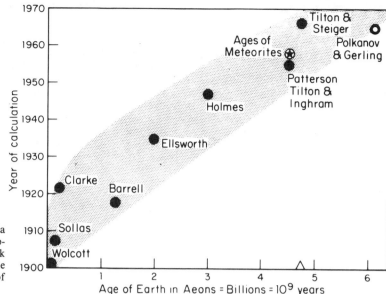

Figure 1. A graph of calculated ages of the earth, plotted as a function of the year the calculation was published. The appended bibliography for Figure 1 includes citations of the work of the individuals named on the graph. The triangle on the horizontal coordinate represents the presently accepted age of the earth.

air, life, and continents have enormously decreased. And we know that in almost any study of time, people, and processes, rates may be as, or more, critical than dates. The history of man is an obvious example. The accepted rate of man's evolution from less "advanced" primates has decreased by an order of magnitude in recent years as his origins have been traced back some 2 or more millions of years. Figure 1 also raises many questions: for example, how are relative and absolute concepts of time and the earth obtained? Just what are the first-order events in this almost incomprehensible span of geologic time? Can we, as environmental scientists, learn from the earth's past how to predict its future in a meaningful way?

Before discussing these points, we may draw several conclusions, or, more honestly, we may formulate other questions couched as assertions.

(1) The accepted age of the earth probably will not continue to increase as in the past, and the implied rates of major processes will not slow enormously with each new decade of investigation. The earth will remain, for man, roughly 5 aeons old. Its crust, atmospheres, oceans, and life began their evolution essentially as the earth formed. These assertions are largely derived from our awareness and use of the radioactivity of the earth, and of meteorites, to tell time on earth.

(2) No more than one percent or so of the history of the earth is decipherable. But that one percent is dispersed through a series of events, or episodes, extending back through geologic time. By imaginative manipulation of the evolving data we can reconstruct a magnificent and awesome history of the earth and its life, if only because there is so much time, so much history, and such diverse, evolving life. Our principal problem is the perennial one, of properly formulating the problem—and recognizing where, how, and what to study [6].

(3) The early history of the earth from its origin about 4.5–5 aeons ago, to the age of the oldest decipherable rocks some 3.5 aeons ago, remains a challenging lost weekend in the earth. "Lost" because most of the history

of the earth must be read from its rocks, and any rocks older than 3.5 aeons appear to have been destroyed by succeeding earth events. This is the gap indicated on the lower right in Figure 2. Note in Figure 2 that, as the accepted age of the earth has grown, it is the so-called Precambrian or "early earth" time that has been enormously enlarged. We will speculate a lot about the first aeon or more of earth's history (prior to 3.5×10^9 years) in the next few years; but in the foreseeable future it will be mostly speculation—essentially geopoetry. The problems here are analogous to those confronting most historical reconstructions. The farther back one probes, the dimmer the record, the fewer the old, recognizable rocks, the greater the differences in early earth environments from the recent—hence, the exponentially increased opportunities for error and misunderstanding. Yet things aren't beyond the pale. A controversy exists over who shot President Kennedy in front of thousands of witnesses a few years ago, and over who wrote *Hamlet* several hundred years ago. Geologists know in a crude way who went swimming and where they swam in primeval seas over 3,000,000,000 years ago [7]. Many of us seem able to reconstruct our childhood as clearly as what we did and said last week.

Telling time in the earth

Much, but obviously not all, of the earth's history is read from the more accessible rocks. But the earth's crust is less than 0.5 percent of the earth's mass. Moreover, the exposed crustal rocks are demonstrably very unlike the ninety-nine-plus percent of the mass of the earth. Crustal rocks have a density of about 2.7. The density of the earth is about 5.5 The density of the earth's core is perhaps 12. Analogous increases occur in the elastic constants of rocks at depth. For example, the velocity of the transverse waves (V_p) generated in the earth by earthquakes and nuclear blasts increases irregularly from some 2 or 3 at the surface to more than 8 just below the earth's crust. These and many other changes in the physical properties with depth imply very different substances in the earth's interior than at its surface. We cannot simply collapse, with pressure, the surficial rocks and endow them

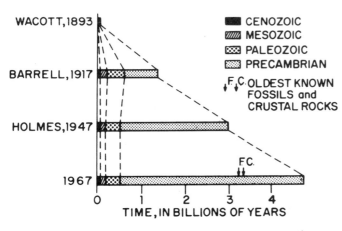

Figure 2. A graphic illustration of changes in concepts of geologic time, indicating the increasing awareness of Precambrian (early earth) time, due to the use of the radioactivity of the earth to tell time. References to the work of Walcott, Barrell, and Holmes are cited in the bibliography for Figure 1. (After James, 1960.)

with the properties of the earth's interior. Crustal rocks appear to be literally a penultimate silicate froth of some of the most uncommon terrestrial elements especially the large cations of the crystal chemist. These large cations, coupled with sodium, aluminum, calcium, iron, and silicon, have differentiated and migrated with the volatile elements from the interior of the earth [8]. In an obvious oversimplification the continental crusts are called "granite," the oceanic crusts basalt. The most time-revealing of the large cations in the crust are the radioactive nuclides noted above, especially potassium, uranium, and lead, with rubidium and carbon. Many crustal rocks are susceptible to radiometric dating because they contain one or more of these long- or short-lived radioactive nuclides. Consequently, their age can be determined rather accurately in years. By comparing the contemporary rock products formed from current earth processes with successively older, dated rock products, we gain considerable insight into earth processes and events of the past. This approach is reflected in the worn expression, "The present is the key to the past." Too frequently, in a dangerous but necessary reversal, the earth scientist must use ancient rock products and the inferred rock-forming processes as a key to the present and the future.

Because the earth's continents—and oceans and air—are extreme distillates and fractionates of the average, original earth, and very unlike it in composition and origin, it is obviously risky, if not misleading, to infer that surficial studies can tell us much that we must know of time and the earth. For the present, however, we must live with the premise that the continents are not only a complex of rocks derived directly, or indirectly, from the earth's interior—but also that their differentiation throughout geologic time should complement and reveal many aspects of the differentiation of the internal core and mantle of the earth. And because the crust remains the most accessible—hence the most revealing—earth's clock, we continue to use it to reconstruct earth history. Obviously, the attempts of earth scientists to drill deep holes in the earth clearly indicate we are dying to go underground for further data. From the history of the ill-fated Mohole, coupled with the difficulties of drilling in hard rock, it looks like most of us will be dead before the

earth is probed very deeply [9]. Space spectaculars are now presumed to be popular with "the people." Moreover, we must outrace the Russians into a vacuum. Hence we are urged to gaze up, or into the "Boob-tube," and not look around, or down, while the pickpockets who over-procreate, pollute, pillage, and procrastinate take their fateful toll on earth.

Relative geologic time

Relative sequences of events on earth were constructed by observant naturalists such as Herodotus long before the birth of Christ. It seems tragic there was so little inter-communication. The scientific methods rely upon the geometrical relations between rocks, the concept of "super-position," the presence in many sedimentary layers of so-called index fossils, as well as the existence of what we call time stratigraphic units. Figures 3 and 4 indicate superficially how the game is played. Quite obviously, the lowermost bed in this sequence of sedimentary rocks in Figure 3 had to be deposited before the "superimposed" overlying sedimentary beds. Note that the fossilized organisms entombed in the rocks are of marine animals which had hard parts. Hence both sequences were deposited in the oceans. Moreover, it is clear to the field geologist who has mapped the region that the lower stratigraphic sequence of marine "beds" was deposited before it was tilted, for the layers are uniform in thickness and of regional extent. During or after the tilting of the lower sequence, moving water, wind, and more locally in time and space, glaciers, tended to level or buff off the irregularities in the tilted sequence to produce a fairly flat, planar surface. A subsequent, "superposed" sequence of still horizontal, marine sedimentary rocks was deposited on the beveled surface of the first. The fact that the marine fossils in these rocks are Metazoan creatures which had evolved hard exo-skeletons enabled them to be conveniently preserved for our study. Some of these organisms happen to be types which evolved relatively rapidly into diverse genera and species, compared with rates of sedimentation and with the length of geologic time. They also were types of organisms that could and did migrate readily or were dispersed widely in oceans by currents. Accordingly, the rate of their dispersal in worldwide seas was infinitely

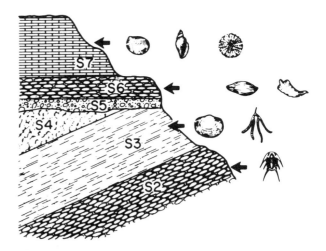

Figure 3. Sketch of a cliffed exposure of two fossiliferous, sedimentary rock sequences separated by an angular unconformity. (From Spencer, 1962.)

rapid relative to the length of geologic time. Even before the sun was over the yard-arm, early oceanographers littered the seas with buoyant objects such as empty bottles, corks, and so on. Their more curious colleagues followed these objects as they circumnavigated the earth via oceanic currents. These observations indicate that floating or swimming organisms can be dispersed throughout the oceans of the world in a decade. Unfortunately, well preserved index fossils of this type have lived in the seas in only the last 10 to 15 percent of recognizable geologic time. Hence, good index fossils serve as tools in the worldwide correlation of the sedimentary rocks deposited only in this last 500 to 600 million years. Sedimentary rocks formed in the preceding 85 percent of geologic time contain fossils of Protozoan life, but there are essentially no good index fossils unless the older reef-building algae prove useful [10].

Most index fossils also are indicative of the environments in which they lived, and thus reveal a great deal about ancient climates and the evolution of the oceans. We have said that the earthly existence of a good index organism must be limited to a relatively small period of time on earth. The presence of such relatively short-lived species or genera in any particular bed or sequence of beds is assurance that they were deposited at about the same time as any bed or sequence bearing this index fossil in any other part of the earth [6].

If we think of one of the rock beds, say in Figure 3, as volcanic ash, it is obvious that it, too, is formed essentially synchronously in geologic time for the explosion of a volcano and the dispersal in the air and water of its fragmental material is an enormously rapid process relative to most rates of erosion and sedimentation and relative to geologic time. There are several such means of time correlation that enable geologists to establish either the contemporaneity or the relative ages of rocks and events in one part of the world, and to make temporal comparisons with rocks and events in another.

Absolute geologic time and rates of earth processes

But what about the absolute age of each event, or the rates at which these sedimentary deposits were formed? What of the rates of evolution of the organisms, or the dates, as well as rates of evolution of other phenomena of the earth such as the differentiation of the earth, the intrusion of magmas in the earth's crust, the quasi-solid state recrystallization of rocks at depth, and the time correlations of the non-fossiliferous rocks? What about the vast expanses of terrestrial time prior to the evolution of Metazoan index fossils? We have said that the determination of absolute time is possible because of the awareness and use of the radioactivity of the earth. Figure 4 indicates how radiometric age measurements have defined and refined our concepts of time and the earth. The same sequence of rocks shown in Figure 3 also appears in Figure 4. In Figure 4, however, there are superimposed several other igneous and metamorphic events that offer powerful clues to our understanding of the absolute time at which some of the geologic events occurred. What we are saying is that the sequences of sedimentary rocks shown in Figure 3 have been traced by the field geologist into an area of Figure 4 that exposes parts of the ancient

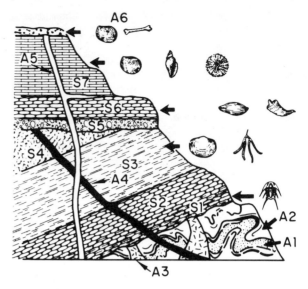

Figure 4. The same sequences of fossiliferous sedimentary rocks (S1 to S7) shown in Figure 3, but here found resting upon an older igneous and metamorphic floor (A1, A2, and A3), variously injected by igneous dikes and sills (A4 and A5), and overlain by a midden-bearing soil.

sea floor upon which the sediments were deposited. This old sea floor, labelled A1 and A2, was once buried deeply in the earth. It consists largely of metamorphic rocks, that is, rocks recrystallized in the quasi-solid state in the roots of a mountain belt, at temperatures of some 300 to 700 °C, and at great depths, hence under pressures of 2000 to perhaps 10,000 atmospheres. The sea floor also includes igneous rocks (A3); that is, rocks intruded in the molten state as mutual solutions of silicates at temperatures as high as 1200 °C. Many of the minerals in both the igneous and metamorphic rocks crystallized as essentially equilibrium products of the igneous and metamorphic events. Deep erosion has removed the original mountains that covered these sea-floor rocks. The critical points with regard to age studies are that (1) the minerals in the sea floor were equilibrated in specific ancient physical-chemical events deep in the earth, and (2) they have remained as essentially closed "systems" despite the deep erosion and succeeding marine environments that followed the time of each of these events. This means that, although many of these minerals have been exposed by erosion at or near the surface of the earth for millions or billions of years as metastable products, they have resisted alteration in the new physical and chemical environments in which they are not in equilibrium. Their radioactivity may be determined, and the ratio of daughter to parent radioactive element permits a determination of the absolute age of the initial, deep-seated igneous or metamorphic event in which they were formed [5].

In the example illustrated in Figure 4 it is possible to date the age of the metamorphic process that formed the basement of this sedimentary sequence. It may even be possible to probe into some of the older metamorphosed sediments (A1) to find the very refractory mineral, zircon, a zirconium silicate. Zircon often is not recrystallized during metamorphic processes which occur at lower temperatures than its igneous origin. This means that the zircon may not have undergone extensive isotopic exchange during the metamorphic epic recorded here. Its radioactive clock has not been completely reset. Dating these zircons gives a "residual" age which may approach the age of some previous high-temperature igneous episode that took place before the metamorphism, and before the evolution of the mountain belt. We may thus bracket the age of the basement complex in Figure 4 using the zircon in A1 to give us a mineral age of a preexisting igneous event, and the metamorphic minerals in A2 to tell us the age of the metamorphism. Finally, we may use the minerals A3 in the younger crosscutting granites to tell us the age of the last igneous event in the evolution of the mountain belt. These data also give us a maximum age for the oldest sedimentary rock (S1) in our lowermost sequence.

The several dikes that have cut the sequence at varying intervals of time (A4 and A5) enable us to bracket the epochs, and to determine the rates of sedimentation, and the rates at which the several forms of fossil life have evolved. For example, dike A4 cuts the lower sequence but quite obviously was emplaced prior to the erosive processes that occurred during or after the tilting of the basal sequence. This means that by dating A4, a maximum age limit is established for the overlying sedimentary sequence and a maximum age for the time when the index fossils in the beds S1, S2, S3, and S4 lived. The absolute ages of the emplacement A3, A4, and A5 also indicate time limits for rates of both sedimentation and of evolution of associated fossil life. The age of emplacement of the youngest dike, A5, indicates the minimum possible age of the second sequence including S5, S6, and S7. Accordingly, the ages of the dikes A4 and A5 bracket in time the age of the second sequence.

The weathered debris at the surface (A6) may include a pile of shells discarded after a dinner by early "prehistoric" man. Radiocarbon measurements on these shells (A6) tell approximately the year at which early man had his banquet, and when he prowled and fished the shores of the nearby lake or sea in which the bivalves lived.

The interrelations of rocks sketched in Figures 3 and 4 are somewhat oversimplified and highly idealized. But innumerable rock complexes in the world, of diverse ages back to 3.5 aeons ago, approach or even exceed these in their potential for telling time. A team of imaginative and clever geologists and geochronologists can extract enormous quantities of information about time and the earth from reconstructions of such age relations of rocks using index fossils, the time stratigraphic units such as S5, and dating each of the several igneous and metamorphic events by radiometric methods. From such studies, especially in the last decade or two, our awareness of time and the earth has grown explosively. And we have only scratched the surface.

The radioactive elements occur in minerals crystallized and recrystallized from molten magmas, saline seas, in sediments, as well as in meteorites possibly formed in the asteroidal belt of the solar system. The half-lives of the long-lived radioactive nuclides potassium, uranium, and lead are on the order of the age of the earth, or some appreciable fraction of that age. Fortunately, many minerals containing these radioactive nuclides have been frozen as metastable relics since the event or events that formed them. Either the lack of alteration, or the

Radiometric 'Absolute' Ages of Geologic Events

Derived from radioactive minerals and rocks that are essentially unaltered, although metastable at and near the earth's surface since the geologic event in which they were formed.

'Time — Stratigraphic markers'

Widely recognizable rock beds indicating a 'point' in geologic time.

Wind blown Volcanic tuffs, and water dispersed fossiliferous beds with one or more fossils whose identifiable features evolved and terminated in times infinitely short relative to geologic time and many geologic processes.

Figure 5. A graphic illustration indicating why dates of geologic events can be fixed in an absolute way, with precision, on a worldwide scale. Please see text for discussion.

relatively slow rate of alteration, of these rocks and mineral products to configurations of matter more stable at the surface is complemented by the relatively rapid rates of evolution of life and its dispersal relative to geologic time. These relationships are suggested in Figure 5. They permit numerous insights into time and the earth.

Paleoecology and historical concepts

Our conception of time and the earth may be further expanded if we interrelate our knowledge of animal physiology and ecology with radioactivity and isotope chemistry and a study of the earth. For example, Urey and his colleagues have measured the oxygen $\delta O_{18}/O_{16}$ frozen in the annular layers of carbonate in the shell of a 100 million year old fossil squid. From their data they have suggested the temperatures of the waters in which the squid lived and the length in absolute years of its life. They conclude from their studies of the isotopic composition of oxygen that the squid was born in the spring, lived 3 years, and died in the fall [11].

Wells of Cornell and a group of enthusiasts at Newcastle in England suggest that the complex and intricate growth patterns of a 300 million year old fossil coral indicate that the Devonian day was only about 12 hours long [12]. The Devonian is the segment of geologic time about 300 to 400 million years ago. The studies have led Runcorn to relate this variation in length of day to the rate of the earth's rotation, its moment of inertia, and to the rate of growth of the earth's core, mantle, and crust [13]. Others

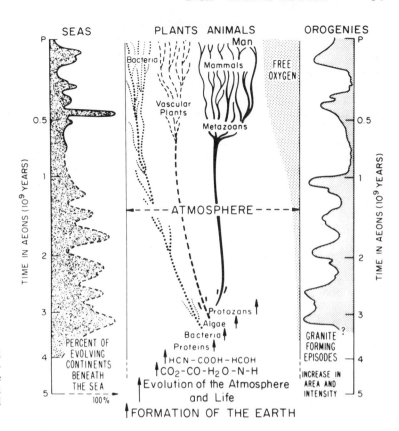

Figure 6. Interrelations, with geological time, of the fossil record of life, the evolution of the atmosphere, mountain-building, granite-forming events (orogenies), and the advance and retreat of seas across the continents. Modified from Cloud, 1968; Engel, 1964; Fisher, 1965; Holland, 1965; and many others.

have suggested that just prior to this time the moon was either derived from, or had an orbit very close to, the earth. Research now in progress suggests that all of these interpretations are not entirely correct. The moon, for example, is probably as old as the earth. Clearly, however, the potential exists to interrelate time, the earth, and the moon in these and related imaginative and revealing ways. Many more and new insights into the history of the earth are on the horizon if the smog, atomic holocausts, urban sprawl, and continental concretion do not obscure the view or terminate the game.

Historical narrative

So much for the means to the end. What are some of the first-order events on earth that we infer from our studies of absolute and relative time? Studies of amounts and ratios of volatile elements, of radioactivity inferred in the earth from its heat flow, and from the composition of the crust, from meteorites, and cosmic abundances of elements suggest that the earth began as a relatively cool agglomerate [8]. Despite this, the radioactive decay of the long-lived nuclides seems to have heated up the earth in its first 500 million years of history to a state of at least partial melting [14]. Much of the abundant, heavy element iron in the primordial earth fragments melted and began to sink into the interior at rates such that an iron-rich protocore and magnesium-rich mantle formed in the first aeon of the earth's history. The volatile elements and the large cations such as potassium, uranium, thorium, barium, and lead began to rise with the more volatile substances to the surface as the iron sank. The oceans, atmosphere, and crust began to evolve [8]. Possibly the first, primordial atmosphere was swept from the earth by solar winds. These first-order differentiative events in the earth are implicit in much of our data, but there is no highly definitive record of the initial differentiation processes,

rates, or dates. They occurred prior to the origin of the oldest crustal rocks, known to be some 3.5 aeons old. In the last 3.5 aeons, the data derived largely from crustal rocks tell us the earth continued to differentiate in a series of perhaps a half dozen second-order episodes as shown in Figure 6. Large quantities of molten rock were emplaced on or near the earth's surface, along with hydrogen, oxygen, helium, and other volatile elements. The seas, air, and the continents continued to grow in volume, to change in composition (Fig. 6). The atmosphere was enriched in oxygen; life evolved and flourished. The so-called orogenies indicated on the right side of Figure 6 are approximate periods in the earth's calendar at which large masses of granitic rock have been emplaced in unstable belts at and near the surface of the earth [15]. The major upward pulses of granite are reflected in the peaks in this graph. In general, these masses of granites have been emplaced in the roots of large evolving mountain belts. Successive orogenies furnish the means by which our continental rafts have grown in volume and in stability [15, 16]. The ages of these granitic regions are determined by complementary field mapping and radiometric dating. One of the most intriguing prospects of the future is to drill these areas of granite to determine their total volume as a function of their age [16]. By the use of radioactive tracers such as the ratio of strontium 87/86, Rb/Sr, K/A, and the array of Pb, U, Th data, it may be possible to decide the percent of this granitic material derived from pre-existing granitic crust versus the percent of the granite derived directly or not too indirectly from the interior of the earth. Preliminary studies using radioactive tracers and geologic mapping suggest that the amount of granite that has arrived from the interior of the earth and been emplaced in the crust may not have decreased appreciably from 4 aeons ago to the present. Thus the rate of differentiation of the earth's crust may be, very crudely, a linear function of geologic time [15, 17].

The squiggly line on the left side of Figure 6 represents an estimate of the extent to which shallow epicontinental seas have covered the evolving and existing continents throughout geologic time. These areas covered by seas are reconstructed from studies of the relic, patchy distribution of marine sediments and sub-aqueous lavas of any given age. Maps of the areas of distribution of the marine sediments, of sub-aqueous lavas, and so on, show that only 70 million years ago, for example, almost half of present-day North America was covered by seas (Fig. 7). About 500 million years ago some three-fourths of North America was covered by somewhat analogous shallow seas (Fig. 6). About 1.8 aeons ago much of what is now North America seems to have been under water; and relic pillow lavas and sediments 3.2-3.4 aeons old suggest that most of the oldest known continental nuclei were evolving from and engulfing oceanic crust. These early continents were periodically inundated. In fact, all of the oldest continental nuclei closely resemble island arcs. But much of this part of Figure 6 is extremely conjectural. The geologic history of the earth does indicate, however, that continental seas will advance across parts of North America and other continents in the future. A note for the Noahs—boats must be tarred and tied at hand. Fossil Miami, Los Angeles, and New York will be properly interred, middens of man's monuments to Minerva.

The center of Figure 6 is an attempt to show crudely the interrelations of the evolution of the atmosphere and of life throughout geologic time. Note that the earliest atmosphere was not enriched in oxygen but rather impoverished in this vital element [8]. Life seems to have evolved in this low oxygen or oxygen-free atmosphere from proteinoids and the first organisms lived in a watery (marine) environment [8, 10]. Here proteinoids, shielded from extreme ultraviolet radiation, could have evolved into bacteria, algae, and protozoan creatures. Algae existed at this time, because we find their fossil remains entombed in rocks over 3 billion years old. Recently, the casts or husks of what seem to be both algae and possibly bacteria have been discovered in African rocks that appear to be as old as 3.4 billion years [7]. The origins of life are rooted in the oldest known crustal rocks. And it is very possible that a proto-continental crust, seas, and organisms existed 4 billion years ago.

Inspection of Figure 6 also indicates that in the period of time between 3 or 3.5 billion years ago and 700 million years ago there are no unequivocal fossil remains of Metazoan life, only Protozoan life, especially algae and bacteria. Consequently, the appearance of Metazoans may have been a relatively rapid event in terms of geologic time, at or about some 600–700 million years ago. But almost monthly, a possible old Metazoan invertebrate is unearthed [18]. Curiously, the now accepted, geologically abrupt appearance of Metazoans was during a period in which there are serious gaps in the geological record; that is, crustal rocks of this period, especially datable igneous and metamorphic rocks, are rare. Thus, the right and left sides of Figure 6 indicate there were no major granite-forming, mountain-building events during this period of almost 600 million years prior to and during the evolution of Metazoan life. This interval from approximately 1.2 aeons to 600 million years ago is about as long as the succeeding period of geologic time up to the present in which the Metazoans have flourished. Thus the evolution of the more complicated animal phyla seems to have occurred during times when as yet there are few datable rocks and innumerable degrees of freedom. These are optimal conditions for the imaginative megathinker and ad hoc hypothecator.

Each of the granite-forming episodes shown on the right of Figure 6 seems to reflect the buildup of radioactive heat in the earth's interior, a partial melting of the mantle, and upward convection of the radioactive heat with molten magmas which contained the volatile elements that form the atmospheres and oceans. A schematic model of such a convective movement of the earth's interior during the contemporary, major orogenic episode is diagrammed in Figure 8. Each major orogenic episode of this type in the earth appears to have been followed by a comparative lull—after the molten rock, volatiles, and heat had moved to the surface, lost extensive heat, and after any remaining heavier elements, especially iron, had migrated toward its core. The historic lulls would seem to reflect periods of reheating prior to the onset of another major convective and orogenic episode. We may infer the periods of heating took place over some 200–600 million years. Large convective cells in the earth's mantle caused by terrestrial reheating have been invoked to move the continents, make the mountains, spread the oceans' floor, and create the tremendous secular changes that we see manifest in the surficial complexity of the earth's crust [1, 19]. The mechanism of large convective cells as agents of these several major dynamic events in the earth's crust is a most functional, if untestable hypothesis. Belief in cells is akin to belief in an orthodox God. Few earth scientists, I suspect, really believe in either completely; yet none is clever enough to invent working hypotheses less testable, more tantalizing, self-sufficing, and all-encompassing as God or the convective cell.

Recent earth history

The clues in earth's crust indicate that vast changes in and on the earth continue in very recent times. Continents still appear to be wandering about, and sea floors spreading at rates of 0.2 to 10 centimeters per year [1]. As the continents wander, they warp. Seas invaded much of North America along shallow depressions as recently as 70 million years ago (Fig. 7). The black patches in Figure 7 mark the distribution of the marine rocks infested with marine fossils at that time. The shaded area is the readily inferred distribution of these recent seas based upon the distribution of relic sedimentary layers formed in them. The widespread and well-known distribution of continental and mountain glaciers that covered the greater portion of North America only about 50–75 thousand years ago is shown in Figure 9. The evidence for the existence of these glaciers, their repeated advance, melting, advance, melting, over a series of 6 to 12 episodes is unequivocal [6]. The distribution of the major lakes of the southwestern United States formed in large part by the melting of glacial ice is indicated in Figure 10. These lakes were in existence as recently as 12,000 years ago. Their shores were thronged with exciting forms of life we no longer see in this presently arid and desolate country. Ancient man camped on the banks of these lakes, fished in them, and left his artifacts on their shores [6, 20]. Judging

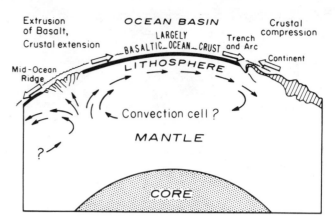

Figure 8. A schematic diagram of heat convective cells in the earth's mantle. The cells are started and stalled by the thermal regimen of the earth. Such cells may move the oceanic crust, and make the mountains. (Revised from Strahler, 1963, and many others.)

Figure 7. Inferred distribution of land and seas in the region of North America about 70 million years ago during the Cretaceous period. The extent of the seas (shaded) over the continent is reconstructed from the known distribution of fossiliferous marine sedimentary rocks of this age (black) in North America. (From Spencer, 1962.)

from the piles of shells left, some of his camps were almost as smelly as Los Angeles smog and those parts of Chicago downwind from the stockyards.

The future

Episodes involving dynamic aspects of the earth will continue into the future. But the rates, dates, and causes will remain speculative until we have thoughtfully collected and evaluated much more data about the earth. This conclusion may be illustrated by some attempts by man to use the past and present to predict the future in smoggy and unstable California.

California has earthquakes. Not many compared to, say, Chile, Indonesia, or some other parts of the Circumpacific circle of seismicity and fire, but very many compared to, let us say, Florida [21]. A strain release curve from earthquakes of Southern California in the period of 1938–1963 is plotted in Figure 11 after Allen *et al.* [22]. It is notable that the data earlier than 1938 are too inaccurate and incomplete to allow this curve to be plotted farther backward in time. This is invariably a major limitation of much of our knowledge. Instruments adequate to record critical events and appropriate field observations have been made only in recent years. Almost surely, subsequent generations in science will bemoan our present feeble and crude attempts to record events on the earth. But note, anyway, in Figure 11 the seeming quasi-periodicity in the times of relatively large strain release. If the curve in Figure 11 could have been

extended up to last week, its slope would be without any of the larger preceding steps or jumps in magnitude. The conclusion seems obvious: a major earthquake is not inevitable in populous California, but a disastrous one would not be entirely unexpected by seismologists. Almost surely, accumulations of this and other geological data through time may permit something in the way of predictions about when the earth will quake, erupt, and sleep.

There are, of course, formidable problems inherent in the prediction of the future of the earth and its life. The most obvious have been noted. Man's limited tenure on earth is virtually momentary relative to geologic time. The period over which we have taken even approximate measurements or attempted to simulate earth processes is infinitely small relative to the rate of many of these processes. This relationship produces its own special anxiety neuroses in Earth Scientists. Some, in some universities, have the curious illusion they must publish or perish. They have the vague feeling their chairmen and their deans are proud and happy only if papers appear at least every 6 months. How does this faculty member, under this illusion, thoughtfully design an earth-time experiment, simulate millions or billions of years in this experiment, perform it, write it up, and publish it in time to get his promotion? It takes most of us, except the odd, errant genius, at least 6 months to think of a good geologic problem and to design the experiment. It takes infinitely longer to make many of the most meaningful observations. Obviously, the coming generation would be wise to work in fields of fast reactions—studies of the earth take time.

Of course, man can and has induced earthquakes, sometimes unintentionally by merely dumping waste water or gas into holes in the ground. Similarly, man has contrived to dry up and salt the seas. The shrinking size of the Caspian Sea in the period 1930–1970 is indicated in Figure 12. This shrinkage is actually the result of the collaboration of nature and man. The Soviets have been using the waters of the Volga River upstream for irrigation. Concurrently, rainfall has diminished. So has the Caspian.

Just suppose one's comrade had built a dacha on the

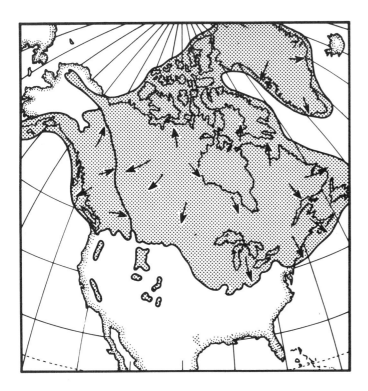

Figure 9. Map of North America showing the regions covered by Pleistocene glaciers some 25,000 to 100,000 years ago. The areas indicate the direction of ice flow as inferred from striae and gouges on bedrock and from the distribution of typical glacial debris. (From Flint, 1957; and Bates and Sweet, 1966.)

shore of the Caspian in 1930. Today he would have to walk or ride about 100 kilometers to go swimming or fishing. But then, only field geologists and other minority groups walk 100 km these days.

Apropos this question of time and walking, we may note the evolution, with time, of the kinds of students of time

and earth. A few decades ago the earth sciences consisted almost solely of those who made up for deficiencies in their heads by having and using their good eyes and feet. Geologists mapped the earth's surface in their attempt to understand its history. Earth scientists are now in their own special rococo phase. The current majority have a method involving a beautiful "black box," and are in

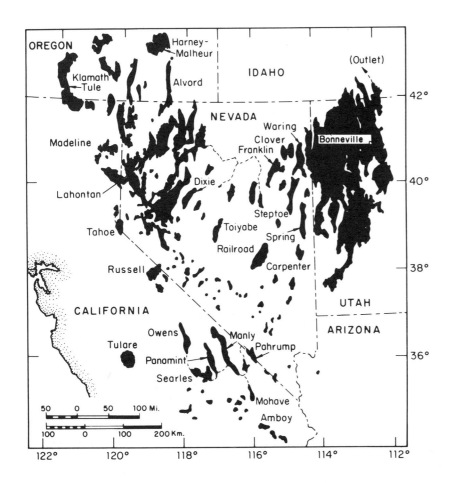

Figure 10. Distribution and size of lakes formed by rainfall and the melting of geologically recent glacial ice in the western United States. Almost all of these lakes have dried up in the last 10,000 years. (Based on maps by Flint, 1957; and Strahler, 1963.)

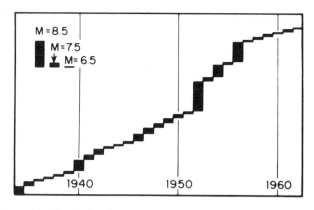

Figure 11. Graph showing the cumulative strain release in Southern California as a function of time in the period 1934–1963. The bars in the upper left of the graph show equivalent strain of single earthquakes of respective magnitudes 8.5, 7.5, and 6.5. (From Allen *et al.*, 1965.)

search of money, a sample, and, too often, an idea. There is evolving, however, a much more sophisticated group. They much prefer to calculate than to be right. From time to time these several types intercommunicate and learn more of time and earth.

Man's ability to manipulate his environment is painfully obvious. The problem has been recognized, stated, and restated in various forms, beginning a thousand years before Malthus. Many men, paradoxically, can grasp the key to our understanding of time and the earth. But the majority are too irrationally compassionate, too preoccupied with procreation, provocation, pollution, recreation, or proclamations of the resurrection.

The most honest concede freely we know almost nothing of the earth or its processes. Yet we have managed to discover enough uranium in a few years to destroy essentially all life in a few seconds. We now attempt to bury deadly radioactive and biological wastes by techniques we are quite unsure of. We pour millions of tons of carbon dioxide into the air from the irresponsible destruction of fossil fuels. The rate of proliferation of this and other smog producing gases is so rapid that we will almost surely effect major and unknown changes in the earth's climate. Because we procreate ourselves so rapidly, our grandchildren may have standing room only, on night soil instead of earth soil. Because we pollute the earth, air, the waters, and the minds of our children, we are surprised at the problems each confronts us with. We hastily and unwisely harvest and dissipate the mineral crops from the earth, that she can replace in at best millions of years. In our Great Society we destroy an average of one species of animal each year. We proclaim that if you have seen any one of the few remaining thousand-year-old redwood trees, you have seen them all. "To hell with posterity." And then we explosively head for the moon, or southeast Asia. Man is the only domesticated animal not bred for his attributes.

Almost anyone can suggest a partial solution for the future of the earth and its life. If man, who so ignorantly controls and modifies his environment on earth really wishes to have a pleasant home here, he probably should reduce his numbers by at least 30 percent in the next ten to twenty generations. Synchronously, man might try to understand better what effect his diverse actions will have upon his earthly environment and himself. He must, as

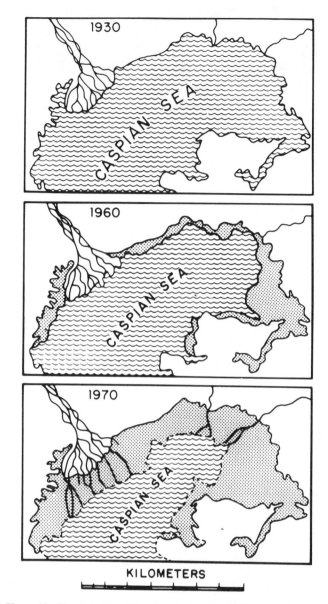

Figure 12. The shrinking of the Caspian Sea since 1930 due to reduced rainfall and in-flow of fresh water from streams.

Lynn White has said so well, reject the Christian axiom that nature has no reason save to serve man [23].

But perverse and mercurial mankind almost surely will not do these things. The history of the earth, which is man's environment, and the history of man is therefore indicated in a general way. Both will change, enormously, episodically, and catastrophically with time. Perverse, independent, and ignorant, man will continue to accelerate these changes, none of which he will have or take the time to understand. Ultimately, man's and nature's capabilities to terminate his adventure here will allow the earth a little peace, if not quiet. These are perhaps two of the greatest luxuries most of us cannot now enjoy. Time alone will tell when man will depart and the earth may continue to be turbulent in its natural disquiet and in a more peaceful way.

References

1. Heirtzler, J. R., 1968, Sea-floor spreading: *Sci. American, 219,* 60–70.

2. Dobzhansky, Theodosius, 1967, Changing man: *Science, 155,* 409–414. Iltis, Hugh H., 1967, A plea for man and nature: *Science, 156,* 581.

3. Polkanov, A. A., and Gerling, E. K., 1961, Geochronology and geological evolution of the Baltic Shield and its folded margins: *T. Lab. Geol. Doekembria. Akad. Nauk., S.S.S.R., 15,* 7–102.

4. Hart, S. R., and Dodd, R. T., Jr., 1962, Excess argon in pyroxenes: *Jour. Geophys. Research, 67,* 2998–2999. Tilton, G. R., and Steiger, R. H., 1965, Lead isotopes and the age of the earth: *Science, 150,* 1805–1808.

5. Faul, Henry, 1966, Ages of rocks, planets and stars: Earth and Planetary Science Series, McGraw-Hill Book Co., New York, 109 pp. Tilton, G. R., and Hart, S. R., 1963, Geochronology: *Science, 140,* 357–366.

6. Woodford, A. O., 1965, Historical geology: W. H. Freeman and Co., San Francisco, 512 p.

7. Engel, A.E.J., *et al.,* 1968, Alga-like forms in Onverwacht Series, South Africa: Oldest recognized lifelike forms on earth: *Science, 161,* 1005–1008. Schopf, J., and Barghoorn, E. S., 1967, Alga-like fossils from the early Precambrain of South Africa: *Science, 156,* 508–511.

8. Ringwood, A. E., 1966, The chemical composition and origin of the earth, p. 287–356 *in* Hurley, P. M., Editor, Advances in earth sciences: The M.I.T. Press, Cambridge, Mass., 502 p. Cloud, P. E., *et al.,* 1965, Symposium on the evolution of the earth's atmosphere: *National Academy of Sciences Proc., 53,* 1169–1226.

9. Hess, H. H., 1962, the AMSOC hole to the earth's mantle, p. 78–88 *in* White, J. F., Editor, Study of the earth: Prentice Hall, Inc., Englewood Cliffs, New Jersey. Anonymous editorial, 1966, Closing the Mohole: *Nature, 210,* 662. Abelson, P. H., 1966, Penny wise, pound foolish: *Science, 152,* 1. Joides, 1967, The deep sea drilling project: *Amer. Geophys. Union Trans., 48,* 817–832.

10. Cloud, P. E., 1968, Pre-metazoan evolution and the origin of the Metazoa, p. 1–72, *in* Drake, Ellen T., Editor, Evolution and environment; Symposium on Centennial of Peabody Museum, 1966: Yale University Press, 470 p.

11. Urey, H. C., *et al.,* 1951, Measurement of paleo-temperatures and temperatures of the Upper Cretaceous of England, Denmark and the Southwestern United States: *Geol. Soc. America Bull., 62,* 399–416.

12. Emiliani, Cesare, 1966, Isotopic paleotemperatures: *Science, 154,* 851–856. Wells, J. W., 1963, Coral growth and geochronometry: *Nature, 197,* 948–950. Runcorn, S. K., 1964, Changes in the earth's moment of inertia: *Nature, 204,* 823–825.

13. Runcorn, S. K., 1965, Changes in the convection pattern in the earth's mantle and continental drift: Evidence for a cold origin of the earth: *Royal Soc. London Philos. Trans.,* Ser. A., *258,* 228–251.

14. Birch, Francis, 1965, Speculations on the earth's thermal history: *Geol. Soc. America Bull., 76,* 133–154.

15. Engel. A.E.J., 1963, Geologic evolution of North America: *Science, 140,* 143–152.

16. Engel, A.E.J., 1966, The Barberton Mountain Land: Clues to the differentiation of the earth: *Econ. Geol. Research Unit of South Africa,* Circular no. 27, p. 1–17.

17. Hurley, P. M., Hughes, H., Faure, G., Fairbairn, H. W., and Pinson, W. H., 1962, Radiogenic strontium-87 model of continent formation: *Jour. Geophys. Research, 67,* 5315–5334.

18. Hoffman, H. J., 1967, Precambrain fossils (?) near Elliot Lake, Ontario: *Science, 156,* 500–504.

19. Hess, H. H., 1962, History of ocean basins, p. 599–620 *in* Engel, A.E.J., *et al.,* Editors, Petrologic studies: A volume to honor A. F. Buddington: *Geol. Soc. America,* 660 p. Griggs, D., 1939, A theory of mountain building: *Am. Jour. Sci., 237,* 611–650.

20. Johnson, Frederick, 1967, Radiocarbon dating and archaeology in North America: *Science, 155,* 165–169.

21. Gutenberg, B., and Richter, C. F., 1954, Seismicity of the earth, 2nd edition: Princeton University Press, Princeton, New Jersey, 310 p.

22. Allen, C. R., St. Amand, P., Richter, C. F., and Nordquist, J. M., 1965, Relationship between seismicity and geologic structure in the Southern California Region: *Seis. Soc. America, 55,* 753–797.

23. White, Lynn, Jr., 1967, The historical roots of our ecologic crisis: *Science, 155,* 1203–1207.

Figure References

Figure 1

Barrell, J., 1917, Rhythms and the measurement of geologic time: *Geol. Soc. America Bull., 28,* 745–904.

Clarke, F. W., 1924, Data of geochemistry: *U.S. Geol. Survey Bull.,* 770, 841 p.

Ellsworth, H. V., 1932, Rare-element minerals in Canada: *Geol. Survey Canada Geol. Ser.* n. 11, 272 p.

Holmes, A., 1947, The construction of a geological time scale: *Geol. Soc. Glasgow Trans., 21,* 117–152.

Patterson, C., Tilton, G., and Inghram, M., 1955, Age of the earth: *Science, 121,* 69–75.

Sollas, W. J., 1909, The anniversary address of the president: *Geol. Soc. London Quart. Journ., 65,* 1–72.

Tilton, G. R., and Steiger, R. H., 1965, Lead isotopes and the age of the earth: *Science, 150,* 1805–1808.

Polkanov, A. A., and Gerling, E. K., 1961, Geochronology and geological evolution of the Baltic Shield and its folded margins: *Tr. Lab. Geol. Doekembria, Akad. Nauk, S.S.S.R.,* no. 15, p. 7–102.

Figure 2

James, H. L., 1960, Problems of stratigraphy and correlation of Precambrain rocks with particular reference to the Lake Superior region: *Am. Jour. Sci., 258-A,* 104–114.

Figure 3

Spencer, E. W., 1962, Basic concepts of historical geology: Thomas Crowell Co., New York, 504 p.

Figure 4

Modified from Figure 3 by the author.

Figure 6

See following references from text: Cloud, 1968 [10]; Engel, 1963 [15]; Fisher, A. F., in Cloud, *et al.* [8]; Holland, H. D., in Cloud, *et al.* [8]; and many others.

Figure 7

Spencer, 1962; see reference for Figure 3.

Figure 8

Strahler, A. N., 1963, The earth sciences: Harper Row Inc., New York, N.Y., 681 p.

Figure 9

Flint, R. F., 1957, Glacial and Pleistocene geology: John Wiley and Sons, New York, N.Y., 553 p.

Bates, R. L., and Sweet, W. C., 1966, Geology, an introduction: D. C. Heath and Company, Boston, Mass., 367 p.

Figure 10

Flint, 1957. See reference for Figure 9.

Strahler, 1963. See reference for Figure 8.

Figure 11

Allen, *et al.,* 1965. Reference [22] for text.

Robert L. Fleischer

Where Do Nuclear Tracks Lead?

Damage trails induced by irradiation and chemical etching offer a new tool for investigations in many diverse fields

To respond to the title of this article with a second question, what do the following subjects have in common?

flight patterns of birds
particle reactions in fusion research
the action of a fossil fission reactor that operated 2,000 million years ago
size and mobility of sperm
exploration for uranium ores
identification of radioactivity on aerosol particles in the earth's atmosphere
distribution of transuranic elements in living tissue
exposure of minute dust particles in primordial space near the time the earth was born
the slow, steady accumulation of sediment at the depths of the ocean

These subjects range from biology to mineral exploration, from the depths of the earth into distant space, from the small to the large, and from the ancient to the recent. The thread that binds them together consists of variations on a single method that can be used to study them—the nuclear-track technique.

The striking sunburst pattern in Figure 1 illustrates the means and the power of nuclear-track revelation in solids. A small, uranium-rich particle was placed on a piece of the layer-

Robert L. Fleischer is staff physicist at the General Electric Research and Development Center in Schenectady, NY. He received his Ph.D. from Harvard in 1956 and was assistant professor of metallurgy at M.I.T. before moving to General Electric in 1960. His interests were in the mechanical properties of crystals before he began to work on the development of nuclear-track technique in 1962. This technique and its uses are described in Nuclear Tracks in Solids (Univ. of California Press, 1975), which he wrote with P. Buford Price and Robert M. Walker. Address: General Electric Research Laboratory, Schenectady, NY 12301.

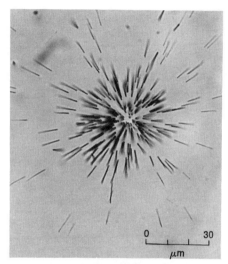

Figure 1. A piece of mica with a uranium-rich dust speck was neutron-irradiated to induce fission and then chemically etched to reveal the uranium distribution. (From Price and Walker 1963a.)

mineral mica (inadvertently, I should note). The mica was then inserted into a nuclear reactor to expose it to slow neutrons, later immersed in hydrofluoric acid, and finally cleaned. The neutrons induced some of the uranium nuclei to split, producing a spray of pairs of heavy particles (fission fragments), the downward-moving half of which entered the mica and created damage trails. The hydrofluoric acid preferentially dissolved the damaged mica and then more slowly opened the originally narrow tunnels into holes that are large enough to be seen clearly through an ordinary microscope.

The two special features of the nuclear-track technique are, first, that properties and positions of nuclear particles can be identified and, second, that holes can be created—holes, as it turns out, whose size, shape, number, and position can be controlled and used. In this article I shall

describe briefly the principles of track etching and how it allows nuclear particles to be identified. Then the broad areas in which track etching has proved to be useful will be outlined and, finally, as examples of recent advances in the use of etched particle tracks, I will point to ways in which the subjects listed above have been approached.

I first became aware of tracks in solids through studies of micas by my associates P. B. Price and R. M. Walker (1962), who were following up the direct electron microscopic observations of tracks by Silk and Barnes (1959). Figure 2 shows the appearance of tracks in untreated mica and the transformation of crystal damage and distortion into holes in chemically etched mica. Price and I later found that the other major classes of solids—glasses and plastics—recorded such tracks and that in fact the uniform bore of etched tracks was atypical of solids in general. In the usual case, also illustrated in Figure 2, etched tracks are tapered with a cone angle θ.

As shown in Figure 3, the value of the angle θ [= $\sin^{-1}(v_G/v_T)$] is a purely geometrical result of chemical attack at a linear velocity of v_T along the track and of simultaneous general, i.e. bulk, etching proceeding at a velocity v_G on the undamaged solid. After these discoveries we learned of Young's (1958) study of track etching in lithium fluoride. His work preceded that of Silk and Barnes as well as our own. Subsequent work has shown that tracks can be observed in virtually any insulating solid and that the chemical etching technique can be used for a vast assortment of jobs in various fields. (See Fleischer et al. 1975 for a detailed treatment; applications are reviewed concisely in Fleischer 1977.)

Most of the uses of particle tracks that were first recognized depend on a well-defined threshold for registration: particles that produce damage below a particular intensity are not revealed by etching, whereas those above that value are recorded with unit efficiency, one track resulting from each incident particle (see Fig. 4). The damage that contributes to accelerated etching is for other scientific fields ranging from biology to oceanography, and for various technological applications.

Major areas of track applications

If we go about deriving quantitative numbers that are pertinent to Figure 1, we quickly recognize three fields for which tracks provide us with new tools. The number of tracks in the sunburst is proportional to three quantities—σ, ϕ, and N_U, where N_U mation to be derived. The spatial resolution that is available for both chemical analysis and dosimetric applications is clearly shown by Figure 1. There is little question as to where the dust speck being analyzed was located. It is also possible by using different nuclear reactions to analyze for an assortment of elements. With projectiles of higher energy than thermal neutrons, other heavy elements besides uranium—thorium, bismuth, lead, and gold, for example—can be made to fission.

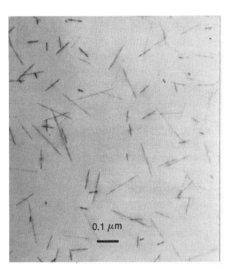 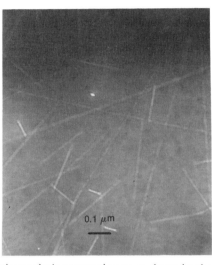 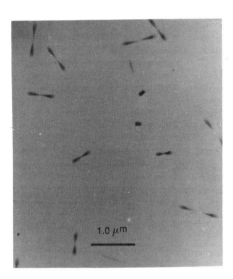

Figure 2. The strain fields around unetched tracks in mica viewed in the electron microscope (*left*) are relieved and holes are produced by etching (*center*). The more usual shape of the tracks is a tapered cone, as shown by the fission tracks in a sheet of phosphate glass at the right.

contained in a narrow cylinder of ionization concentrated within 5 to 10 atomic distances of the particle trajectory, and can be measured radially outward for not more than 40 atomic distances (C. P. Bean, in Fleischer et al. 1975).

To a very good approximation, track-detecting solids behave like threshold detectors that record all highly ionizing particles and ignore all lightly ionizing ones. Thus most natural minerals record fission fragments and very heavy particles, such as the iron nuclei that are found in cosmic radiation, but they are essentially unaffected in any direct way by immense doses of electrons, protons, and even alpha particles. This property of seeing certain particles well and others not at all is a highly valued quality in a detector. The detection qualities of track etching have provided a valuable tool for several branches of physics and astronomy, is the number of uranium atoms in the speck of dust, ϕ is the integrated flux of neutrons to which it was exposed, and σ is the probability of fission (fission cross section). If we know any two of the three quantities, a count of the tracks tells us the magnitude of the third. In the example of Figure 1, we knew the neutron dose and the cross section for fission of the uranium, and therefore we could infer how many uranium atoms were present. But alternatively we could have specified the neutron dose and the number of uranium atoms, and used such an experiment to measure the cross section. Or knowing σ and N_U, we could measure the dose of neutrons. And so merely by counting tracks we can do chemical analysis, nuclear physics, and neutron dosimetry.

In each case track detectors have unique combinations of properties that allow qualitatively new infor- Slow neutrons induce alpha-particle emission from isotopes of certain light elements, most notably ^{6}Li and ^{10}B. The reaction products can be recorded in many of the most sensitive class of solid track detectors—the polymeric materials.

In nuclear physics the sharp threshold of track detectors allows bombardment with a flood of projectiles that do not themselves leave tracks but do produce occasional reactions that release heavily ionizing, track-producing particles. In such experiments remarkably low cross sections for interaction ($\sim 10^{-36}$ cm) can be measured, as for example is typical with electron-induced fission or particle-induced fission at very low energy. Other sorts of nuclear phenomena that have been studied include fission reactions, proof of synthesis of new elements, lifetimes of spontaneously fissioning isomers, nuclear lifetime measurement by blocking

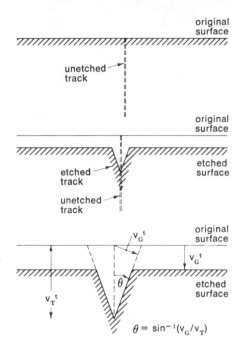

Figure 3. The cone angle θ of a track results from the competition of the general rate of attack v_G and the chemical, or track, attack rate v_T. t= time.

and channeling in crystals, and intensive searches for superheavy elements and for the elusive (and possibly imaginary) magnetic monopole.

Dosimetry can be done for a vast variety of particles in addition to neutrons: electrons, photons, protons, and heavier ions. The energy distributions of neutrons can be inferred. For the lighter particles detection is by means of reactions they induce; for the heavier ones direct detection of individual particles makes it possible to measure minute doses.

As an example of direct particle dosimetry, Figure 5 shows four individual tracks of heavy cosmic-ray nuclei that penetrated the Apollo 14 spacecraft and entered an experimental device that happened to include one of the better particle track-recording solids—Lexan® polycarbonate. The abundance of tracks seen in that apparatus indicates the dose of heavy, cytologically damaging particles to which the astronauts were exposed within their spacecraft (Fleischer et al. 1973), while the distinctive shapes of these millimeter-long tracks contain precise information about the nature of the particles.

Tracks of time

If minerals from a rock are etched, they are usually found to contain particle tracks that have the characteristic length and random orientation of fission tracks. Since these crystals have not been irradiated, the tracks must be natural. Careful analysis of the possible origins indicates that on earth, in all but rather bizarre circumstances, these tracks are from the spontaneous fission of uranium, primarily the major isotope ^{238}U (Price and Walker 1963b). If these tracks remain without fading, their abundance will depend on the product of the uranium concentration in the particular crystal or glass being examined and of the time since it began to record the fission events. Since we can measure uranium content by inducing new fissions by thermal neutrons, the density of the natural tracks gives an age for the sample.

The two steps in the procedure are pictured in Figure 6. The promise of this simple technique (Fleischer and Price 1964a, b; Maurette et al. 1964) has been borne out by the accumulated results, summarized in Figure 7, which shows the usual agreement between fission track dates and those known by other means. Occasional deviations are significant and often result from natural heating, which can "reset" the fission-track clock by removing the tracks that existed at the time of the rise in temperature. This effect makes possible the dating of metamorphic periods and the times of lava intrusions and of orogenic uplift. Because track dating is generally more sensitive to heat than other age methods, it often allows chronology that would otherwise be unobtainable to be measured.

The irradiation in space encountered by extraterrestrial objects such as meteorites, lunar-surface material, and (possibly) tektites leaves a much more complicated track record (Fleischer et al. 1967a). Identified contributions from objects that have been in space include tracks of heavy cosmic rays, fission and spallation events induced by light cosmic-ray nuclei, low-energy solar-flare nuclei, and another variety of spontaneous fission—from ^{244}Pu. Spontaneous fission from ^{244}Pu is not inherently associated with space exposure but occurs because some of the extraterrestrial matter has been kept cool enough to retain tracks since very close to the beginning of the solar system, when natural ^{244}Pu was present. The tracks that record its

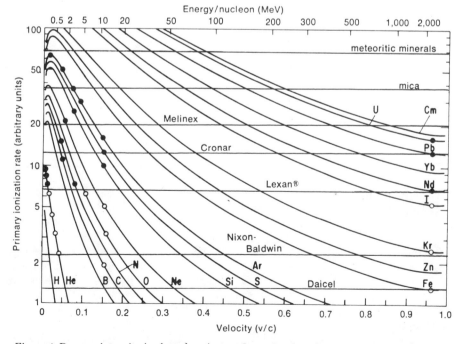

Figure 4. Damage intensity is plotted against velocity for various charged particles. Each detector has an ionization level below which no tracks are etched and above which all particles create tracks. The experimental points are for Lexan® polycarbonate; open circles indicate zero registration; closed circles unit registration. Thresholds are given for several other detectors.

former presence are evidence for stellar nucleosynthetic processes that occurred $\sim 4.6 \times 10^9$ years ago. Table 1 summarizes the new information that can be derived from tracks in objects from space.

How particles are identified

The striking shapes of the tracks shown in Figure 5 contain information that can be used to identify the particles. The simple picture of track etching in Figure 3 assumed that the attack rate v_T along a track is constant, an assumption that is quantitatively wrong and usually is a good approximation only if a short length of track is considered. In fact, the etching rate increases with the ionization rate of the particle, and that rate in turn is a function of the velocity and nuclear charge. As an energetic particle slows, its ionization rate and therefore v_T increase, and consequently θ—the degree of taper—decreases in a manner characteristic of the particle. The curvature of the sides of a track is a quantitative measure of how v_T varies along the path of the particle and allows v_T to be determined.

A simple model for describing how different track shapes arise pictures the track profile as the bow wave from a boat. If the velocity of the boat is constant, the wave makes a simple **V**, whereas a boat that is slowing down produces a convex outward shape (as in the upper portion of the argon track in Fig. 5), and an accelerating boat gives a concave outward shape (lower part of the argon track, which was etched in the opposite direction from the lower surface).

The ability to identify tracks in solids has had its major impact in studying cosmic-ray nuclei and in recognizing and measuring the presence and the abundance of exceptionally heavy nuclei. For nearly twenty years following the nuclear emulsion studies of Frier and colleagues in 1948, iron nuclei were the heaviest known cosmic-ray particles. This "iron curtain" in the search for rarer, heavier particles was broken by studies of fossil tracks in meteorites (Fleischer et al. 1967b), in which the discovery of long tracks amidst the many short tracks of iron-group nuclei allowed the abundance of trans-iron nuclei to be measured. In 1967 Fowler and co-

workers used nuclear emulsions to identify the still-heavier nuclei uranium and thorium.

Since that time most data have been collected in stacks of plastic track detectors, which have greater accuracy in identifying heavy particles, are less sensitive to low pressures and elevated temperatures, and are much less expensive—a consideration that looms large when many square meters of detectors are to be exposed. A recent summary of the results makes evident the significant presence of

elements heavier than lead (Israel et al. 1975). Although new ideas on how elements are synthesized may well develop, the only astrophysical settings where such nuclei are expected to be produced and accelerated are supernova explosions. And because heavy nuclei would be fragmented into lighter particles if they were forced to traverse through more than a few grams per square centimeter of interstellar hydrogen, we infer that these present-day nuclei were accelerated relatively recently, i.e. within the last few million years.

Table 1. Information derived from tracks in extraterrestrial materials

Information	Type of tracks used	Basis for information
Lunar samples		
Surface-exposure ages (rocks, soil grains, soil layers)	Solar flare iron tracks	Known flux of particles stops in surface layers
Near-surface exposure ages of rocks	Recoil nuclei from proton-induced spallation	Known flux of cosmic-ray protons penetrates ~ 1 m into rock
Erosion rates	Cosmic-ray iron nuclei	Track numbers vs. depth in rock are changed by erosion
Soil-accretion rates	Cosmic-ray iron nuclei	Surface-exposure ages of sequence of soil layers are added
Crater ages	Cosmic-ray iron nuclei	Near-surface exposure ages of rocks ejected from crater are measured
Shock-event ages	Cosmic-ray iron nuclei	Tracks in samples where earlier tracks were removed by shock deformation are counted
Ancient cosmic-ray flux	Cosmic-ray iron nuclei	Tracks in soil sample covered at a known accretion rate are counted
Micrometeorite flux	Solar flare iron tracks	Tracks and "zap" pits on clean surfaces are counted
Cosmic-ray abundances	Cosmic-ray tracks	Longer tracks are caused by heavier nuclei
Early chronology	^{244}Pu fission tracks	Tracks formed near the beginning of the solar system by now-extinct ^{244}Pu
Meteorites		
Early chronology	^{244}Pu fission tracks	Tracks formed near the beginning of the solar system by now-extinct ^{244}Pu
Early ^{244}Pu abundance	Fission tracks	Ratio of ^{244}Pu to ^{238}U fission tracks
Cosmic-ray abundances	Cosmic-ray tracks	Longer tracks are caused by heavier nuclei
Pre-atmospheric sizes	Cosmic-ray iron tracks	Track abundance at a given depth is a function of exposure time
Erosion rates in space	Cosmic-ray iron tracks	A limit is derived from the pre-atmospheric size plus the exposure time
Grain history prior to meteorite formation	Solar flare iron tracks	As for surface-exposure ages: abundant tracks are found in surfaces of grains
Tektites*		
Formation age	^{238}U fission tracks	Fission track dates center of tektite
Age since fall on earth	^{238}U fission tracks	Date rim that melted during fall through the earth's atmosphere
Time in space (upper limits)	Induced fission tracks	Absence of these distinctive tracks sets a limit

*There is no agreement as to whether tektites are extraterrestrial.

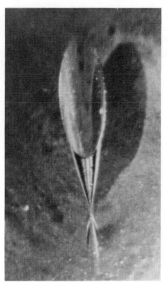

Figure 5. Four heavy cosmic-ray ions penetrated the Apollo 14 command module and entered a piece of track detector, leaving the etched tracks shown in the electron micrographs above. The nuclei can be identified by their shapes. *Left to right:* an argon ion, iron, titanium, and calcium. Tracks are approximately 1 mm in length. (From Fleischer et al. 1975.)

Some examples of recent advances

In recent years two significant advances—one toward increasing the sensitivity of detection, the other toward improving the precision and reproducibility—have extended the uses of track etching. Lück (1974) has described the high sensitivity of a new "highly esterified," additive-free, high-molecular-weight cellulose nitrate, which has a potential resolution of ~5%. He has directly observed protons of energies up to 1 MeV and alpha particles up to 8.5 MeV. His energy resolution for protons is 50 kilovolts. By fitting his results to a primary ionization model he infers that protons of 1.2 MeV and alphas of 22 MeV would be detectible with $v_T/v_G = 1.2$ (a cone angle of 56°). Comparison with the thresholds for different materials given in Figure 4 shows that this material is considerably more sensitive than the substance previously believed to be the most sensitive—Daicel cellulose nitrate. The new registration level lies just below the bottom of the graph and permits iron nuclei to register even at relativistic velocities. This property is a great aid in studying heavy cosmic-ray nuclei, in which the iron group can be used as an internal calibration, since the flux of these particles is well measured.

A slightly less sensitive detector but one of greater resolving power, allyl diglycol carbonate, has been described by Cartwright and co-workers (1978). This material is thought to be of greater submicroscopic homogeneity because it is a thermosetting plastic. Possible resolution is estimated to be ~0.5%, provided that problems of macroscopic inhomogeneity across the cast cross section can be solved. The potential resolution is to be compared to ~5% for the cellulose nitrate. In previous cosmic-ray work with track detectors there

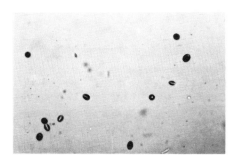

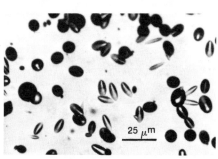

Figure 6. The ages of minerals can be measured from natural tracks caused by the spontaneous fission of trace amounts of ^{238}U in materials such as obsidian (*top micrograph*). The uranium content is measured by inducing further fission by a neutron irradiation, as shown in the bottom photograph. Magnification is 50% greater in the bottom picture.

has been just a hint that isotopic resolution is possible for the iron group, a mass resolution of about one mass unit (±2%) (Bartholoma et al. 1975; Henke and Benton 1975). The new material should allow more decisive recognition of adjacent iron isotopes.

Microchemical analysis for elements that were previously difficult to distinguish by track etching is possible because of the ready identification of protons. For example, two elements of geochemical importance, boron and lithium, undergo (n,α) reactions with thermal neutrons, the boron giving short-range ^{4}He and ^{7}Li particles and the lithium a short-range ^{4}He and a longer-range ^{3}H. When a detector is pressed against the substance being analyzed and the assembly is exposed to neutrons, the tracks are distinctive, as pictured in Figure 8. Hence the proton component, which represents the lithium content, can be recognized. Alternatively, an absorber can be used to stop the short-range alphas and ^{7}Li nuclei, so that the detector reveals only the more penetrating ^{3}H. Because of background tracks from the ^{14}N(n,p)^{14}C reaction, it is important to exclude air from the space adjacent to the detector.

Other elements that can now be mapped by track techniques are deuterium, lead, and bismuth. The last two can be distinguished if they are bombarded with 30 MeV alpha particles to produce, respectively,

[210]Po and [211]At, at each of which is an alpha-emitter. Because the half-life of the [211]At is considerably less than that of the [210]Po, two exposures of track detectors to the sample after irradiation allow the two parent elements to be measured and mapped (Woolum et al. 1976).

Deuterium is of special interest because it can be used as a tracer in biological studies. The first technique used was an irradiation with photons produced by bombarding a target with 3 MeV electrons (Ettinger and Malik 1975). The bremsstrahlung photons that exceed 2.23 MeV produce neutrons ($^2H(\gamma,n)^1H$): these in turn are detected by a plastic track detector with a ^{235}U fission foil using the reaction ^{235}U (n, fission). A second technique is to utilize a resonance in the fusion reaction $^2H(^3H,n)^4He$: the deuterium (^2H) is bombarded with 160 KeV tritium (^3H), and the emitted alpha particles are recorded (Geisler et al. 1974). Because of the low energy of the inducing particles, the background of other reactions is negligible. The authors use this process to detect individual labeled cells.

Because the first tracks observed in solids were from fission fragments, to many people the nuclear track method has often been known as the fission track method. It can now also be known as the fusion track method. The deuterium-tritium reaction used by Geisler and co-workers (1974) has been studied in experiments on laser-induced fusion (Ceglio and Coleman 1977). By means of a novel holographic technique, the image of the tracks from the ^4He reaction products viewed through a Fresnel-zone plate was used to show that the reactions occurred in a region of diameter 7 μm, i.e. about one-sixth the diameter of the target. The temperature of fusion reactions has been measured in other experiments, by Bogomolov and colleagues (1977), using the other fusion product, the neutrons, to induce fission in separate elements that have different energy responses, and recording the events in glass track detectors.

Radiobiological effects of ^{239}Pu have become a subject of scientific and public interest because increasing amounts of the alpha-active nuclide ^{239}Pu are being produced by nuclear reactors and are used in both reactor and weapons operations. Thus it is

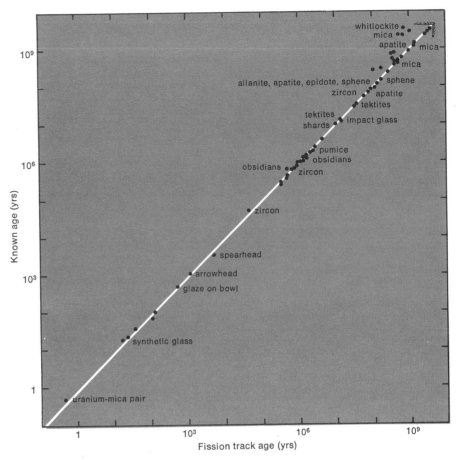

Figure 7. Fission-track dating has been shown to give correct ages over time spans ranging from one-half year to a few billion years. The graph compares fission-track ages, measured by counting natural tracks formed by spontaneous fission of ^{238}U, with ages known by other means, either documented ages for man-made samples, which appear in the lower left, or ages measured by other radioactive-decay techniques for geological samples, indicated at the upper right. Occasionally, fission-track ages of natural samples are lowered by thermal effects, allowing the fission-track technique to serve as a thermometer for past events. Cross-hatching at upper right indicates the age of the earth, the moon, and meteorites. (After Fleischer et al. 1975.)

not surprising that mapping the locations where plutonium concentrates in living matter has become a significant area of track studies.

The most direct method of mapping—autoradiographic recording of the spontaneous alpha decays (Polig, 1975, in tissue and Levy et al., 1978, in coral from the Bikini nuclear test site)—is much less sensitive and considerably slower than the procedure of inducing fission and recording plutonium fission tracks (Bleaney 1969). Tissue sections are mounted against plastic detectors, and an irradiation gives graphic results, such as those shown in Figure 9. Rabbits (Bleaney 1969), beagles (Jee 1972), rats (Storr et al. 1975), and one human subject (Schlenker et al. 1976) have been studied by this procedure. Papers have described improved techniques for image quality (Fellows et al. 1975), sample stability and ease

of readout (Storr et al. 1975), and image resolution (Becker and Beach 1975), which is normally ~10^{-3} cm.

Recoils, sputtering, and isotopic disequilibrium

Atomic collisions within but near the surfaces of solids eject atoms in a process that is usually called sputtering. If the solid is composed of an element that can be induced to emit charged particles, the ejected atoms can be caught and later irradiated and their abundance and distribution measured by the track technique.

The collisions that give rise to the sputtering process are usually caused by irradiations from outside the solid but can also be caused by particles released inside. The first study of sputtering, by Verghese and Piascik (1969), in fact examined self-sput-

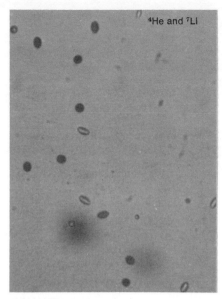
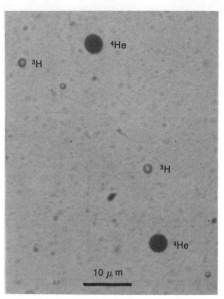
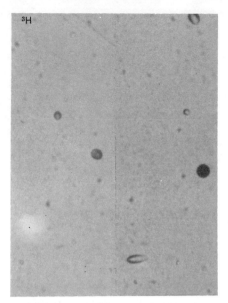

Figure 8. Neutron-induced alpha-particle reactions can be used to measure lithium and boron contents of materials. Because the $^{10}B(n,\alpha)^7Li$ products ^4He and ^7Li are short-ranged, they can be stopped by a thin absorber so that only the longer-range ^3H is seen from the $^6Li(n,\alpha)^3H$ reaction. *Left:* ^4He and ^7Li tracks from boron; *center:* ^4He and ^3H tracks from $^6Li (n,\alpha)^3H$ reaction are clearly different in size; *right:* when 1.4 mg/cm² of polycarbonate foil is interposed between the lithium and the detector, the ^4He particles are stopped, and only ^3H particles are seen—enlarged relative to the view in the center since they have been slowed by the foil.

tering induced in ^{235}U-foil by the fission fragments released during irradiation with thermal neutrons. Ejected ^{235}U atoms were caught on an aluminum substrate, which later was placed next to a glass detector and reirradiated with neutrons to induce fission of the sputtered atoms. According to their paper, it appears that a number of uranium atoms were ejected by each fission fragment that exited the uranium foil, and an occasional fragment, perhaps one of each 10^4, ejected clusters with $>10^6$ atoms. These were detected by means of sunburst patterns of tracks similar to that shown in Figure 1. More recently Biersack and colleagues (1974) have described similar but more complete experiments in which ~200 atoms were ejected per fission fragment that exited, and large chunks were produced only at high neutron doses and were associated with crater formation in the sputtered substrate.

Gregg and Tombrello (1978) and Weller and Tombrello (1978) have given other quantitative data in related experiments. Instead of internal bombardment from fission fragments they used various ions that were accelerated to energies ~100 KeV. They found a systematic increase in the number of uranium atoms ejected per incident projectile ion, from 2×10^{-4} for hydrogen to 2 for argon. For each of the ions tested, chunks of several million atoms were released, but only ~10^{-9} to 10^{-10} of the projectiles produced such chunks. Sputtering can be of importance in exotic places, as Tombrello (1977) emphasized in discussing the erosion of interstellar dust particles and of lunar soil grains exposed to solar wind ions.

Internal radioactivity appears to be intimately associated with the solution properties of plutonium dioxide, a substance of importance and concern in relation to nuclear reactors, energy production, and health and safety. A striking isotopic effect seen by Raabe and co-workers (1973) was that $^{238}PuO_2$ "dissolved" at 200 times the rate for $^{239}PuO_2$—an effect that could not be chemical but was inferred to be caused by radiation damage induced by the ^{235}U recoils that result from alpha decay (Fleischer 1975). Fleischer and Raabe (1977) have shown, however, that the effect is not true solution since the plutonium is present in the liquid as minute subparticles containing up to 10^4 atoms. This effect is not caused by sputtering, as was demonstrated by catching the ejected ^{235}U and ^{239}Pu atoms on quartz crystals and then inducing fission with neutrons (Fleischer and Raabe 1978a). Since the upper limit on the number of sputtered ^{239}Pu atoms turned out to be orders of magnitude less than the observed "dissolution" in water, it was suggested that tracks produced by alpha recoil were being attacked by the water, releasing plutonium into the liquid.

The preferential dissolution of material damaged by alpha recoils has been suggested to explain two significant geochemical phenomena—the copious natural release of the gas radon from rocks and soil and the preferential release of ^{234}U relative to ^{238}U (Fleischer and Raabe 1978b). If, for example, a recoiling ^{222}Rn atom produced by alpha decay of ^{226}Ra moves from one grain into another, it lies at the end of a damage track that crosses the surface, and therefore it can potentially be released by etching. If, instead, the recoiling atom had been a ^{234}U atom from the decay of ^{238}U, isotopic separation would have resulted—with the solid depleted preferentially in ^{234}U and the liquid enriched, a common observation in nature (Thurber 1962).

The atmosphere, the oceans, the earth

Mapping by particle tracks is useful in the air to identify radionuclides in atmospheric particles, at the earth's surface to measure the geochemically interesting radionuclides radon, thorium, and uranium, and beneath the oceans to measure the accumulation of sediments at the ocean floors.

Aerosols that contain alpha-emitting

nuclides are potential health hazards if inhaled, and if the alpha activity can be identified, the source of the input to the atmosphere can be recognized and monitored. Aerosols may be readily identified as containing transuranics if they are placed against a track detector and fission is induced by particle irradiation. Identification of the specific nuclide involved requires further information, which, as Center and Ruddy (1976) noted, can be provided by measuring the alpha-decay rate for the same aerosol particle.

For example, the ratio of the number of spontaneous tracks formed in a plastic detector of alpha tracks to the number of fission tracks induced by thermal neutrons allows the ratio λ/σ of the decay constant to the fission cross section to be measured. Although occasional ambiguities will arise, for most single nuclides widely different values of λ/σ are obtained—for example $1.5 \times 10^{12}/cm^2$-s for ^{238}Pu, 1.2×10^8 for ^{239}Pu, and 1.2×10^6 for natural uranium.

Independently, Fisher (1977) recognized that the same principle applied to measuring thorium and uranium contents of rocks, these being two elements of widespread geochemical interest. Because the detector records all the alpha-active nuclides in the respective decay series, the results are directly valid only if both decay chains are in equilibrium. But if they are not, other opportunities arise.

For example, in the deep ocean, sedimentation rates can be measured by means of ^{230}Th that is formed from ^{238}U dissolved in the seawater. Because of the low solubility of thorium, it precipitates and thus is present in the sediments in excess. As it is progressively covered by sediment, the ^{230}Th gradually decays, with a half-life of 77,000 years. A profile of ^{230}Th with depth, such as the one shown in Figure 10, therefore gives a measure of sedimentation rate. The background that limits such measurements, given by the uranium in the sediments themselves, is measured by inducing fission tracks with neutrons to allow point-by-point determination of uranium. Anderson and Macdougall (1977) have done similar measurements and have extended the technique to manganese nodules, which require readings over much smaller distances, since typical

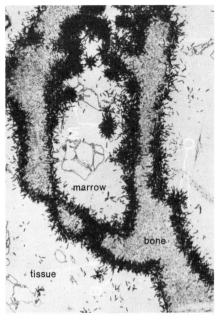

Figure 9. This fission radiograph shows the location of plutonium in the lumbar vertebra of a dog. The dog was given 2.7 μCi of $^{239}Pu/kg$ and killed 28 days later. A Lexan® polycarbonate detector was pressed against a section from the lumbar vertebra, neutron irradiated, and then etched to produce this image, which shows the plutonium concentrated on bone surfaces and in star-producing aggregates within the marrow. Typical tracks are ~10 μm in length. Many minute alpha tracks from boron produce the dark image of the bone. (Courtesy of W. S. S. Jee.)

growth rates are smaller by a factor of 1,000 than the usual sedimentation rates in the deep ocean.

For measuring more rapid sedimentation rates, ^{230}Th would not show significant changes over reasonable lengths of sediment core, and therefore shorter-lived nuclides such as ^{226}Ra (1,620 years) and ^{210}Pb (22 years) are of interest. Jensen and coworkers (1977) have reported that preliminary studies of ^{210}Pb in the North Sea are consistent with a rate of 2 mm/yr, a factor of ~1,000 times more rapid than typical deep-ocean results.

Radon in the earth is a potential tracer for locating subsurface uranium ore (Fleischer et al. 1972; Gingrich and Fisher 1976). Ore has been discovered by drilling where the highest near-surface signal is found in arrays of shallow test holes. The measurement is most commonly made by placing plastic track detectors sensitive to alpha particles into the air in the holes and integrating, over a month's time, the signal of alphas emitted by ^{222}Rn and its prompt

daughter alpha-emitters ^{218}Po and ^{214}Po. The integration of the signal has the virtue of minimizing the unpredictable transient effects of atmospheric pressure, temperature, and precipitation; other noise—signals from ^{220}Rn (a decay product of thorium) and effects of moisture (Fleischer and Mogro-Campero 1978)—is eliminated by use of a semipermeable membrane and a drying agent.

How do signals propagate over distances as large as 100 m? Since the short 5.5-day mean life of radon implies that the average atom could move at most 1.5 m by diffusion before it decays, some transport mode must be active. One possibility has been suggested—convective flow of fluids in the earth (Mogro-Campero and Fleischer 1977)—but firm evidence has yet to be acquired for present-day, long-distance transport of radon in the earth. Near-surface changes in radon concentrations are being sought as precursors of earthquakes, but again there are no clearly definitive results.

In some unusual environments fossil tracks have yielded interesting chronological information. Most natural tracks in terrestrial samples are solely from spontaneous fission, but the site of the natural nuclear reactor at the Oklo uranium mine in Gabon is an exception. The surprise finding that the natural ore there was depleted in ^{235}U below the usual present-day value of 0.72% implied that many neutrons had been present and in fact that the natural deposit had acted as a water-moderated nuclear reactor at or near the time the ore was formed 1.8×10^9 years ago. High-quality ore with natural water to slow down the neutrons from fission can sustain a chain reaction if the $^{235}U/^{238}U$ abundance exceeds 3%, as it did at that time.

Small quartz grains in and around the Oklo reactor zone have monitored the generation of neutrons by the reactor (Dran et al. 1977). When the quartz is not too close to the "core," the tracks around minute inclusions result only from spontaneous fission and give the age of the bedrock beneath the uranium-rich zone as 2.6×10^9 years. Closer to the core, there is an excess of fission tracks, induced by the neutrons from the reactor. This excess gives, for example, a total neutron dose of $10^{17}/cm^2$ at a position of about 20 to 30 cm from the core, a value well below that which must have obtained within the ore. This technique, when extended to measure the gradients in and near the core, will aid in deciphering the operation of this fossil reactor.

Irradiation effects on interplanetary matter store a wealth of information about the history of small solid grains—both in free space and partially exposed to space on unshielded planetary bodies such as the moon and meteoroids (see reviews by Walker, 1975, and Maurette and Price, 1975). The high flux of low-energy ions that make up the solar wind can in a short time ($\sim$2,000 yr) convert the outer $\sim$1000 Å of crystalline grains into amorphous material. The flux of the more penetrating, somewhat higher energy ions is much diminished, yet micron-sized grains are sometimes found to be amorphous also, thus giving evidence of much longer space exposures. When this destruction of the crystal structure is seen in some individual grains of a meteorite but not in all, it tells us that the irradiation preceded the assemblage of the grains into a solid body. Since these meteorites were formed near the beginning of the solar system 4.6×10^9 years ago, the tracks were formed then, and their numbers measure the time of exposure to space during that distant period. Thus Macdougall (1975) determined that some meteoritic grains had been exposed for 10 to 100 years, 4.6×10^9 years ago.

Technological uses

Many technological uses of tracks have been developed, and new companies have been formed solely to utilize track techniques for making filter membranes of highly uniform hole size (Nuclepore Corporation) and for locating high-radon sites at which to drill for uranium (Terradex Corporation). Nuclepore Corporation provides filters made by irradiating thin plastic sheets and then etching holes to the desired size. Such holes have been put to uses that range from measuring the size distribution of aerosols, by forcing air through a graded set of filters (Twomey 1976), to study of conduction and electro-osmosis in microcapillaries that can be made small relative to the screening distance in electrolytes (Koh and

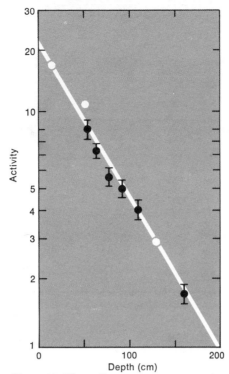

Figure 10. The sedimentation rate near the East Pacific Rise has been shown to be about 6 mm/1,000 years by measuring ^{230}Th, which forms from ^{238}U dissolved in the seawater. Alpha-active ^{230}Th precipitated from the seawater is recorded on plastic detectors as a function of depth in the core. The white dots are comparison determinations made by laborious radiochemical techniques. (After Fisher 1977.)

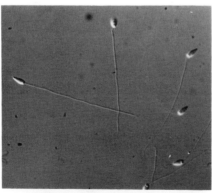

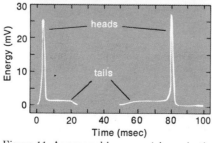

Figure 11. As sea urchin sperm (*shown in the electron micrograph*) move through a hole in a plastic sheet, the resistive pulse tells the size and shape of the sperm. The two pulses represent two sperms—one entering head first, the other tail first. (Electron micrograph courtesy of C. P. Bean; resistive pulse courtesy of R. W. DeBlois.)

Anderson 1975), to checking theories of thermal conductance by providing known scatterers of phonons (Zaitlin and Anderson 1975), to creating quantized vortex rings in a superfluid by forcing it through 5 μm holes at the critical velocity (Gamota 1973), to measuring small objects with high precision by monitoring the electrical resistance of a hole as the object passes through it (DeBlois et al. 1977).

As only one example, the last procedure, the resistive pulse technique, or "DeBlois-Bean counter," named after its inventors, is able to deal with odd shapes. Figure 11 shows sea urchin sperm and the resistive trace as two sperms pass through the hole—one head first, the other tail first. The large part of the pulse is from the head passing through the hole and the longer "tail" on the pulse is from the time when only the tail of the sperm is in the hole. From these measurements DeBlois (1978) could assert that the somewhat variable volume of the head is 5.5($\pm$.5) μm^3 and that the tail has a highly uniform diameter of 0.26($\pm$.01) μm.

Even in a seemingly unrelated field such as bird altimetry, nuclear tracks have provided a valuable tool for research. Kristiansson and colleagues (1977) have built a simple integrating barometer and used it to measure the distributions of flight altitudes of birds (Gustafson et al. 1977). Assembled in a housing, the 1 gm, 5 cm-long system is attached to the back of the bird with biodegradable paste; for more "acrobatic" birds, such as the swifts, a smaller (0.4 g, 2 cm-long) version is used. A source of alpha particles and a plastic alpha detector are positioned so that the alphas reach the nearest portion of the detector when the bird is at ground level. At higher altitudes the lower atmospheric density allows the particles to make tracks farther along the detector, and thus the distribution of the distance of tracks along the detector gives the distribution of times spent at different altitudes. It was found that in homing flights over long distances these birds flew at heights of up to 3,600 m in clear weather and at lower altitudes during cloudy periods. Individuals tended to resume the same altitude on flights under similar weather conditions, but different individuals chose different altitudes.

An occasional, but happily atypical, response to learning about an unexpected use of track etching such as bird altimetry is the question (really a comment) "So what?" The answer lies at the heart of why science is done—first, to gain knowledge for its own sake and, second, to learn on occasion how to do something of economic use. The case of track etching is an unusual one in physics not primarily because of the direct contribution in that field—in understanding a solid state phenomenon—but because of the tools it has provided in other branches of physics—nuclear, cosmic-ray, and elementary particle physics—and in other scientific fields that range from geology and geochemistry to virology and ornithology, as well as in technological areas that include nuclear engineering, filtration, and uranium exploration.

References

Andersen, M. E., and J. D. Macdougall. 1977. Accumulation rates of manganese nodules and sediments: An alpha track method. *Geophys. Res. Lett.* 4:351–53.

Bartholoma, K.-P., G. Siegmon, and W. Enge. 1975. Isotopic composition of the cosmic ray iron nuclei. *Proc. 14th Intl. Conf. on Cosmic Rays,* 14:384–88.

Becker, K., and J. L. Beach. 1975. Improvement in spatial resolution of track-etching radiography. *Nucl. Instr. Meth.* 130:499–506.

Biersack, J. P., D. Fink, and P. Mertens. 1974. Sputtering and chunk ejection from UO_2 and metallic layers deposited on UO_2. *J. Nucl. Materials* 53:194–200.

Bleaney, B. 1969. The radiation dose-rates near bone surfaces in rabbits after intravenous or intramuscular injection of plutonium 239. *Brit. J. Radiol.* 42:51–56.

Bogomolov, A. M., A. D. Molodtsov, and L. Ya. Tikhonov. 1977. Measuring the temperature of the neutron gas with solid-state detectors of fission fragments. *Atomnaya Energiya* 42:228–29.

Cartwright, B. G., E. K. Shirk, and P. B. Price. 1978. CR-39: A nuclear-track-recording polymer of unique sensitivity and resolution. *Nucl. Instr. Meth.* 153:457–60.

Ceglio, N. M., and L. W. Coleman. 1977. Spatially resolved α-emission from laser fusion targets. *Phys. Rev. Lett.* 39:20–24.

Center, B., and F. H. Ruddy. 1976. Detection and characterization of aerosols containing transuranic elements with the nuclear track technique. *Anal. Chem.* 48:2135–39.

DeBlois, R. W. 1978. Analysis of the morphology of sperm cells using a resistive-pulse technique. *Biophys. J.* 24:149a.

DeBlois, R. W., C. P. Bean, and R. K. A. Wesley. 1977. Electrokenetic measurements with submicron particles and pores by the resistive pulse technique. *J. Colloid Interface Sci.* 61:323–35.

Dran, J. C., J. P. Duraud, Y. Langevin, and J. C. Petit. 1977. Fission track dating of quartz grains from the Oklo uranium ore deposit. *Nucl. Inst. Meth.* 147:101–03.

Ettinger, K. V., and S. R. Malik. 1975. Solid state track detectors for the detection of deuterium in water. *Health Phys.* 28:75.

Fellows, M. H., L. Clark, Jr., J. J. O'Toole, D. B. Kimmel, and W. S. S. Jee. 1975. An improved technique for neutron-induced autoradiography of bone containing plutonium. *Health Phys.* 29:97–101.

Fisher, D. E. 1977. f/α particle track analysis: A new geological technique for the measurement of uranium, thorium, and istopic disequilibria. *J. Radioanalytical Chem.* 38:477–90.

Fleischer, R. L. 1975. On the "dissolution" of respirable PuO_2 particles. *Health Phys.* 29:69–73.

———. 1977. The past and future roles of solid state nuclear track detectors. *Nucl. Inst. Meth.* 147:1–10.

Fleischer, R. L., H. W. Alter, S. C. Furman, P. B. Price, and R. M. Walker. 1972. Particle track etching. *Science* 178:255–63.

Fleischer, R. L., H. R. Hart, Jr., G. M. Comstock, M. Carter, A. Renshaw, and A Hardy. 1973. Apollo 14 and 16 heavy particle dosimetry experiments. *Science* 181:436–38.

Fleischer, R. L., and A. Mogro-Campero. 1978. Mapping of integrated radon emanation for detection of long-distance migration of gases within the earth: Techniques and principles. *J. Geophys. Res.* 83:3539–49.

Fleischer, R. L., and P. B. Price. 1964a. Glass dating by fission fragment tracks. *J. Geophys. Res.* 69:331–39.

———. 1964b. Techniques for geological dating of minerals by chemical etching of fission fragment tracks. *Geochim. Cosmochim. Acta* 28:1705–14.

Fleischer, R. L., P. B. Price, and R. M. Walker. 1975. *Nuclear Tracks in Solids.* Univ. of Calif. Press, Berkeley.

Fleischer, R. L., P. B. Price, R. M. Walker, and M. Maurette. 1967a. Origins of fossil charged particle tracks in meteorites. *J. Geophys. Res.* 72:333–53.

Fleischer, R. L., P. B. Price, R. M. Walker, M. Maurette, and G. Morgan. 1967b. Tracks of heavy primary cosmic rays in meteorites. *J. Geophys. Res.* 72:355–66.

Fleischer, R. L., and O. G. Raabe. 1977. Fragmentation of respirable PuO_2 particles in water by alpha decay—a mode of "dissolution." *Health Phys.* 32:253–57.

———. 1978a. On the mechanism of "dissolution" in liquids of PuO_2 by alpha decay. *Health Phys.* 35:545–49.

———. 1978b. Recoiling alpha emitting nuclei—mechanisms for uranium-series disequilibrium. *Geochim. Cosmochim. Acta* 42:973–8.

Fowler, P. H., R. A. Adams, V. G. Cowen, and J. M. Kidd. 1967. The charge spectrum of very heavy cosmic ray nuclei. *Proc. Roy. Soc. A* 301:39–45.

Frier, P. S., E. J. Lofgren, E. P. Ney, F. Oppenheimer, H. L. Bradt, and B. Peters. 1948. Evidence for heavy nuclei in the primary cosmic radiation. *Phys. Rev.* 74:213–17.

Gamota, G. 1973. Creation of quantized vortex rings in superfluid helium. *Phys. Rev. Lett.* 31:517–20.

Geisler, F. H., K. W. Jones, J. S. Fowler, H. W. Kraner, A. P. Wolf, E. P. Cronkite, and D. N. Slatkin. 1974. Deuterium micromapping of biological samples using the $D(T,n)^4He$ reaction and plastic track detectors. *Science* 186:361–63.

Gingrich, J. E., and J. C. Fisher. 1976. Uranium exploration using the track-etch method. *IAEA Intl. Symp. on Exploration of Uranium Ore Deposits,* pp. 213–25.

Gregg, R., and T. A. Tombrello. 1978. Sputtering of uranium. *Rad. Effects* 35:243–54.

Gustafson, T., B. Lindkvist, L. Gotborn, and R. Gyllin. 1977. Altitudes and flight times for swifts *Apus apus* L. *Ornis Scand.* 8:87–95.

Henke, R. P., and E. V. Benton. 1975. Isotopic resolution of the iron peak. *Proc. 14th Intl. Conf. on Cosmic Rays* 14:395–99.

Israel, M., P. B. Price, and C. J. Waddington. 1975. Ultraheavy cosmic rays. *Phys. Today* 28:23–31.

Jee, W. S. S. 1972. ^{239}Pu in bones as visualized by photographic and neutron-induced autogradiography. In *Radiobiology of Plutonium,* ed. B. J. Stover and W. S. S. Jee. Salt Lake City: J. W. Press, pp. 171–93.

Jensen, J. M., W. Enge, H. Erlenkeuser, and H. Willkomm. 1977. Age determination of sediments by ^{210}Pb using a plastic detector technique. *Nucl. Inst. Meth.* 147:97–99.

Koh, W.-H., and J. L. Anderson. 1975. Electroösmosis and electrolyte conductance in charged microcapillaries. *AIChE J.* 21:1176–88.

Kristiansson, K., B. Lindkvist, and T. Gustafson. 1977. An altimeter for birds and its use. *Ornis Scand.* 8:79–86.

Levy, Y., D. S. Miller, G. M. Friedman, and V. E. Noshkin. 1978. Analysis of alpha emitters in the coral, *Favites virens,* from Bikini Lagoon by solid-state track detection. *Health Phys.* 34:209–17.

Lück, H. B. 1974. A plastic track detector with high sensitivity. *Nucl. Instr. Meth.* 114:139–40.

Macdougall, J. D. 1975. Precompaction irradiation of mafic silicate in C1 and C2 chondrites. *Meteoritics* 10:449–50.

Maurette, M., P. Pellas, and R. M. Walker. 1964. Etude des traces de fission fossiles dans le mica. *Bull. Soc. Franc. Min. Cryst.* 87:6–17.

Maurette, M., and P. B. Price. 1975. Electron microscopy of irradiation effects in space. *Science* 187:121–29.

Mogro-Campero, A., and R. L. Fleischer. 1977. Subterrestrial fluid convection: A hypothesis for long-distance migration of radon within the earth. *Earth Planet. Sci. Lett.* 34:321–25.

Polig, E. 1975. α-Microdosimetry in bone sections by means of dielectric track detectors and electronic image analysis. *J. Appl. Rad. Isotopes* 26:471–79.

Price, P. B., and R. W. Walker. 1962. Chemical etching of charged particle tracks. *J. Appl. Phys.* 33:3407–12.

———. 1963a. A simple method of measuring low uranium concentrations in crystals. *Appl. Phys. Lett.* 2:23–25.

———. 1963b. Fossil tracks of charged particles and the age of minerals. *J. Geophys. Res.* 68:4847–62.

Raabe, O. G., G. M. Kanapilly, and H. A. Boyd. 1973. Studies of the *in vitro* solubility of respirable particles of ^{238}Pu and ^{239}Pu oxides and on accidentally released aerosol containing ^{239}Pu. *Inhalation Toxicology Res. Inst. Ann. Rept. Lovelace Found.* Report LF-46, UC-48, pp. 24–30.

Schlenker, R. A., B. G. Oltman, and H. T. Cummins. 1976. Microscopic distribution of ^{239}Pu deposited in bone from a human injection case. In *The Health Effects of Plutonium,* Salt Lake City: J. W. Press, pp. 321–28.

Silk, E. C. H., and R. S. Barnes. 1959. Examination of fission fragment tracks with an electron microscope. *Phil. Mag.* 4:970–71.

Storr, M. C., J. G. Hollins, and R. L. Clark. 1975. An improved method of fission-fragment radiography of plutonium in rat bone. *Intl. J. Appl. Rad. Isotopes* 26:708–13.

Thurber, D. L. 1962. Anomalous $^{234}U/^{238}U$ in nature. *J. Geophys. Res.* 67:4518–20.

Tombrello, T. A. 1977. Sputtering in astrophysics and planetary science. *Cal. Inst. Tech.* Preprint OAP-501/LiAP, Sept.

Twomey, S. 1976. Aerosol size distributions by multiple filter measurements. *J. Atmos. Sci.* 33:1073–79.

Verghese, K., and R. S. Piascik. 1969. Ejection of uranium by fission fragments—evidence of clusters and knock-ons. *J. Appl. Phys.* 40:1976–77.

Walker, R. M. 1975. Interaction of energetic nuclear particles in space with the lunar surface. *Ann. Rev. Earth Planet. Sci.* 3:99–128.

Weller, R. A., and T. A. Tombrello. 1978. Energy spectrum of sputtered uranium—a new technique. *Rad. Effects* 37:83–92.

Woolum, D. S., D. S. Burnett, and L. S. August. 1976. Lead-bismuth radiography. *Nucl. Instr. Meth.* 138:655–62.

Young, D. A. 1958. Etching of radiation damage in lithium fluoride. *Nature* 182:375–77.

Zaitlin, M. P., and A. C. Anderson. 1975. Phonon thermal transport in noncrystalline materials. *Phys. Rev.* B12:4475–86.

Irving Friedman
Fred W. Trembour

Obsidian: The Dating Stone

The depth of water penetration on a surface of this stone reveals the dates of geologic and archaeologic events

In the seventeen years since obsidian hydration measurement as a dating tool first became known, the method has drawn increasing attention from a variety of disciplines. No other stone lends itself so well to the age determination of events as varied as those in, for example, late quaternary geology and anthropology. Wherever prehistoric peoples and obsidian—or volcanic glass—occurred together in nature, the people made use of it as a raw material for manufacture, and the evidence remains in the form of worked fragments in the vicinity of present-day deposits of eruptive rocks in much of the world.

From the earliest cultural use of obsidian, the Acheulian hand axes of East Africa, almost half a million years ago, to the end of the stone age in most of the New World (Fig. 1), less than half a thousand years in the past, traits of mankind's developing ways

Fred Trembour, employed by the U.S. Geological Survey in Denver in laboratory research on obsidian properties, concurrently pursues field and university studies on archaeological lithic tool materials. He earned a Metallurgical Engineering degree from Carnegie Institute of Technology in 1940 and worked in the tool steel industry and in several departments of the U.S. Government into the late 1960s. Irving Friedman received his Ph.D. from the University of Chicago, followed by two years as a research associate at the Enrico Fermi Institute for Nuclear Studies before joining the U.S. Geological Survey in 1952. At present he is a research geochemist at the U.S.G.S. and an adjunct professor in the Department of Geology, University of Pennsylvania. His primary research activity has been concerned with the application of light stable isotope studies to problems in the earth sciences. However, he has carried on intermittent research on volcanic products for the past 23 years as a subsidiary interest. Address: Dr. Friedman, U.S. Geological Survey, Box 25046, Denver Federal Center, Denver, CO 80225.

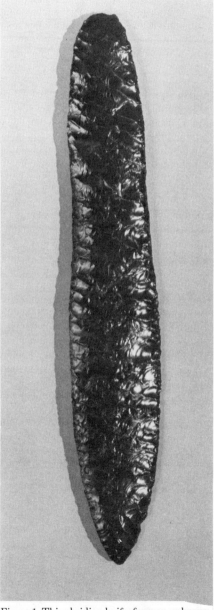

Figure 1. This obsidian knife, from an unknown source in the Western United States, is 31 cm long (12″). Elaborately chipped bifacial forms of this kind have been found in several cultural areas of the New World and may well have had a ceremonial function. Smaller obsidian knives were used by the Aztecs in conquest times for sacrificial purposes. (Photograph by the authors.)

of life have been reflected in the use of this stone. Thus technological and psychological preferences alike are recorded in the archaeological residue of obsidian manufacture—from the common chipped flakes (Fig. 2) and weapon points, the edged blades of composite sickles from the Near East, and the sacrificial knives of the Mayas, to the Maya's ground and polished ear spools and mirrors and ceremonial objects, the drilled beads and pendants of neolithic Anatolians, and the carved vessels of Pharaonic Egyptians as well as sacred objects inlayed with obsidian (Fig. 3). A remarkable example of the obsidian knapper's flaking skill is the 18-inch-long ceremonial weapon blade of the Ohio Hopewell culture of about 2,000 years ago that is pictured in Figure 4.

Although the use of other stone materials far outweighed that of obsidian in human prehistory and later, obsidian nevertheless is found scattered through the traces of all cultures that could acquire it locally or by trade. Human or other transport often carried the natural glass far from its source. Wherever it was deposited, the intrinsic chemistry of the hygroscopic material led to moisture absorption and formation of a growing superficial hydration rind whose ultimate thickness was a function of elapsed time and the temperature conditions of the immediate surroundings. Laboratory techniques for hydration depth measurement and the method of converting depth to age values were first described by Friedman and Smith in 1960.

The work of many investigators over the succeeding years has brought information from a growing number of archaeological sites as well as geologic

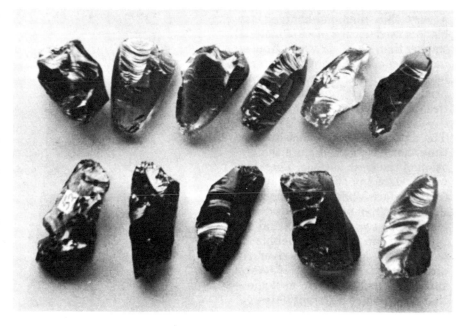

Figure 2. This group of small, bifacially worked tools of nearly clear obsidian, possibly chisels, shows percussion use at the ends. The piece in the lower right corner is 22 mm (⅞″) long. These stones come from the surface of farmland in southern Colombia. (Photograph by the authors.)

sources. Laboratory test results have been added to field findings to determine hydration rates of different obsidians and for different site conditions. Progress has been achieved in the mathematical formulation of the relationship between the chief chemical and environmental determinants of the hydration rate. In this way, the method has come closer to realizing its potential of becoming one of the few reliable direct dating means for obsidian sites that sometimes involve crucial natural phenomena or human activities of the past 300,000 years or more.

The nature of obsidian

What is this material, where is it found, and how can it be used as a dating tool? These are some of the questions that we will try to answer in this article.

Magmas of granitic composition frequently are erupted on the earth's surface as viscous fluids. The interiors of these flows or domes cool slowly and have time to crystallize as a fine-grained volcanic rock called rhyolite. The exterior cools rapidly, and the lava chills to a glass, which may be as much as 10 or 20 meters thick. It is this volcanic glass—often called obsidian—upon which people and surficial geologic processes leave their marks. Obsidian occurs in almost all lava flows of rhyolite composition, and its distribution is therefore associated with the volcanic belts that circle the earth.

Obsidian varies in appearance from clear colorless to pinkish glass through gray-green, brown, red, and black. It is usually transparent in thin pieces but may appear opaque due to the presence of microscopic crystals of various minerals that scatter the transmitted light. Its usefulness as a tool-making material derives from the ease with which it furnishes the sharp edges required for cutting or scraping tools and projectile points. Obsidian can be readily flaked by percussion and pressure techniques similar to

those used for working flint. The shiny, vitreous appearance of the shaped object as well as the mirrorlike surface of polished pieces make it attractive for decorative purposes. Naturally occurring boulders and cobbles of obsidian often have characteristic bright facets that shine in the sun and draw the attention of inquisitive eyes.

Obsidian dating

The obsidian dating method depends on the fact that while the glass, when formed, contains about 0.1 to 0.3% water by weight, it absorbs additional water from the atmosphere, which slowly diffuses into the body of the glass. The depth of penetration of the water tells us how long it has been since the surface of the piece of ob-

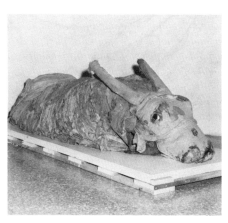

Figure 3.   This mummified sacred bull from Dahshur, Egypt, 200 B.C. to A.D. 200, has obsidian eyes. (Courtesy of Smithsonian Institution, photo #44042C.)

sidian was created or when the stone was last cracked or chipped.

To measure the depth of water penetration, a small piece of the object, removed by cuts made perpendicular to the surface in question, is mounted on a glass slide, ground down to a thickness of about 0.08 mm (0.003″), and examined under the microscope by transmitted light. The glass that has been penetrated by water (hydrated) has a higher refractive index than the pristine glass, and the interface between the hydrated and nonhydrated glass can be seen in the microscope as a thin line (Fig. 5). The uptake of water also changes the volume of the affected glass and causes mechanical strain in the hydrated rind which renders it visible as a bright band under polarized light due to an effect called strain birefringence.

The thickness of the rind may vary from less than 1 μm (0.00004″) to more than 50 μm and can be measured to a few tenths of a μm by the use of a filar or image-splitting eyepiece attached to the microscope. Measurements are usually made at × 500 to 1000 magnification. The cutting and grinding of an obsidian sample requires about ten minutes, and the measurement of hydration thickness about five minutes. The relative rapidity and simplicity of this dating technique make it possible to examine hundreds of artifacts or geologic samples in a few days, and at low cost. To minimize damage to a valuable object the test piece that is removed can be small. In such a case, a very thin diamond-charged saw blade is used to cut a piece of material smaller than a grain of wheat from an inconspicuous spot that may later be backfilled with a bit of wax of appropriate color, rendering the cut unnoticeable.

The dating technique measures the time that has elapsed since the creation of the fresh surface. Therefore we can date any event that causes a fresh surface—be it the purposeful flaking of a tool by an ancient craftsman or the casual chipping and abrading of an obsidian pebble in transport by a glacier or a river. An obsidian artifact may exhibit several surfaces—man-made or geologic—that may involve different dates.

At present, the visual measuring method using the optical microscope is the only one that is generally practiced. A nuclear reaction technique for hydration profile measurements developed in 1974 holds promise for the detection of especially thin hydration layers (Lee et al. 1974).

In order to convert the thickness measurement of a hydrated rind of obsidian into age, the rate of penetration (diffusion) of water must be known. During the development of the obsidian dating method, suitable artifacts of known age or those excavated along with material that could be dated by the method radiocarbon (^{14}C) were used to derive rates of hydration. It was found that the hydra-

Figure 4. The source of the obsidian for this 18″ long ceremonial weapon blade, of the Ohio Hopewell culture of about 2,000 years ago, was the Yellowstone region of Wyoming. (Photograph courtesy of the Ohio Historical Society.)

tion proceeded faster in tropical climates than in cooler zones. It was also found that the hydration rate was independent of the relative humidity of the environment—explained by the fact that there is always enough moisture even in the driest places to completely saturate the surface and

form a layer of water one molecule thick. Experiments have shown that the rate of penetration of the water into the body of the glass is not linear but proceeds as the square root of time (Fig. 6). Since the water molecules must travel through the already hydrated rind to reach fresh glass, the thicker the hydrated rind, the longer it takes for the water to reach the interface, and therefore the process slows down as it proceeds.

Intermittently through the years of applied hydration dating, investigators have come out with different formulas to express the depth–time relationship. Departures from the originally employed equation (depth $\approx$ time$^{1/2}$) have usually been in the direction of a faster rate of penetration (Table 1). Various possible reasons for the divergencies have been given. Kimberlin suggested that previous investigators might have used obsidian of varying compositions which hydrated at differing rates. Others have cited stratigraphic mixing of excavated obsidian artifacts as an obvious possible cause of distorted rate formulas. Friedman et al. (1968) and Meighan (1976) see the problem as lying in the rather short spans of archaeological time usually involved in actual studies, which would cause the depth-to-age conversion to appear to be linear, whereas on a much longer time scale the true curve form of the relationship would show up clearly.

An additional consideration is the modification that must be made in those plots of hydration depth/age that were drawn up before the recently available corrections to the ^{14}C dating technique. In order to take account of variations in the ^{14}C content of the atmosphere through time, the ^{14}C in accurately dated tree rings has been measured. The published data by Suess (1970) go back to about 6,500 years B.P. The amount of correction needed tends to increase with time and can be as much as 600 years for samples about 5,000 years old. In most cases, the corrected ages are older than the uncorrected ages. Another correction that will increase the ^{14}C ages is the conversion of the old half-life ^{14}C constant to the new half-life constant: the ^{14}C dates must be multiplied by 1.03.

A further likely source of data distortion in rate calculations may par-

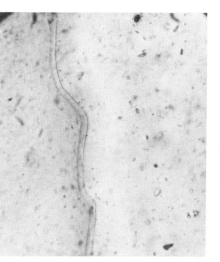

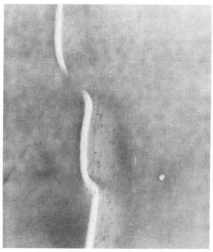

Figure 5. (left) This photomicrograph of a thin section of an obsidian artifact shows the hydrated rind, or the interface between the hydrated and the nonhydrated glass. The rind is about 6 μm thick, and the depth of water penetration, from the surface to the hydrated rind, reveals the length of time since the surface was created—by man or by nature. (right) The same hydrated rind under crossed nicols is bright because the rind is under strain.

ticularly apply to certain stratified archaeological deposits. At many site locations the mean annual soil temperature just beneath the surface is higher than that of either the air above or the deeper soil levels below (Friedman 1976). Thus chemically identical obsidian flakes in the deposit will have hydrated at different rates while the strata built up with time. Furthermore, this distortion of the thermal history can be compounded by the effect of human presence in a series of living floors that spanned long periods of habitation in a cold climate where fire was used for warmth. Shallowly buried obsidian under these floors would have been subject to accelerated hydration; a 1° C change in the effective hydration temperature produces a 10% change in the hydration rate. In general, to correct for this kind of higher temperature episode in the underground life of a buried artifact requires subtraction of some constant amount from the observed hydration depths of the pieces from all levels before the age is converted.

Another effect that can create problems in dating stratified deposits is the vertical movement of artifacts caused by disturbance of the deposit by man and animals as well as by the natural movement of large objects in the soil caused by alternate wetting and drying (Springer 1958). This effect will be especially important in deposits where the rate of accumulation with time is slow or variable, and a shift of a few centimeters in stratigraphic position may represent a thousand years or more.

Hydration rates, usually expressed in $\mu m^2/1,000$ yrs, vary from less than one to as much as twenty $\mu m^2/1,000$ yrs. As an example, if the rate is 5 $\mu m^2/1,000$ yrs, the layer will be $\sqrt{5} = 2.24$ μm thick after 1,000 years, $\sqrt{5 \times 2} = 3.16$ μm thick after 2,000 years, and $\sqrt{5 \times 3} = 3.87$ μm thick after 3,000 years. In an attempt to measure hydration rates experimentally, obsidian of different chemical compositions (rhyolite magmas can vary, for example, from 70 to 78% silica by weight) was exposed to water vapor at elevated temperatures in order to accelerate the hydration process (Friedman & Long 1976). Rates of hydration were then derived from the data as a function of the chemical composition as well as of the hydration temperature.

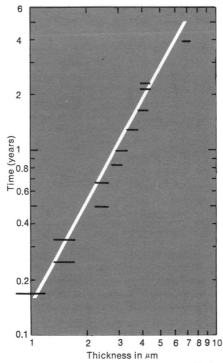

Figure 6. In this plot of the experimentally induced accelerated hydration of obsidian at 100°C, each point represents a measurement of the hydration thickness of a piece of obsidian hydrated for the time shown. (The measurement errors are given by the length of the dash passing through each point.) The slope of the line indicates that the hydration proceeds at a rate proportional to the square root of time.

Obsidian is a glass of complex chemical composition, containing about 12 major components that include silicon (Si), sodium (Na), potassium (K), calcium (Ca), magnesium (Mg), aluminum (Al), hydrogen (H), oxygen (O), phosphorus (P), titanium (Ti), iron (Fe), and manganese (Mn) in addition to many other elements present in minor amounts. It has been found that the hydration rate is controlled mainly by the amounts of Si, Ca, Mg, and H_2O; curves that estimate the hydration rate as a function of the temperature as well as the chemical composition expressed as $(SiO_2)-45(CaO + MgO)-20(H_2O+)$ are given in Figure 7. If the chemical composition of the obsidian is not known, the refractive index—a more quickly determined property that can be measured under the microscope—can be used as a substitute.

To apply the technique shown in Figure 7 to a practical problem, we monitored the ground temperature at an open site about 1 km from the rock shelter called Mummy Cave in Wyoming (Wedel et al. 1968). The

integrated temperature over a period of 365 days at 1 m below the surface was 10.3°C. Unlike the measured site, however, the floor in Mummy Cave was not exposed to the sun; thus, from measurements on sunny and shady sites made elsewhere, we applied a temperature about 2.3° cooler to the shaded site. Using a temperature of 8°C, the rate of hydration for obsidian quarried from Obsidian Cliff (as were most of the artifacts) would be about $4m\mu^2/1,000$ years. In Figure 8 we have plotted the corrected ^{14}C age of the dated strata in which artifacts made from Obsidian Cliff obsidian were found against the square of the hydration thickness that was measured on these artifacts. The hydration rates given by the solid lines are 4 and 4.5 $\mu m^2/1,000$ years—in good agreement with the rate calculated from temperature data.

Geological uses

For many years the main use that was made of obsidian hydration dating was for archaeological purposes. However, the technique is now being applied to the dating of geologic events.

The dating of volcanic phenomena was initiated by a study of very recent obsidian flows datable by the ^{14}C method. In order to determine the amount of hydration that might develop during the time that a lava flow

Table 1. Published hydration depth vs. time formulas to determine age of volcanic glasses (obsidian except as noted) grouped in the order of faster penetration reading downward

Formula type	Investigators
depth = K × time$^{1/3}$	Kimberlin (1976)
depth = K × time$^{1/2}$	Friedman and Smith (1960)
	Friedman et al. (1966)
	Michels (1967)
	Johnson (1969)
	Suzuki (1973)
	Friedman and Long (1976)
depth = K × time$^{1/2}$ + K′	Findlow et al. (1975)
depth = K × time$^{3/4}$	Clark (1961)
depth = K × time	Meighan et al. (1968)
	Layton (1972b)
	Ericson (1975)
(for basaltic glass)	Barrera and Kirch (1973)
(for basaltic glass)	Morgenstein and Riley (1975)

cooled to surficial temperature, we examined flows whose ages were thought, on the basis of their morphology, to be very young. These included Big Obsidian Flow in Newberry Craters, Oregon, which overlies pumice that contains wood dated by ^{14}C as 1400 ± 400 yrs old, as well as Glass Mountain in the Medicine Lake highlands of northern California. During a visit to the Glass Mountain flow by Friedman a tree was found in place but partially pushed over by the slowly moving lava. The outer part of this partially burned tree was ^{14}C dated at 380 yrs $\pm$ 200 yrs B.P. The obsidian from the two ^{14}C flows contained a minor thickness of hydration: 1.3 μm for Glass Mountain and 0.8 μm for Big Obsidian Flow. Assuming a reasonable rate of hydration, much or most of the observed amounts should have formed since cooling of the rock, and therefore only a negligible thickness of hydration could have developed while the material was still hot. This conclusion indicates that the outer surfaces of the volcanic glass were cooled through the temperature range of rapid hydration (ca. 150° to 70°C) in times of less than a few months. Rapid cooling of the exposed surfaces of flows would have been hastened by precipitation, which can be expected in both central Oregon and northern California at least once a month, even in the dry summer season.

In attempting to date obsidian-containing flows and domes older than the examples given above, a complication has been observed. When thin sections for microscopic examination

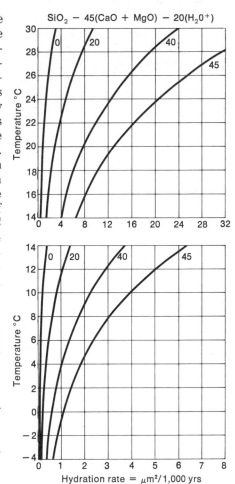

Figure 7. The relationship between the chemical composition of the obsidian—expressed as $SiO_2 - 45(CaO + MgO) - 20(H_2O+)$ where the various oxides are expressed in weight percent and where H_2O+ is water that is expelled at temperatures in excess of 110°C (bound water)—and the hydration rate—expressed in μm^2 per 100 years—is given as a function of temperature. These data were derived by experimentally hydrating obsidians of different chemical composition at elevated temperatures to speed up the hydration, and extrapolating the data so obtained to lower temperatures.

were cut from several samples of a single flow, it was found that these sections contained hydration of different thicknesses. For example, Figure 9 gives the thickness of hydration measured on 85 thin sections taken from specimens of seven different eruptive events in the Newberry Craters area. In all cases the thickest hydration is believed to represent the surface created by the eruptive event that originally emplaced the material. Subsequent cracking that provided new surfaces for hydration is assumed to have been due to earthquakes or renewed movement of the flow or dome caused by nearby eruptions. It should be noted that all the eruptive events were very close together geographically: all occurred within a square of about 5 × 5 km.

The sequence of events at Newberry is shown in Figure 10. In assigning ages to the hydration thicknesses, a rate of hydration of 3.0 μm^2/1,000 yrs was used for all the samples except those from Big Obsidian Flow, where a rate of 0.8 μm^2/1,000 yrs was employed. These values were derived by (1) measuring the effective hydration temperature in the soil at a site close to Newberry Craters and (2) using the experimental data on rate of hydration as a function of temperature for obsidian of different compositions as given in Figure 7.

Inasmuch as obsidian hydration dating is a measure of the time since a particular surface came into being, the method can be used to date erosional events that cleaned off or fractured the obsidian, thereby generating a new surface. The fact that obsidian deposited by a stream or by a glacier will have surfaces or cracks formed during transport has been used in dating the last two glaciations that occurred in and near Yellowstone National Park, Wyoming. Except for the final deglaciation about 13,000 years ago, glacial history in this part of the world is beyond the reliable 40,000-year range of ^{14}C dating.

In order to date these glaciations, obsidian was collected from the terminal moraines that were deposited by the glaciers. The hydration thickness was measured on deep cracks that were formed in the rock by abrasion during glacial transport. To estimate the rate of hydration of the

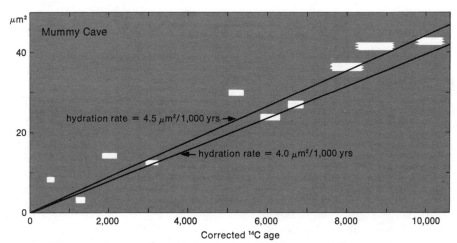

Figure 8. Hydration thickness is expressed as μm^2 for obsidian from corrected ^{14}C dated horizons from Mummy Cave rock shelter, Wyoming. The corrections to the ^{14}C dates for the three oldest samples are estimated from an extrapolation of graphs of the existing ^{14}C corrections which only extend to 7,150 years B.P. (From data by Davis 1972.)

glass concerned, the hydration thicknesses of two nearby obsidian-containing flows, both dated by K-Ar, were also measured. The hydration on all the material deposited by the Pinedale, or last, glaciation is thinner than the hydration on the adjacent West Yellowstone rhyolite flow dated by K-Ar as 115,000 years old. The hydration thickness of the previous, Bull Lake, glaciation is thicker than that on the West Yellowstone flow, but thinner than the hydration thickness measured on the Obsidian Cliff flow, K-Ar dated as 179,000 years B.P. When the hydration rind thicknesses are plotted as a function of age, (time)$^{1/2}$, we see that the end of the Pinedale came about 13,000 years ago, and the Pinedale terminal moraines, which were deposited at the farthest extent of the ice, are about 30,000 years old (Fig. 11). The Bull Lake moraines were deposited about 140,000 years ago.

The evidence of high stands of sea level at about 125,000 years B.P. that mark the culmination of the last interglacial (we are at present again in an interglacial period) indicates that the Bull Lake glaciation occurred before the last interglacial and is pre-Wisconsin in age, and not an early Wisconsin glaciation as had been previously believed.

Applications in archaeology

Even while hydration measurement as a dating tool was in its infancy in the early 1960s, archaeologists were using the method to analyze their field findings. In his review of progress in obsidian dating, Michels (1967) described a number of achieved and potential archaeological applications of the technique. He stressed the relative dating approach made possible by the abundance of man-made obsidian flakes at many sites and the low cost and speedy nature of the method. About the same time other investigators were pressing for direct dating of their excavated obsidian finds, depending for calibration of the hydration scale on the empirical application of associated radiocarbon data and other observations. Archaeologists have learned to apply hydration measurement in several ways, some of which are listed below.

To reconstruct the relative chronol-

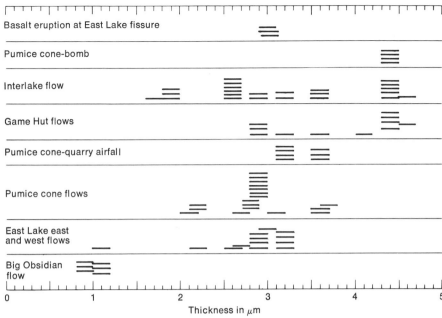

Figure 9. In this plot of measured hydration thickness of 85 samples from Newberry Craters, Oregon, the position of each horizontal line represents the thickness of hydration along a crack or edge. (The length of the line segments represents the estimated measuring error.) For all samples, the thickest hydration—indicating the oldest date—is found on the surface created by the eruption that originally emplaced the obsidian.

ogy of artifact style changes, a class of archaeological objects can be arranged in order of decreasing hydration depths. Thereby, for example, the total time range of use or manufacture of weapon points at a location can be established, and within that, the temporal sequence and duration of different styles of points. The application yields valuable information for both stratified and mixed collections. An example of this use is reported for a site in northwest Nevada by Layton (1973).

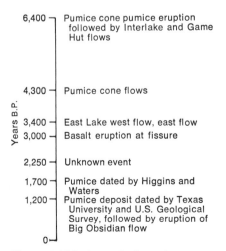

Figure 10. This time scale shows the sequence of major eruptive events during the past 6,700 years at Newberry Craters.

In testing site stratigraphy, we can look for discrepancies in the vertical hydration sequence of the excavation, i.e. for indications of possible disturbance or mixing of the contents of apparently separate geological layers. Such evidence warns against too ready acceptance of primary data. Intrusion into older strata by human action has often been demonstrated in archaeology, but displacement by subtle natural processes, for example the previously mentioned progressive upward travel of large objects imbedded in soil particles consequent to alternate wetting and drying, has been little heeded by excavators. Here the application of large hydration samplings may help resolve stratigraphic contradictions that at times puzzle archaeologists.

The phenomenon of reuse of the same artifact by peoples centuries apart in antiquity has been detected by investigators by means of hydration studies (e.g. Katsui and Kondo 1965). In this event one or more facets of the obsidian object show a markedly different depth of hydration than the rest. The unique capability of the hydration method, when statistically applied to a large collection of flakes, can illuminate our understanding of the lithic technology and economy of the cultural period concerned.

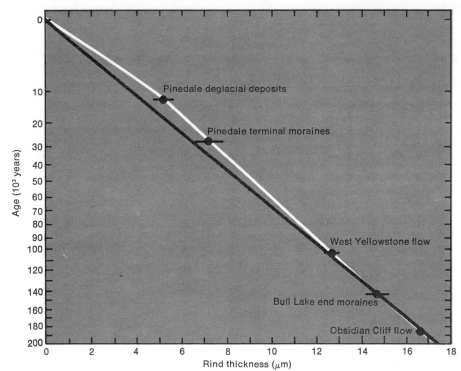

Figure 11. The rate of obsidian hydration for the West Yellowstone Basin is determined by hydration rinds on cooling cracks of the West Yellowstone flow, K-Ar dated as 179,000 ± 3,000 years old. The black line gives the average rate based on the two dated flows. The white line is drawn to account for variation in hydration rate due to climatic change during the late Quaternary. (From data by Pierce et al. 1976.)

Hydration analysis can be used as a first step in planning further field work and excavation (Meighan et al. 1968). This application has been advocated as a boon to the economy of archaeological operations at new site complexes where surface obsidian is relatively abundant and other age indicators, such as pottery sherds, are scarce or unevaluated.

Obsidian hydration rates have been derived (usually in terms of hydration amount per thousand years) from artifacts and associated radiocarbon dates for circumscribed regions and time spans, such as for Central California (Clark 1961), Western Mexico (Meighan et al. 1968), the Klamath Basin of California and Oregon (Johnson 1969), and Government Mountain–Sitgreaves Peak area of Arizona (Findlow et al. 1975). Ericson (1975) in his work on "source-specific" hydration rates for numerous obsidian sources of California and western Nevada attempted to assign unique hydration rates to obsidians from different sources.

The dating of ancient heating events has become possible by measuring the amount of hydration (rehydration) of burnt obsidian. The technique depends on the onset of superficial thermal crazing, at sufficiently high temperature in the firing, and the subsequent hydration of the newly opened crack walls. This approach has been applied to estimating the date and duration of prehistoric agricultural burning practices, to which the high prevalence of burnt flakes in surface collections from farmland in South America has been attributed (Trembour 1976). Suzuki (1973) took advantage of datable fission tracks in burnt obsidian from prehistoric hearths of Kanto, Japan, and correlated the results with artifact hydration from 33 archaeological sites of the district to derive hydration rates for a number of obsidian sources.

Analyzing gaps in the hydration sequence pattern of a large collection of artifacts, Layton (1972a, b) was able to associate apparent prehistoric abandonment at two northwest Nevada sites with the climatic Altithermal maximum of about 7,600 years B.P. and thus to time-calibrate all hydration depth measurements for cultural obsidian of the area.

Determining the antiquity—or recency—of obsidian artifacts that are presented for dating without authenticated archaeological context, but presumed to be of ancient origin, is of obvious service to museum curators and others who wish to determine the age of objects of interest that appear on the market.

Composite artifacts containing obsidian can be used to correlate directly the hydration time scale with other time scales such as one based on the thermoluminescence of pottery. Some candidate artifacts of this nature are fired pottery statuary with decorative obsidian inlays from the Near East and ceramic food rasps with obsidian teeth that have been found in South America. The ancient Egyptian cloth-wrapped mummified bodies of animals and humans that occasionally contain attached pieces of obsidian offer a similar opportunity for correlating obsidian and radiocarbon ages.

Dating basaltic glass

About seven years ago hydration measurement was extended to artifacts made from the basaltic glass of the Hawaiian Islands (Morgenstein and Rosendahl 1976). Although differing appreciably from the more acidic (higher SiO_2 content) natural glasses grouped under obsidian, the basaltic variety was found to hydrate, or alter, at a predictable rate. Fortunately, in view of the short human record of the Islands, the alteration rate is so rapid that useful layers accrue in a relatively short time. Calibration of the hydration or alteration rate of basaltic glass—about 12 μm per 1,000 years—has proved possible from the study of historically dated lava flows as well as from radiocarbon dating of charcoal and artifacts excavated from the same levels.

The advantages cited for basaltic glass hydration dating (Barrera and Kirch 1973) reaffirm those recognized elsewhere for obsidian. The Hawaii experience emphasizes several points of superiority of the hydration method over radiocarbon, especially more refined chronological control, the wide prevalence and relative ease of collection of the samples, and the low test cost per sample, which is less than one-tenth that for radiocarbon determinations.

Future research

Further growth in the application of the hydration dating method would benefit greatly from an assured way of assigning an intrinsic hydration rate to any piece of obsidian without recourse to source identification. This is true because in regions where eruptive rocks are prevalent, all the prehistoric cultural obsidian sources—with their unique hydration rates—may not be discoverable. Also, obsidian pebbles and cobbles often occur in mixed secondary deposits, and if they have been used by ancient peoples, it is not readily feasible to attribute them to a single primary source from which a hydration rate can be derived. The path via the laboratory can get around these difficulties.

Future research on hydration dating should focus on a rapid, simple laboratory technique for determining the hydration rate as a function of temperature on small samples of natural glass. Ambrose (1976) has suggested measuring the change in weight of a powdered obsidian sample that is exposed to a moist atmosphere at different but surficial temperatures for reasonably short periods of time, and this may prove to be the best direction for research. We also need a quick and simple method for establishing the concentrations of the major elements present in a sample. As a likely prospect, the electron microprobe provides a rapid method for the chemical analysis of selected areas as small as a few square micrometers, permitting avoidance of the small crystals usually present in the glass. The crystals, which differ markedly in composition from the glassy matrix, are not involved in the hydration process. Generally, the analytical methods used in the past, including wet chemistry, neutron activation, and X-ray fluorescence, have required a small sample in which there is no discrimination between the glass and the enclosed crystals.

Once the rapid processes for hydration rate measurements and chemical analyses have been established, additional data to relate chemical and physical properties of the glasses to hydration rate can be researched and acquired. At that point in the development, a key chemical or physical attribute, or a combination of them, can be used to ascertain a hydration

rate on any archaeological or geological specimen that comes along. The idea emerging is that simple tests of this kind should allow quick placement of any obsidian specimens into usefully narrow parts of the total range of the hydration rates of all obsidians, which stretches from about $1 \ \mu m^2$ to about $20 \ \mu m^2$ per 1,000 years. Furthermore, as not all natural glasses are of rhyolitic composition (as are obsidians), the products of this line of research could well lead to the reliable use in dating of these other types of volcanic glasses, particularly dacites and possibly basalts, whose hydration or alteration modes and rates are still uncertain.

Since the discovery of radioactivity in the late nineteenth century, methods of dating have been dominated by techniques that make use of radioactive disintegration clocks. There are, however, other processes that lend themselves to the determination of time intervals. The diffusion of water into natural materials is a time-dependent process that shows great promise for dating. No tool will work in every situation, and hydration dating is no exception. It must be used with discretion and with an appreciation of its limitations. When so applied, it is capable of giving valuable information with a minimum of effort and expense in studying materials that can be dated by no other method.

References

Ambrose, W. R. 1976. Intrinsic hydration rate determinations. In *Advances in Obsidian Glass Studies,* ed. R. E. Taylor. Noyes Press.

Barrera, W. M., Jr., and P. V. Kirch. 1973. Basaltic-glass artifacts from Hawaii: Their dating and prehistoric uses. *J. Polynesian Soc.* 82(2):176–187.

Clark, D. The application of the obsidian dating method to the archaeology of central California. 1961 diss. Stanford Univ.

Davis, L. B. The prehistoric use of obsidian in the Northwestern plains. 1972 diss. Univ. of Calgary, Alberta.

Ericson, J. E. 1975. New results in obsidian hydration dating. *World Archaeology* 7(2): 151–59.

Findlow, F. J., V. Bennett, J. Ericson, and S. De Atley. 1975. A new obsidian hydration rate for certain obsidians in the American southwest. *Am. Antiquity* 40:344–48.

Friedman, I., and C. Evans. 1968. Obsidian dating revisited. *Science* 162:813–14.

Friedman, I., and W. Long. 1976. Hydration rate of obsidian. *Science* 191:347–52.

Friedman, I., and R. L. Smith. 1960. A new dating method using obsidian, part 1: The

development of the method. *Am. Antiquity* 25:476–93.

Friedman, I., R. L. Smith, and W. Long. 1966. Hydration of natural glass and the formation of perlite. *Bull. Geol. Soc. of Am.* 77:323–28.

Johnson, Le Roy, Jr. 1969. Obsidian hydration rate for the Klamath Basin of California and Oregon. *Science* 165:1354–56.

Katsui, Y., and Y. Kondo. 1965. Dating of stone implements by using hydration layer of obsidian. *Jap. J. of Geol. and Geog.* 36:45–60.

Kimberlin, J. 1976. Obsidian hydration rate determination on chemically characterized samples. In *Advances in Obsidian Glass Studies,* ed. R. E. Taylor. Noyes Press.

Layton, T. N. 1972a. Lithic chronology in the Fort Rock Valley, Oregon. *Tebiwa* 15(2):1–20.

———. 1972b. A 12,000-year obsidian hydration record of occupation, abandonment and lithic change from the northwestern Great Basin. *Tebiwa* 15(2):22–28.

———. 1973. Temporal ordering of surface-collected obsidian artifacts by hydration measurement. *Archaeometry* 15(1):129–32.

Lee, R., D. Leich, T. Tombrello, J. Ericson, and I. Friedman. 1974. Obsidian hydration profile measurements using a nuclear reaction technique. *Nature* 250:44–47.

Meighan, C. W. 1976. Empirical determination of obsidian hydration rates from archaeological evidence. In *Advances in Obsidian Glass Studies,* ed. R. E. Taylor. Noyes Press.

Meighan, C. W., L. Foote, and P. Aiello. 1968. Obsidian dating in West Mexican archaeology. *Science* 160:1069–75.

Michels, J. W. 1967. Archaeology and dating by hydration of obsidian. *Science* 158:211–14.

Morgenstein, M., and P. Rosendahl. 1976. Basaltic glass hydration dating in the Hawaiian Islands. In *Advances in Obsidian Glass Studies,* ed. R. E. Taylor. Noyes Press.

Morgenstein, M., and T. J. Riley. 1975. Hydration-rind dating of basaltic glass: A new method for archaeological chronologies. *Asian Perspectives* 17(2):145–59.

Pierce, K., J. Obradovich, and I. Friedman. 1976. Obsidian hydration dating and correlation of Bull Lake and Pinedale Glaciations near west Yellowstone, Montana. *Bull. Geol. Soc. of Am.* 87:701–10.

Springer, M. E. 1958. Desert pavement and vesicular layer of some soils of the desert of the Lahonton Basin, Nevada. *Soil Sciences Soc. of Am. Proc.* 22:63–66.

Suess, H. E. 1970. Bristlecone-pine calibration of the radiocarbon time-scale 5200 BC to the present. In *Radiocarbon Variations and Absolute Chronology,* ed. I. U. Olsson. Wiley.

Suzuki, M. 1973. Chronology of prehistoric human activity in Kanto, Japan. *J. of the Faculty of Science, Univ. of Tokyo* (5) 4: 241–318.

Trembour, F. W. The continuing problems of hydration dating. 1976 paper del. at symposium on early man in the New World, Univ. of Calif., San Diego, Aug. 7.

Wedel, W. R. et al. 1968. Mummy cave: Prehistoric record from Rocky Mountains of Wyoming. *Science* 160:184–86.

Elizabeth K. Ralph
Henry N. Michael

Twenty-five Years of Radiocarbon Dating

The long-lived bristlecone pines are being used to correct radiocarbon dates

When radiocarbon (C^{14}) dating was introduced in 1949, there was some reluctance to accept the dates; later there seemed to be such general reliance upon them that often they were uncritically quoted without the cautioning remarks of both field investigator and radiocarbon laboratory report. We have now learned that one of the basic implicit assumptions of the method—namely, the constancy of the atmospheric inventory of $C^{14}O_2$—is not strictly correct. This problem and the means of correcting it are the main subjects of this article.

In 1934, A. V. Grosse anticipated the existence of "cosmic radioele-

Elizabeth K. Ralph, Associate Director of the Museum Applied Science Center for Archaeology at the University of Pennsylvania, received her Ph.D. in geology from the University of Pennsylvania, where she has been a member of the staff since 1951. Her research interests have centered around C^{14} dating and, more recently, thermoluminescence dating of pottery and the use and development of instruments, primarily magnetometers, for archaeological prospecting at sites such as Sybaris in southern Italy.
Henry N. Michael, Research Associate in the Radiocarbon Laboratory at the University of Pennsylvania and Professor of Geography at Temple University, received his Ph.D. in anthropology from the University of Pennsylvania in 1954. He joined the faculty at Temple University in 1959 and was Chairman of the Department of Geography there from 1965 to 1973. He has done field work in the Arctic and has edited and translated numerous Russian publications on Siberian anthropology and ethnology.
The authors are grateful to C. Wesley Ferguson, Barbara Lawn, Kathleen Ryan, and William E. Stephens, who have helped in various ways with their dating projects and in writing this article. The dating program is currently supported by NSF grant GA-12572. Address: Museum Applied Science Center for Archaeology, University Museum, University of Pennsylvania, Philadelphia, PA 19174.

ments" (1). More than a decade later, in 1947, W. F. Libby, Grosse, and others (2) proved that natural radiocarbon does indeed exist by enriching methane from sewage gas in a thermal diffusion column. Libby and his students then developed a technique to detect and measure this low-level radioactivity without the cumbersome process of enrichment. They used a screen wall-counter (a Geiger counter with a grid between the center wire and the outer-tubing), which had been developed by Libby (3), and a surrounding ring of counters in anticoincidence. The apparatus was enclosed in a massive shield of steel which, in combination with the anticoincidence ring, reduced the background of unwanted counts below the level of natural C^{14} activity, thus making possible its detection without enrichment.

In the late 1940s and early 1950s, Geiger-counting of solids was in vogue, and it therefore seemed logical to convert samples to solid carbon for lining the counter wall—that is, to have as much carbon as possible per mole. However, two basic weaknesses of this technique were soon discovered: first, self-absorption of β-rays (maximum energies of 160 Kev), which reduced the counting efficiency, and, second, the likelihood of carbon absorption of radioactive contamination, especially from the atomic bombs then being tested. In seeking an alternative, investigators experimented with gas counting methods, and a few developed scintillation detection. Eventually the presently used methods were developed—proportional counting of pure CO_2, CH_4, C_2H_2, and liquid scintillation counting of C_6H_6.

Radiocarbon dating

Radiocarbon is formed in the upper atmosphere by the reaction of ordinary nitrogen (N^{14}) with neutrons (n). Thus $N^{14} + n \rightarrow C^{14} + H^1$. Since nitrogen is abundant in the earth's atmosphere, the constancy of C^{14} production depends upon the supply of neutrons. These are produced by cosmic rays, and therefore their quantity is dependent upon the intensity of cosmic rays. In turn, the intensity of the earth's magnetic field determines the number of cosmic rays which arrive in the upper atmosphere: when the field is strong, fewer cosmic rays reach the upper atmosphere; when it is weak, the inverse takes place.

The C^{14} produced constitutes a very small part of the earth's carbon dioxide—about one part in a million million (1 in 10^{12} parts). The CO_2 is in equilibrium with terrestrial life and with the surface water of the oceans, the largest reservoirs of carbon. For the application of natural C^{14} in radiocarbon dating, it was assumed that the balance among the reservoirs has been constant and that the mixing rate within the atmosphere is rapid—on the order of two or three years, short compared to the 8,268-year average life of nuclei of C^{14}. In 1949 Arnold and Libby (3) demonstrated that these conditions are fulfilled within an accuracy of approximately 10 percent.

Photosynthesis causes all living vegetation to be in equilibrium with atmospheric $C^{14}O_2$, and almost all forms of terrestrial life that consume vegetation contain the same proportion of C^{14}. When an organism dies, of course it does not

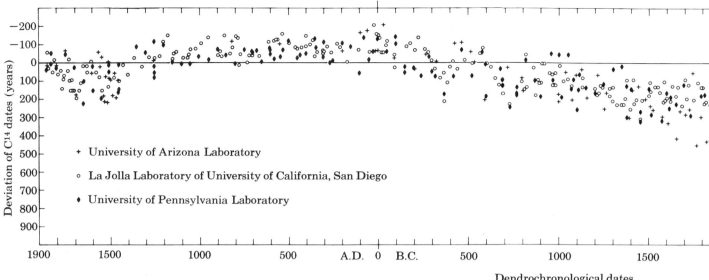

Figure 1. Deviations from the C^{14} dates become progressively greater as the wood material being measured is older. C^{14} dates were obtained by three independent laboratories for dendro-dated sequoias and bristlecone pines and were calculated with the 5,730-year half-life.

continue to recycle carbon from the atmosphere and to replenish C^{14}. The C^{14} contained in the dead organism decays back to N^{14} by emission of a β particle: $C^{14} \rightarrow N^{14} + \beta^-$. The half-life of C^{14} provides a measure of the rate of radioactive decay of C^{14}, and thus the precision of radiocarbon dating depends to a high degree upon the accurate determination of this half-life.

In 1951 the best estimate of the half-life was $5,568 \pm 30$ years (3), and this value continues to be used for all dates published in *Radiocarbon,* although at the Fifth Radiocarbon Dating Conference, held in Cambridge, England, in 1962, the half-life of $5,730 \pm 40$ years was accepted as the most probable one (4). The conversion from the former to the latter half-life is easily achieved, however, by multiplying the Before Present (B.P.) date calculated with the 5,568-year half-life date by a factor of 1.03—that is, a 3 percent increase.

Radiocarbon, an isotope of the very abundant element carbon, provides a tracer for age determination for the past 40,000 years. In most laboratories the limit of detection of C^{14} is reached at 40,000 years—i.e. the counting rate is so low that it cannot be detected above the background rate. Also, for samples this old or older, the problem of contamination with a minute amount of modern carbon is severe. Since radiocarbon dating is independent of other chronological assumptions

or stratigraphic considerations, it is of great use to archaeologists, anthropologists, Pleistocene geologists, glaciologists, and many other scientists.

Age calculation

The familiar exponential decay equation $I = I_o e^{-\lambda t}$ describes the change in activity of a radioactive sample. I is the activity of the sample when measured. I_o is the original standard activity of the sample; this value is obtained from samples of known age (approximately 100 years old but corrected for zero age), and the same volume is measured as that of the unknown sample. For this reason, an absolute calibration of the counter is not required. λ is the decay constant ($= 0.693/t_{1/2}$, with $t_{1/2}$ the half-life), and t is the time elapsed. If we feed the value of the half-life into this equation (using 5,568 years in this case), we have the simple formula

$$t = \log \frac{I_o}{I} \times 18.5 \times 10^3 \text{ years}$$

for routine age calculations.

Radioactive disintegration is a random process. When a particular atom will decay cannot be predicted; we know only that after a certain length of time a certain fraction of the original number of atoms will have disintegrated. The uncertainty in measuring radioactive disintegration is expressed in terms of statistical deviation, and it

has become customary to quote the standard deviation of one sigma with each date to indicate that the date lies within the range quoted in approximately two out of three cases. When the one-sigma tolerance is doubled, the chance that the correct date lies within the doubled range increases to 21 out of 22 cases. The standard deviation represents not only the uncertainty inherent in the counting of an unknown sample; it also includes the uncertainty in radiation background and in the calibration samples. Such basic uncertainties cannot be ignored in the quoting of C^{14} results.

In most laboratories, the standard counting time for radiocarbon samples is usually a period of 1,000 minutes. Counting is normally done overnight, and each sample is counted at least twice. Making the second count about a week after the first provides a check on possible radon contamination. The radon may have originated from traces of uranium and thorium in the soil mixed with or surrounding the sample. In the process of purifying the CO_2 or other gas for counting, radon is one of the most difficult contaminants to remove, but since it has a half-life of only 3.82 days, it will have decayed measurably by the second week. In some laboratories, including the one at the University of Pennsylvania, samples are counted a third time or more if the first two counts are not statistically consistent.

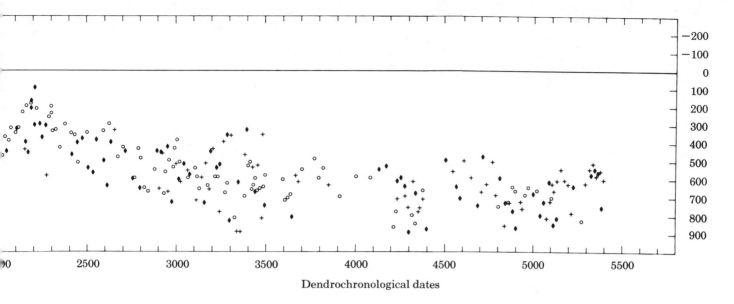

Dendrochronological dates

For interlaboratory calibration, the National Bureau of Standards has provided a batch of oxalic acid in bulk. From comparison of samples from this batch (which is homogeneous in age) with age-corrected wood samples, it has been found that the C^{14} activity is 95 percent of age-corrected "modern-day" wood. (Modern-day wood is wood which has been determined by tree-ring dating to be at least 100 years old; such samples are used to avoid the effects of depletion of $C^{14}O_2$ from the atmosphere due to the combustion of fossil fuels and of enrichment due to nuclear bomb tests.)

When maximum precision of a radiocarbon date is desired (as, for example, in dating the sequoia and bristlecone pine samples to be described later), C^{13}/C^{12} ratios are measured by means of mass spectrometry as a check on natural and/or laboratory fractionation. If the C^{13}/C^{12} ratio differs from that of modern-day, or standard, wood samples, the amount of difference is doubled and applied as a correction to the C^{14} counting rate. The standard wood samples in most laboratories are calibrated against a limestone standard; the difference between the limestone and wood standards is usually 25 mils (1 mil = 1 part in 1,000), the limestone being 25 mils richer in C^{13} than the wood.

We now know that the assumption that the biospheric inventory of C^{14} has remained constant over the past 50,000 years or so is not true. At the Twelfth Nobel Symposium, held in Uppsala in August 1969, many papers (5), as well as formal and informal discussions, were addressed to this problem. As a result of the C^{14} dating of dendro-dated (tree-ring dated) wood samples, disagreements became evident (see Fig. 1).

In 1958 de Vries (6) noted, on the basis of a comparison of radiocarbon dates of precisely dated samples from the A.D. era, that there were differences between the radiocarbon and calendric (tree-ring) dates. After his untimely death in the same year, this type of comparative study was taken up (on a much larger scale) almost simultaneously at three different places—the University of Arizona, the La Jolla laboratory of the University of California at San Diego, and the University of Pennsylvania (see refs. 5 and 7 for results of their studies). Before going into the details and implications of the radiocarbon problems, let us consider the technique of tree-ring dating, or dendrochronology.

Dendrochronology

The concept of dendrochronology is old. There are, for instance, accounts of its use from the fifteenth century, when attempts were made by Leonardo da Vinci to tell the age of trees by simple ring counting and to tell the nature of "past seasons" by the width of the rings (8). In 1880, in Holland, J. C. Kapteyn was apparently among the first to conduct a scientific composite study of the relation of tree rings to climate (9).

The first extensive study of tree rings in the United States was begun in 1901 by the astronomer A. E. Douglass (10, 11), who had hoped to correlate sunspots and climatic changes. His work led to the development of precise tree-ring dating, which eventually was extended by Douglass (12) and his collaborators (13) and successors (14) to 5400 B.C.—a span of almost 7,400 years. This very long tree-ring chronology is based principally upon two long-lived trees—the *Sequoia gigantea* and *Pinus aristata* (bristlecone pine). Some living sequoias reach beyond 1000 B.C., the oldest recorded dating to 1306 B.C. Living bristlecone pines have reached the age of about 4,600 years—that is, their innermost rings date to about 2600 B.C. Obviously, even these extreme ages would not suffice to establish the chronological span of almost 7,400 years.

Chronology building and extending is done by cross-dating wood from living and dead trees. The tree rings from two or more samples must contain signatures—similar sequential arrangements of wide and narrow rings—which can be recognized in all the samples to be correlated. Since wood from long-dead trees must be used, it is fortunate that the high resin content (30 percent) of the bristlecone pine tends to preserve it better than most other woods. The relative dryness and short summers of the ecological zone in which the oldest bristlecone pines are found (at altitudes of 10,000 to 11,000 feet in the White Mountains of California) also help to preserve dead trees.

With accurately dated wood samples available for a long span of time, studies could be undertaken to determine the nature and extent of differences between calendric dates and radiocarbon dates. The published raw data (15, 16, 17) of the three main laboratories conducting this research are assembled in Figure 1. There is excellent agreement among the three on the average, but there is also a significant long-term deviation of C^{14} dates from dendro-dates in the earlier B.C. millennia of the plot and possibly short-term oscillations. The pronounced scatter among the dates exceeds statistical expectations, and we have smoothed the raw data for Figure 2 by employing a simple 9-cell regression, placing the weighted average at its midpoint.

The 9-cell regression, selected as the smoothing medium after much experimentation with other methods, including polynomials, provides a relatively smooth curve, preserving at the same time the major, and most of the minor, oscillations expressed in the raw data. The curve thus derived (Fig. 2) serves as the means for correcting (where necessary) radiocarbon dates as far back as 4760 B.C. The dashed line in the figure represents the following third-order polynomial, which is the best fit on the average:

$$T_{C^{14}} = -43.96 + 0.918 \times T_D +$$
$$7.17 \times 10^{-5} \times T_D{}^2 + 1.18 \times 10^{-8}T_D{}^3$$

where $T_{C^{14}}$ is the C^{14} date, and T_D is the dendro-date. $T_{C^{14}}$ and T_D are positive for A.D. and negative for B.C. Both the curve derived from the polynomial and that from regression averaging were calculated with the 5,730-year half-life. (The calibration curve is divided into six large plots, which can be easily used to correct C^{14} dates, in ref. 18.

For those who prefer tables, they are also given for ten-year intervals in ref. 18.)

To correct a radiocarbon date, one locates the C^{14} date on the left of the graph and proceeds horizontally across the graph until the curve is reached. If the curve is crossed only once, the point of crossing becomes the midpoint of the corrected date, which is read off the dendro-dated scale. It must be kept in mind that the new date still has at least the one-sigma uncertainty. As an example of the correction of the midpoint for a single crossing, take the C^{14} date of A.D. 880, which crosses the curve at A.D. 940; thus correction to the calendar date is minus 60 years. In the case of multiple crossings and/or spans, the situation is more complicated. For example, the midpoint of the C^{14} date of 1950 B.C. crosses the curve at 2190, 2230, and 2290 B.C., and the true date could fall anywhere within this range.

Fortunately, 85 percent of the C^{14} dates within the ranges shown in Figure 2 cross the curve only once —that is, they may be corrected as single crossings. For the remaining 15 percent, three situations arise: (1) the C^{14} date may cross the curve more than once; (2) it may follow the curve for a distance; (3) it may do both.

Validity of the tree-ring calibration

The validity of the bristlecone pine–sequoia time scale has been questioned for a number of reasons. To us, the most serious concern is that for periods earlier than 3,000 years ago, it is based entirely on one species of tree that grows, or grew, under rather atypical conditions—namely, at elevations of 10,000 to 11,000 feet. That the extension of the dendro–time scale during the past fifteen years has been done in a single laboratory and mostly by one person we consider to be a minor point, because of the reputation of the Laboratory of Tree-Ring Research at the University of Arizona and of the extreme care which C. Wesley Ferguson has exercised in building this chronology. Also, independently, La Marche (19) has compiled a chronology of bristlecone pines from locations other than the White Mountains that agrees with Fergu-

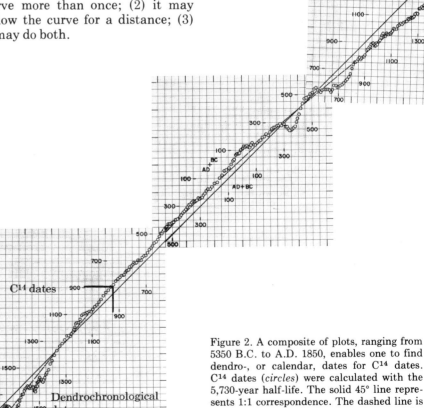

Figure 2. A composite of plots, ranging from 5350 B.C. to A.D. 1850, enables one to find dendro-, or calendar, dates for C^{14} dates. C^{14} dates (circles) were calculated with the 5,730-year half-life. The solid 45° line represents 1:1 correspondence. The dashed line is the best fit for a third-order polynomial for the average of all C^{14} dates.

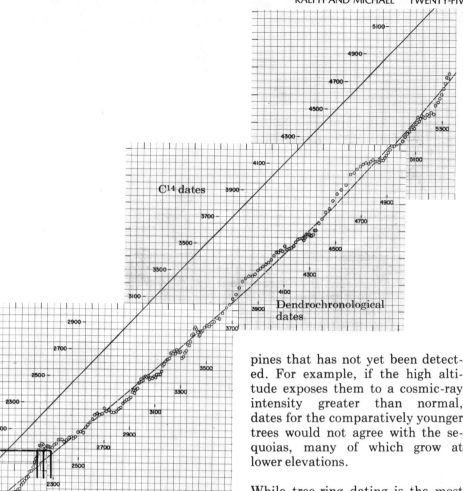

C14 dates

Dendrochronological dates

Other independent confirmation comes from dates of lake-sedimentation and lake-varves in the United States (21). These studies—especially that of Lake of the Clouds in Minnesota—suggest that there may be a discrepancy of 600 years between the De Geer (Swedish) and the U.S. varve chronologies. The U.S. varves do contain datable carbon, but it is not possible to determine the age of the lake water (depleted in C^{14} by carbonates derived from limestone), which adds an uncertainty. Despite this uncertainty, there are indications that the C^{14} dates depart from true ages in a manner similar to that of the bristlecone pines. A study of marine varves from Saanich Inlet, British Columbia, indicates similar differences (22).

For confirmation of the shorter-term oscillations, there is independent evidence from ice cores, especially from the long core from Camp Century, Greenland (23). For this core, correlation of time with depth is unusually precise. Variations in climate for the past 15,000 years have been determined by the measurement of O^{18}/O^{16} ratios, and for the past 8,000 years, short-term oscillations appear to be similar to short-term C^{14} periods. The records from the ice cores show long-term trends that reflect the climatic changes during the past 100,000 years. If climatic changes, as a secondary effect of solar and magnetic variations, did affect the C^{14} inventory, this would tend to confirm Leona Libby's theory (24) that climate acting on the biosphere changed the rates of bacterial decay of organic carbon and thus released more or less stored carbon with lower C^{14} content.

pines that has not yet been detected. For example, if the high altitude exposes them to a cosmic-ray intensity greater than normal, dates for the comparatively younger trees would not agree with the sequoias, many of which grow at lower elevations.

While tree-ring dating is the most precise chronological technique, support for the corrected C^{14} chronology comes from other, completely independent methods. Confirmation of the long-term C^{14} trend is provided by dating layered deposits of annually formed varves, discernible because the spring and summer melt deposit of coarse particles is overlain with the autumnal fine particles from the slowed, receding waters.

Possible causes of radiocarbon variations

We now face the question: What are the possible causes of the deviations in the atmospheric inventory of C^{14}?

Mixing rate. We do know with reasonable certainty that the mixing rate of $C^{14}O_2$ in the atmosphere is rapid. In assessing the mixing rate, the above-ground nuclear bomb tests, which created neutrons and hence excess C^{14}, have been of help. The majority of the first

son's. In our laboratory we have processed some of the younger sections of bristlecone pines which overlap the sequoias, and the agreement is excellent.

Another query frequently heard is whether the results of dating trees from the southwestern United States can be applied on a worldwide basis. Fortunately, the mixing rate of the atmosphere is rapid and the mixing extensive. Therefore, a tree making its annual ring anywhere in the world should have the same C^{14} content (within a two-year period) as the bristlecone pines of the southwestern United States, unless there is something very unusual about the bristlecone

The Scandinavian varve chronology has been studied most intensively; present investigations indicate that the correlation of the various deposits of varves in southern and central Sweden may be uncertain to about 300 years in the 12,000-year run of varve chronology (20). Another problem is that a varve contains a negligible amount of organic carbon, and thus it is difficult to compare the varves with C^{14} dates. A few varve sequences that have been related to radiocarbon dates by means of associated pollen sequences indicate that differences in C^{14} dates become greatest (too young) at about 4000 B.C., in reasonable agreement with the bristlecone pine data.

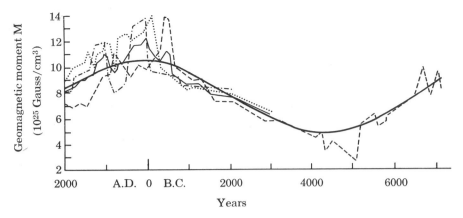

Figure 3. Changes in the intensity of the earth's magnetic field have influenced the atmospheric inventory of C^{14}. The dashed line represents a long series of changes in the geomagnetic moment determined by Bucha at sites in Central Europe; dashed-dotted line is based on data from the U.S. Southwest; dotted line shows results from Japanese sites. The thin solid line expresses the unsmoothed average, the smoothly curved, thick solid line shows the smoothed average. (From ref. 37, with permission of M.I.T. Press.)

bombs were exploded in the Northern Hemisphere, and monitoring of the increase by laboratories in both hemispheres (25) has shown that the mixing rate is on the order of two to three years. Other studies have confirmed this. For example, Tauber (26) collected grasses from latitude 0° to 80° N during the growing season in 1960 and found that there was no significant latitudinal variation in C^{14} content. Haugen (27) compared major trends in tree-ring indices in Scandinavia, the Urals, Alaska, and Labrador, which resulted in highly significant correlations among the trees from these areas from A.D. 1650 on, thus supporting the proposition of a rapid mixing rate.

The turnover in the oceans may be much slower. The atmosphere is in equilibrium with the surface ocean waters, but the lag in the mixing of C^{14} between the oceans and the atmosphere is approximately 10 years (25). The surface water then becomes part of the deep ocean water with an uncertain lag of from 2 (28) to more than 11 years (25). The return from deep ocean to surface may, then, require 500 years or more (25). Numerous models have been proposed to explain and evaluate this complicated mixing process on a worldwide basis (see, for example, refs. 29 and 30).

Cosmic-ray intensity. A precise assessment of the constancy of cosmic-ray intensity is difficult to obtain when one is looking for relatively small variations in the concentrations of atmospheric radiocarbon. Investigations of decay series of other nuclides in meteorites indicate that there have been no major changes in the intensity of cosmic rays during the past 300,000

years (31). This period has been extended by Crevecoeur to five million years (32). The results of studies by Fireman and Goebel (33) on Ar^{37}/Ar^{39} ratios in iron meteorites for recent time ranges (of less than 2,000 years) indicate that the cosmic-ray flux near 1 AU (AU = average distance between earth and sun) did not differ significantly in the spring of 1969 from the cosmic-ray flux of the previous several hundred years.

Explosions of supernovae, suggested by Dergachev and Kocharov (34) as possible causes of the variations in the atmospheric inventory of C^{14}, should also be considered. However, the effect would be due to the generation of excess cosmic rays at the time of the explosion. If we believe the evidence from meteorites, the cosmic-ray intensity has probably not fluctuated more than 5 percent during the past 50,000 years.

Much more work will have to be done, however, before any supportable accuracy can be claimed for the behavior of cosmic radiation. The basic difficulty is that, with few exceptions, the orbits of meteorites are unknown and the cosmic-ray flux received by meteorites differs with the distance of their origin from the sun. One step in overcoming this difficulty was the establishment of seven tracking stations in the U.S. Midwest in 1964. Thus far they have determined the trajectory of a meteorite which fell near Lost City, Oklahoma, in 1970 (35). Another trajectory was previously recorded near Příbram, Czechoslovakia, in 1959 (36). Studies of these and future recordings will improve our knowledge of the constancy of cosmic-ray intensity.

The magnetic effect. The role of terrestrial magnetism in controlling the amount of cosmic radiation which reaches the upper atmosphere is important. From measurements of archaeomagnetic samples of fired clays, Bucha (37) has traced the changes in the intensity of the earth's magnetic field to about 7000 B.C. (Fig. 3). As measured in uncorrected radiocarbon years, the magnetic intensity reached a maximum, probably on a worldwide basis, between 100 B.C. and A.D. 100 and a minimum about 4250 B.C. From Bucha's data we can calculate the effect of these changes upon the content of atmospheric C^{14}. Since we now know that the changes in atmospheric inventory appear to be cyclic, it seemed appropriate to try a sinusoidal equation (38). Using such an equation, we found that these changes account for approximately half of the long-term divergences in C^{14} production and that a lag on the order of 250 years should occur between minimum magnetic intensity and maximum C^{14} production.

Furthermore, long-term changes in magnetic intensity may have been caused by a "recent" pole reversal —that is, a reversal of the earth's main dipole moment. Until a few years ago, it was thought that there had been no pole reversals for the past 700,000 years. There is now evidence that two have occurred during the past 35,000 years but that there was an absence of reversals between about 30,000 and 13,000 years ago (39, 40, 41, 42). Cox (43) has estimated that the duration of the "recent" reversal processes may be on the order of 10^3 to 10^4 years.

If we consider the situation during

a pole reversal as a time of minimum magnetic intensity and therefore of maximum cosmic-ray intensity, the C^{14} production could be evaluated in terms of the present value near the magnetic poles. On the basis of the work of Merker (44), Wada and Inoue (45), and Dergachev (34), reduced magnetic intensity could cause an increase in the C^{14} inventory by a factor of two. To test this assumption, we need to extend the tree-ring scale back at least 13,000 calendric years or to live long enough to witness the next pole reversal.

Earthly magnetic fields may also be influenced by the varying magnetic field of the sun. Here, we are even more stymied in attempting to correlate the effects, because the fields of the sun have only recently been measured (46). The solar wind, caused by the continually expanding solar corona, freezes in the solar magnetic field lines, which are transported outward and travel past and beyond the earth (47). The earth's magnetic field presents an obstacle to the flow of the solar wind. Current and proposed studies of the magnetosphere (48) should allow more quantitative evaluations in the near future.

Sunspot activities and cycles. Numerous investigators have attempted to explain the short-term oscillations of C^{14} in terms of sunspot activities and cycles, and both negative (49, 50) and positive (51) results have been presented. The difficulties connected with this particular problem are that the measurement uncertainties in C^{14} dates are greater than the short-term sunspot cycles and data on sunspot activity prior to A.D. 1750 are scarce and, if available, unreliable. The short-term oscillations in the C^{14} inventory tend to have a periodicity of 400 years, which may reflect the secular variation of the nondipole magnetic field. A harmonic analysis by O'Brien (52) suggests, however, that there is no true sinusoidal periodicity in short-term oscillations.

Confirmation from ancient civilizations

As the bristlecone pine chronology is pushed back in time, many of the questions about the possible causes of the deviations in atmospheric C^{14} may be answered. For the present, however, as differences between radiocarbon and tree-ring data were being measured, it was realized that in order to substantiate the validity of correction factors suggested by the studies, it would be necessary to develop other convincing evidence. At Uppsala in 1969, our own feeling and that of Säve-Söderbergh and Olsson (53) was that Egyptian dynastic chronology, which is better documented archaeologically than that of any other civilization of similar age and extent, would constitute a logical case for testing.

We assembled seventy-four radiocarbon dates of fifty-five Egyptian archaeological samples (some dated more than once and by different laboratories) and divided the samples into three major classes: short-lived (linen, cloth, reed matting, human bones and tissues), long-lived (supporting timbers, parts of wooden coffins or sarcophagi constructed of large pieces of wood with many discernible rings—from at least fifty up to several hundred), and mixed (charcoal samples which may have come from old or young wood) (54). With the short-lived samples we would expect the corrected dates to agree with the archaeological ones; with long-lived samples, the corrected dates should be older than the archaeological dating.

In most cases the corrected radiocarbon dates agreed with the archaeologically determined ones. This was particularly true of the short-lived samples. In several cases it was possible to confirm or deny dates that had been designated as doubtful by the field archaeologist (54). In recent months we have assembled additional radiocarbon dates for Egyptian dynastic samples, bringing the total to 140, and are analyzing these samples; preliminary studies indicate that an overwhelming majority of the corrected radiocarbon dates agree with the archaeologically determined ones.

Taking into account the successful correlation of C^{14} and archaeological dates from Egyptian dynastic samples, Colin Renfrew (55) has argued that if the correction factors apply in a consistent way to dynas-

tic Egypt, they must also apply to the Late Neolithic, Bronze, and Iron Age cultures of Europe—including mainland Greece. Briefly, this makes the Neolithic dates of Europe up to 800 years older than previously determined, and, more fundamentally, it drastically revises the notions of direct Egyptian and Mycenaean influence upon the Bronze and Iron Age cultures of Europe, thus forming a base for the argument that they developed independently, particularly in southeastern Europe.

In central Europe, Neustupný (56) has long argued for just such a chronological shift of the Aeneolithic and earlier Neolithic cultures, as indicated by the recalculated radiocarbon dates. The works of Renfrew and Neustupný are impressive because of their careful documentation (when available) and their cautious statements about the need for additional work—which is forthcoming. In the next few years we should hear further discussion about the revision of European prehistory based on corrected radiocarbon dates. If there is validity to the new calendar—and we strongly believe there is—the outcome should be a better chronology.

References

1. A. V. Grosse. 1934. An unknown radioactivity. *J. Am. Chem. Soc.* 56:1922–23.

2. E. C. Anderson, W. F. Libby, S. Weinhouse, A. F. Reid, A. D. Kirschenbaum, and A. V. Grosse. 1947. Natural radiocarbon from cosmic radiation. *Phys. Rev.* 72:931–36.

3. W. F. Libby. 1952. *Radiocarbon Dating.* Chicago: Univ. of Chicago Press.

4. H. Godwin. 1962. Half-life of radiocarbon. *Nature* 195:984.

5. I. U. Olsson, ed. 1970. *Radiocarbon Variations and Absolute Chronology.* Proc. 12th Nobel Symposium, Uppsala, Sweden, Aug. 1969. New York: John Wiley and Sons.

6. Hl. de Vries. 1958. Variation in concentration of radiocarbon with time and location on earth. *Proc. Kon. Ned. Akad. van Wetenschappen* B 61:1–9.

7. *Proc. 8th International Radiocarbon Dating Conference,* Vol. 1, 1973. Lower Hutt, New Zealand, Oct. 1972.

8. W. S. Stallings. 1937. Some early papers on tree-rings. *Tree-Ring Bulletin* 3:27–29.

9. J. C. Kapteyn. 1914. Tree growth and meteorological factors. *Recueils Trav. Bot. Neerland* 11.

10. A. E. Douglass. 1919. *Climatic Cycles and Tree Growth*, Vol. 1. Washington, D.C.: Carnegie Institution of Washington, publ. no. 289.

11. A. E. Douglass. 1928. *Climatic Cycles and Tree Growth*, Vol. 2. Washington, D.C.: Carnegie Institution of Washington, publ. no. 289.

12. A. E. Douglass. 1935. Dating Pueblo Bonito and other ruins of the Southwest. *Contributed Technical Papers, Pueblo Bonito Series No. 1*. Washington, D.C.: National Geographic Society.

13. E. Schulman. 1956. *Dendroclimatic Changes in Semiarid America*. Tucson: Univ. of Arizona Press.

14. C. W. Ferguson. Dendrochronology of bristlecone pine, *Pinus aristata*. In ref. 5, pp. 237–59.

15. P. E. Damon, A. Lang, and D. C. Grey. Arizona radiocarbon dates for dendrochronologically dated samples. In ref. 5, pp. 615–18.

16. E. K. Ralph and H. N. Michael. MASCA radiocarbon dates for sequoia and bristlecone-pine samples. In ref. 5, pp. 619–23.

17. H. E. Suess. Bristlecone pine calibration of the radiocarbon time-scale 5200 B.C. to the present. In ref. 5, pp. 303–11 and Plate 1.

18. E. K. Ralph, H. N. Michael, and M. C. Han. 1973. Radiocarbon dates and reality. *MASCA Newsletter* 9(1).

19. V. C. La Marche, Jr., and T. P. Harlan. 1973. Accuracy of tree-ring dating of bristlecone pine for calibration of radiocarbon time scale. *J. Geophys. Res.* 78:8849–58.

20. E. Fromm. An estimation of errors in the Swedish varve chronology. In ref. 5, pp. 164–72.

21. M. Stuiver. Long-term C14 variations. In ref. 5, pp. 197–213.

22. A. I. C. Yang. Variations of natural radiocarbon during the last 11 millennia and geophysical mechanisms for producing them. In ref. 7, pp. 60–73.

23. W. Dansgaard, S. J. Johnsen, H. B. Clausen, and C. C. Langway, Jr. 1971. Climatic record revealed by the Camp Century ice core. In Karl K. Turekian, ed., *The Late Cenozoic Glacial Ages*. New Haven: Yale Univ. Press, pp. 37–56.

24. L. M. Libby. 1973. Globally stored organic carbon and radiocarbon dates. *J. Geophys. Res.* 78:7667–70.

25. A Walton, M. Ergin, and D. D. Harkness. 1970. Carbon-14 concentrations in the atmosphere and carbon dioxide exchange rates. *J. Geophys. Res.* 75:3089–98.

26. H. Tauber. 1967. Copenhagen radiocarbon measurements. VIII. Geographic variations in atmospheric C14 activity. *Radiocarbon* 9:246–56.

27. R. K. Haugen. 1967. Tree-ring indices: A circumpolar comparison. *Science* 158:773–75.

28. T. A. Rafter and B. J. O'Brien. 14C measurements in the atmosphere and in the South Pacific Ocean: A recalculation of the exchange rates between the atmosphere and the ocean. In ref. 7, pp. 241–66.

29. S. Gulliksen, R. Nydal, and K. Lovseth. Further calculations on the C14 exchange between the ocean and the atmosphere. In ref. 7, pp. C63–C72.

30. G. M. Woodwell and E. V. Pecan, eds. 1973. *Carbon and the Biosphere*. Proc. 24th Brookhaven Symposium in Biology, Upton, New York, May 1972. Technical Information Center, Office of Information Services, USAEC.

31. D. Heymann and O. A. Schaeffer. 1962. Constancy of cosmic rays in time. *Physica* 28:1318–23.

32. E. Crevecoeur. 1966. Détermination de la constance du rayonnement cosmique et des âges terrestres et cosmiques des météorites ferreuses par la radioactivité de l'aluminium 26 et du béryllium 10. *Bulletin Classe Science Academie Royale Belgique* 52:261–75.

33. E. L. Fireman and R. Goebel. 1970. Argon 37 and argon 39 in recently fallen meteorites and cosmic-ray variations. *J. Geophys. Res.* 75:2115–24.

34. V. A. Dergachev and G. E. Kocharov. 1972. Ob odnoy vozmozhnosti izucheniya variatsiy kosmicheskikh luchey v proshlom (A possible cause of cosmic-ray variation in the past). *Izvestiya Akademii nauk SSSR* 36(11):2312–18.

35. R. E. McCrosky, A. Posen, G. Schwartz, and C.-Y. Shao. 1971. Lost City meteorite: Its recovery and a comparison of other fireballs. *J. Geophys. Res.* 76:4090–4108.

36. Z. Ceplecha. 1961. Multiple fall of Přibram meteorites photographed. *Bull. Astr. Inst. Czech.* 12:21.

37. V. Bucha. 1971. Archaeomagnetic dating. In H. N. Michael and E. K. Ralph, eds., *Dating Techniques for the Archaeologist*. Cambridge: M.I.T. Press, pp. 57–117.

38. E. K. Ralph. A cyclic solution for the relationship between magnetic and atmospheric C14 changes. In ref. 7, pp. 91–98.

39. N. Bonhommet and J. Zähringer. 1969. Paleomagnetism and potassium argon age determinations of the Laschamp geomagnetic polarity event. *Earth and Planetary Science Letters* 6:43–46.

40. N. A. Mörner, J. P. Lanser, and J. Hospers. 1971. Late Weichselian palaeomagnetic reversal. *Nature (Physical Sciences)* 234:173–74.

41. M. Barbetti and M. McElhinny. 1972. Evidence of a geomagnetic excursion 30,000 yr BP. *Nature* 239:327–30.

42. R. J. Blakely and A. Cox. 1972. Evidence for short geomagnetic polarity intervals in the early Cenozoic. *J. Geophys. Res.* 77:7065–72.

43. A. Cox. 1969. Geomagnetic reversals. *Science* 163:237–45.

44. Milton Merker, pers. comm.

45. M. Wada and A. Inoue. 1966. Relation between the carbon 14 production rate and the geomagnetic moment. *J. Geomagnetism and Geoelectricity* 18:485–88.

46. H. W. Babcock and H. D. Babcock. 1955. The sun's magnetic field, 1952–1954. *Astrophys. J.* 121:349–66.

47. J. M. Wilcox. 1973. Solar magnetic fields and their influence on the earth. *Naval Research Reviews* 24:16–25.

48. J. G. Roederer. 1974. The earth's magnetosphere. *Science* 183:37–46.

49. F. Link. 1966. Variation de l'activité solaire et de la production de C14 par les rayons cosmiques. *Bulletin Classe Science Academie Royale Belgique* (Ser. 5) 52:486–89.

50. J. R. Bray. 1967. Variation in atmospheric carbon-14 activity relative to a sunspot-auroral solar index. *Science* 156:640–42.

51. M. Stuiver. 1961. Variations in radiocarbon concentration and sunspot activity. *J. Geophys. Res.* 66:273–76.

52. Douglas O'Brien, pers. comm., July 1973.

53. T. Säve-Söderbergh and I. U. Olsson. C14 dating and the Egyptian chronology. In ref. 5, pp. 35–53.

54. H. N. Michael and E. K. Ralph. Correction factors applied to Egyptian radiocarbon dates from the era before Christ. In ref. 5, pp. 109–20.

55. C. Renfrew. 1973. *Before Civilization: The Radiocarbon Revolution and Prehistoric Europe*. New York: Alfred A. Knopf.

56. E. Neustupný. 1969. Absolute chronology of the Neolithic and Aeneolithic periods in central and south-east Europe II. *Archeologické Rozhledy* 21:783–810.

PART 3 *Plate Tectonics and Drifting Continents*

Dan P. McKenzie, "Plate Tectonics and Sea-Floor Spreading," **60**:425 (1972), *page 64.*

William A. Nierenberg, "The Deep Sea Drilling Project after Ten Years," **66**:20 (1978), *page 75.*

Richard K. Bambach, Christopher R. Scotese and Alfred M. Ziegler, "Before Pangea: The Geographies of the Paleozoic World," **68**:26 (1980), *page 86.*

Bruce D. Marsh, "Island-Arc Volcanism," **67**:161 (1979), *page 99.*

G. Brent Dalrymple, Eli A. Silver and Everett D. Jackson, "Origin of the Hawaiian Islands," **61**:294 (1973), *page 112.*

Dan P. McKenzie

Plate Tectonics and Sea-Floor Spreading

Simple geometric ideas have led to a profound understanding of continental motions and the creation and destruction of sea floor

Until about 1960 the possibility that the continents might have drifted huge distances in the course of geological time excited all earth scientists, but produced more imaginative speculation and bitter controversy than sound scientific argument. In the last ten years, however, two instrumental developments have completely changed this situation, and it is now possible to describe the motions of the continents and the evolution of the ocean basins between them in the most remarkable detail. The two new developments were the installation of standardized seismic stations on a global scale to record earthquake waves and the systematic measurement of the earth's magnetic field at the sea surface using proton magnetometers. Both were closely connected with, and to a considerable extent paid for by, the armed forces of the United States, and without this data the advances of the last ten years could not have taken place. Like all major scientific advances, the history of the discoveries is unrelated to the logical development of the theory, and there now seems little purpose in following the various attempts that were made to produce a consistent framework for understanding the tectonics of the earth before plate tectonics was formulated.

The essential idea of plate tectonics is that the surface of the earth may be divided into rigid undeformable

Dan P. McKenzie received his B.A. and Ph.D. from the University of Cambridge and is now on the staff of its Department of Geodesy and Geophysics. His research during the last five years has been on plate tectonics and associated problems, and he is now working on the problem of the driving mechanism and on the deformation of continents. Address: Department of Geodesy and Geophysics, Madingley Rise, Madingley Road, Cambridge CB3 0EZ, England

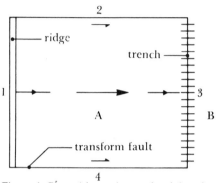

Figure 1. Plate *A* is moving to the right relative to plate *B*. It is being formed along the ridge (1), sliding on transform faults (2 and 4), and being destroyed along the trench (3). (The same symbols are used in following figures.)

spherical caps in relative motion, and that the boundaries of these caps are the belts of earthquakes which cross the earth's surface. The consequences of the rigidity of the caps, or plates, are easier to understand on a plane than on the spherical earth, and Figure 1 shows two flat plates, *A* and *B*, in relative motion. Since there is no way of defining an absolute frame of reference to which we can fix ourselves to measure the absolute motion of any plate, the whole theory must be constructed in terms of the relative motion between two plates. A statement such as "America is moving westward" is meaningless unless an absolute frame of reference can be defined: it is only possible to say that "America is moving westward relative to Eurasia."

In Figure 1 the motion of *A* relative to *B* taken as fixed is shown by the solid arrow. There is no reason why *B* rather than *A* is taken to be fixed, but it is generally easier to understand an argument if one or the other plate is used as a frame of reference. Since *A* is a rigid plate, the motion of all points relative to *B* must be in the

same direction as the heavy arrow, and, in particular, on the boundary marked with a double line, *A* is moving away from *B*. This is only possible if more rigid plate is being created along the boundary 1. Such boundaries are known as ridges because they form shallow linear structures crossing the major ocean basins (see below). This name does, however, sometimes cause confusion since there are many shallow linear structures that are not producing plates; also some boundaries now producing plate are in fact troughs.

One of the surprising features of ridges is that they spread symmetrically, adding to the plates on either side at the same rate. No convincing explanation is known for this behavior, and no good example of an asymmetric ridge has yet been found. Also, ridges are commonly at right angles to the motion direction between the two plates on either side. Though several exceptions to this rule are known, it is obeyed sufficiently often to form a useful guide. There is, of course, no geometric reason why ridges should spread symmetrically or lie at right angles to the motion direction; it is merely an observed fact that they generally do.

In Figure 1, the motion of *A* is parallel to the boundaries 2 and 4, and therefore no new plate can be formed or existing plate destroyed along them. Such boundaries are known as transform faults and are lines of pure slip. They are important because they must by definition be parallel to the direction of motion between the two plates, and are also often easy to recognize on the sea floor and on land. In Figure 1 both transform faults join a ridge to a trench, but they can also join two ridge segments or two trench

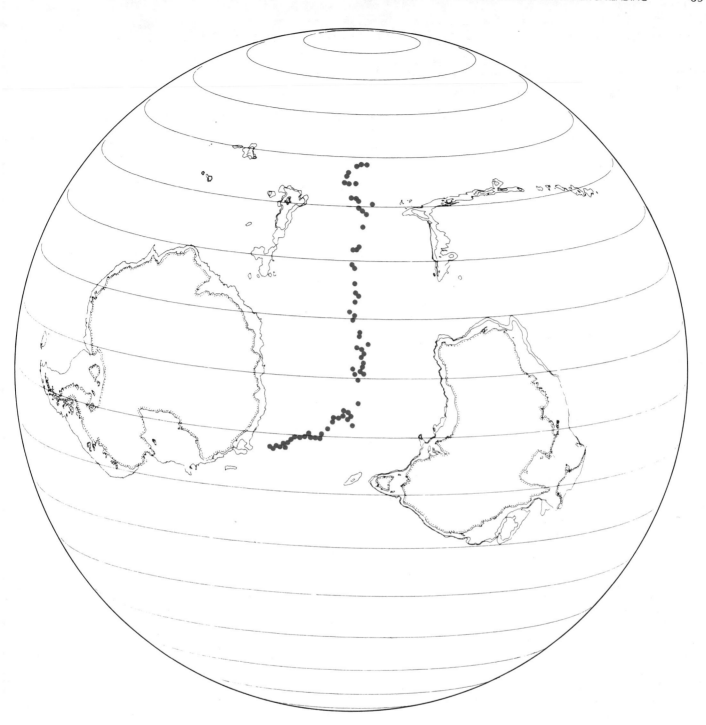

Figure 2. The present position of Australia and Antarctica, with the earthquakes marking the ridge axis.

segments, as shown in Figure 4 (1).

Boundary 3, between *A* and *B* in Figure 1, is a boundary along which one plate moves under the other and is consumed. Such boundaries are called trenches because they are often deep linear slots in the ocean floor. Although there is no geometric reason why trenches should not consume plate on each side, all well-developed ones are asymmetric and consume one or the other plate. An arrow is often used to indicate which plate is being consumed, to show the direction of relative motion between the plates on each side of the boundary, as in Fig-

ure 1 where *A* is being consumed. Unlike ridges, trenches show no tendency to lie at right angles to the motion direction between the plates. They are, indeed, often curved, as shown in Figure 4, with the plate that is not being consumed having a convex plate boundary.

The motion in Figure 1 is a translation of *A* relative to *B*, and therefore all parts of *A* must move in a direction parallel to the heavy arrow. This type of motion is particularly simple to understand, but is not sufficiently general to describe all rigid motions of *A* with respect to *B*, because *A* can

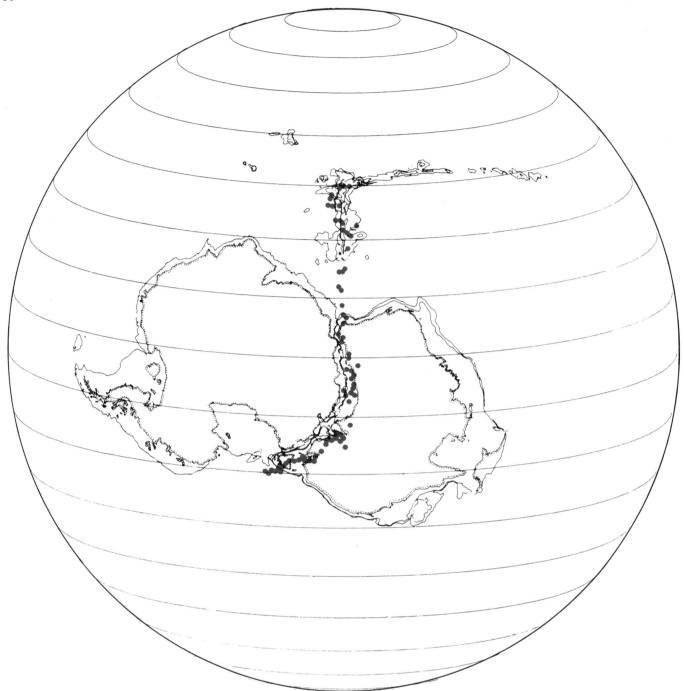

Figure 3. The position of Australia and Antarctica before their separation. Each has been rotated through half the angle of separation, and the earthquakes have not been moved. The initial break is marked by the shocks because the ridge has spread symmetrically.

rotate as well as translate. Figure 5 shows a particular case of the more general motion, with A rotating about a point with respect to B. The transform faults must now be circles centered on the point since the motion must everywhere be parallel to the boundaries, and the motion of all points of A must be along concentric circles centered on the point of rotation. An essential result for plate tectonics is that any motion of A with respect to B may be described by a rotation of A about some point on the plane. This theorem is known as the fixed point theorem, since if A and B are two rigid sheets one on top of the other and each covering the whole plane, any motion of A with respect to B leaves one point on A unmoved with respect to B. This point is therefore at the center of rotation. The special case of a translation can be regarded as a rotation about a point at infinity.

Plate tectonics on the earth is not quite so simple as the examples in Figures 1, 4, and 5 because the earth is a sphere. Fortunately the fixed point theorem also applies to spherical shells, and therefore any relative motion between rigid caps on a sphere can be described by a rotation about some axis. It is easy to see that if the plate is to remain on the surface of the sphere, this axis of rotation must pass through the center of the sphere. This theorem, known as Euler's theorem (2), is fundamental to plate tectonics since it permits any rigid motion between two plates to be described by three parameters: the latitude and longitude of the intersection of the rotation axis with the earth's surface, and the angle through which the plate must be rotated. These three parameters describe the motion and also the fact that both plates move rigidly.

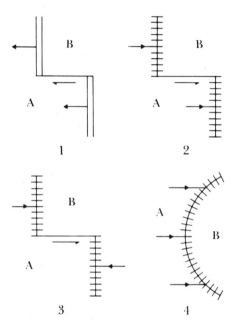

Figure 4. 1–3. Ridges and trenches offset by transform faults, showing the direction of motion between the two sides of the fault. 4. The plate on the side of a trench that is not being consumed is generally convex.

It is obvious that translation on a sphere is impossible to define, since there is no spherical equivalent of the point at infinity. Every axis through the center of the sphere must intersect the surface at some point, and therefore only rotations are possible. On a sphere, as on a plane, each point on *A* moves on a circle with the rotation axis at the center. The lines of latitude on an earth globe are a set of such circles, since each point on the surface of the globe moves along the lines of latitude when the globe is rotated about the earth's rotational axis. Such circles are known as small circles, and transform faults must therefore be parts of small circles about the relative rotational pole (3, 4).

Figure 2 shows the present positions of Australia and Antarctica, and the points at which earthquakes have

taken place on the ridge between them are shown as solid dots. The curved lines are small circles with the axis of relative motion of Australia with respect to Antarctica as their center. Figure 3 shows the relative position of Antarctica and Australia before they separated, and has been produced by rotating each continent through an equal angle toward the ridge axis, which was not moved. Notice that the continents have not moved at right angles to the small circles, because these circles mark the path of points on each continent. Also, note that the earthquake positions lie on the line of the original break, because the ridge between the two continents had spread symmetrically and therefore now lies midway between them.

The part of the ridge axis that was originally between Tasmania and Antarctica is almost parallel to the small circles and now consists of a series of short ridge segments joined by several long transform faults. The original break was very oblique to the direction of motion, and the present arrangement of ridges and transform faults is a consequence of a change from an oblique ridge, shown as a solid double line in Figure 6, to a series of offset ridges at right angles to the direction of motion joined by transform faults, shown as the dashed boundary. The ridge between Australia and Antarctica therefore shows how faults or ridges can be the result of the shape of the initial break. This is a common way in which such offsets are formed.

The discussion of how plates move on a plane and on a sphere has so far been entirely two dimensional. This restriction is artificial, since new plate can only be formed on ridges if material is supplied from below. Also, if the plates are to move rigidly, they must be sufficiently thick to transmit stresses over large horizontal distances without buckling. However, it is advantageous to restrict the theory to two-dimensional motions on a spherical surface and to regard ridges as sources of plate and trenches as sinks because the theory is then closely related to what can be directly observed at the earth's surface.

A three-dimensional picture of planar plate motions is shown in Figure 7 (5). The motions in the hot material beneath the black plates are unknown and are not shown. Along the ridges

this hot material wells up to the surface, but it does not become part of the plate until it has lost heat by conduction to the sea floor and hence has become cool enough to have some mechanical strength. The rigid plates are therefore very thin on the ridge axes and gradually thicken as they move away from the active plate boundary. Plates are destroyed beneath trenches by being pushed back into the hot material that forms the earth's mantle, and because they are very poor conductors of heat, they sink several hundred kilometers be-

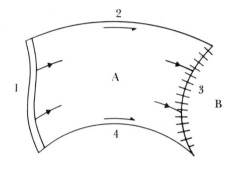

• Point of rotation

Figure 5. *A* is rotating relative to *B* about the point of rotation. Transform faults 2 and 4 form concentric circles about the point of rotations, but there are no restrictions on the shape of 1 and 3.

fore they reach approximately the temperature of the surrounding mantle. These cold slabs beneath the trenches are shown as black tongues sinking into the mantle material in Figure 7, and it is within these tongues that all earthquakes deeper than about 70 km take place.

The remarkable success of these simple geometric ideas when used as a framework for understanding the tectonics of the earth is surprising and impressive. The first striking observation is the narrowness of the plate boundaries when they are defined as the belts in which earthquakes occur. A large number of earthquake positions are plotted in Figure 8 (6), and in oceanic areas they form narrow linear belts separating the aseismic plates. Earthquakes are now known to be the direct result of the relative movement between the two plates on each side of a plate boundary. For some reason not yet properly understood this motion often takes place in jerks, each of which may produce relative motion, from millimeters to meters, on the fault that forms the plate boundary, and each of which causes an earthquake.

The accurate location of large numbers of earthquakes was a consequence of the installation of a world-wide seismic network, which for the first time showed how narrow and well defined are the oceanic plate boundaries. Earthquakes on continents are not confined to such narrow belts, and for this reason there are

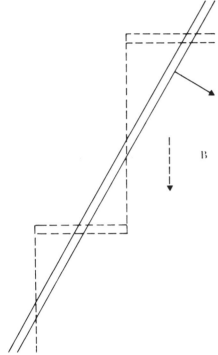

Figure 6. The ridge shown by the solid lines marks the shape of the plate boundary when the relative motion between *A* and *B* was in the direction of the solid arrow. When the direction changed to that shown by the dashed arrow, the shape of the boundary changed to the dashed ridges and transform faults.

still doubts about how useful plate tectonics will be for understanding continental as distinct from oceanic tectonics. The names of seven of the largest plates are shown in Figure 8, together with the direction of relative motion of the plates on each side of a boundary. It is not easy to see that the arrows correspond to rigid relative rotations because Figure 8 is a Mercator projection of the earth's spherical shape onto a plane.

Ridges and magnetic anomalies

For various reasons much more is known about the structure and evolu-

tion of ridges than of trenches. Since we believe that the floor of all deep ocean basins has been produced by active ridges at some time in the past, many inactive structures on the floors of these basins must have been formed when the sea floor was part of the active plate boundary. Therefore a detailed record of past plate motions can often be disentangled from a careful study of sea-floor structures.

Ridges were the first of the three types of plate boundary to be recognized. Hess (7) pointed out that many features of an ocean basin could be understood if new sea floor was produced only along a narrow strip on the crest of the active ridge. Vine and Matthews (8) then used this concept to account for the remarkable magnetic lineations that had been discovered in the northeast Pacific and in the Indian Ocean. Their suggestion may be understood with the help of Figure 9 (9).

Along the ridge axis hot mantle material wells up, and as it does it partially melts to produce basalt, which is extruded at the surface as pillow lavas and intruded at depth as dikes and huge bodies of liquid basalt. The pillow lavas are particularly important because the basalt in them becomes strongly magnetized as the lava is suddenly cooled by contact with the sea water. The magnetization is in the direction of the earth's magnetic field at the time it was cooled. The dikes and larger bodies also become magnetized as they cool, but their magnetization is not so stable and probably does not remain constant for millions of years.

The reason this original magnetization of the pillow lavas is important is that the main magnetic field of the earth

reverses at long and irregular intervals. Such a reversal would cause all compass needles to point south instead of north, and in the past has caused the new sea floor formed on the ridge to be magnetized in the opposite direction. Since the ridges are long linear features, the effect of this process is to magnetize the sea floor in long linear strips of normal and reverse polarity. In areas where the plates are separating at a constant velocity this process is like that of a tape recorder: the active ridge acts as the recorder head and the history of the reversals of the earth's magnetic field is recorded by the magnetization of the lavas on the sea floor (10).

To many people the most puzzling part of the whole process is why the direction of the main field of the earth should change, and this question cannot at present be properly answered. Reversals are not, however, peculiar to the earth; the sun's main magnetic field changes direction every eleven years, at the start of each new sunspot cycle. Also, there are good reasons to believe that the earth's magnetic field is influenced by very complicated movements in the fluid motions of the earth's liquid core. Since studies of much simpler systems have shown reversals, it is not particularly surprising that the earth's magnetic field reverses also. There is therefore no reason to believe that a reversal is the result of any event outside the core. This fascinating problem is the subject of much research at present, but it has little bearing on plate tectonics. What is important, however, is that the reversals of the earth's field are not periodic (nor are they truly random). If they were periodic, the magnetic blocks from a pair of plates separating at a constant velocity would all have the same width and

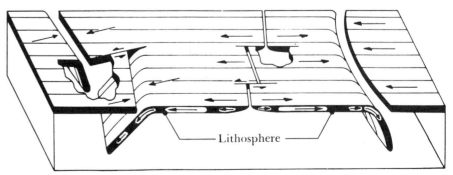

Figure 7. Three-dimensional plate motions, showing the generation of plates along the ridges by slow cooling and the penetration of the cold sinking slabs to great depths beneath the trenches (adapted from Isacks et al., *J. Geophys. Res.* 73:5855, 1968).

Lithosphere

could not easily be used for dating the sea floor.

It is fortunately not necessary to drill a hole in the sea floor to determine the magnetic polarity of the blocks. The magnetic field at the sea surface is principally caused by processes in the earth's core, but it also contains a small component produced by the magnetization of the sea floor. These two parts are easily separated because the core component varies slowly over large distances, whereas the magnetic anomalies produced by the blocks cause rapid variations in the magnetic

have been formed during the last 76 million years. The parts of the ocean where anomalies between 1 and 32 have been recognized, shown in Figure 11, cover somewhat more than half the area of the deep oceans. Furthermore, the oldest rocks yet recovered from the oceans are only between 150 and 200 million years old. Many continental rocks are much older than this—the oldest rocks known at present are 3,900 million years old. This difference in age is one of the most striking discoveries made during the exploration of the oceans and is believed to be a consequence of the low

anomalies have been numbered, according to the scheme described above).

The age of the sea floor decreases northward and eastward, and the anomalies are offset by a series of huge faults—the remains of the transform faults that offset the original ridge axis. These faults and their offsets were discovered before the concepts of sea-floor spreading and plate tectonics were proposed, but they were very difficult to understand, because the continental margin of the western United States showed no correspond-

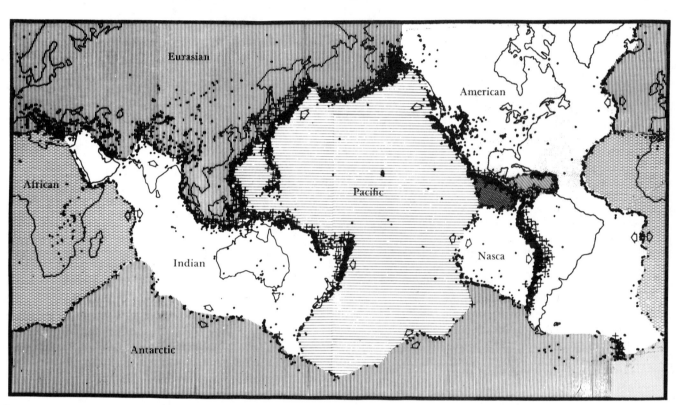

Figure 8. Positions of earthquakes with depths of less than 100 km, Note the difference in their distribution between continents and oceans (6).

field over tens of kilometers. The magnetic anomalies produced at certain times are very distinctive and can often be recognized immediately. The anomalies have quite different shapes because the reversals of the main magnetic field are not periodic. The most obvious anomalies are identified by number (11), from 1 to 33 in order of increasing age (see, e.g., Fig. 10).

The ages of several anomalies have now been directly determined by drilling through the sediment to the pillow lavas beneath the deep ocean, and these studies have shown that all anomalies younger than anomaly 32

density of continental rocks, which prevents them from sinking into the mantle beneath trenches (12).

As well as giving the age of some part of the ocean floor, the magnetic anomalies also mark the shape of the ridge that formed them and are offset by faults that were originally the transform faults between ridge segments. Therefore, if an accurate two-dimensional survey of the magnetic anomalies is carried out, the shape of the lineations can be used to understand the complete tectonic evolution of the ocean floor. Such surveys and interpretations have only been carried out in a few parts of the oceans, of which the northeast Pacific is perhaps the best example. Figure 10 (taken from Atwater, 13) shows the anomalies as straight lines (some of the

ing features. There is no equivalent difficulty of interpretation in terms of plate tectonics, as shown in Figure 12, and this view is now generally accepted. The transform faults offset the ridge axis, whereas the western margin of North America was a trench that consumed the sea floor east of the ridge axis.

This consumption produced geological structures which still exist along the western coast of the United States, and because they are especially striking near San Francisco, the rocks formed during this period are known as the Franciscan formation. They have been contorted by the huge motions between the adjacent plates. Although a small portion of both the ridge (shown dotted in Fig. 10) and the trench remains off the coast of

Oregon and Washington, and indeed the spectacular volcanoes of the Cascade Range are a consequence of the continued consumption of the Pacific sea floor, the plate boundary is now in most places a large transform fault—the San Andreas fault (*13, 14*). The existence of this plate boundary has been known since 1906, when San Francisco was destroyed, but its general significance has only recently become clear. Now that we understand the nature of this fault, we can make some progress toward earthquake prediction, since the relative motion between two plates must be the same everywhere along a transform fault. We therefore expect an earthquake similar in size to the 1906 San Francisco shock to occur along a section of the San Andreas fault north and east of Los Angeles. It is likely to be much more destructive over a much greater area than the 1971 San Fernando earthquake, but unfortunately we have no idea when it will take place.

Another important feature of the magnetic lineations in Figure 10 is the sharp bend in anomalies 25 to 32 in the Gulf of Alaska. The origin of this structure was mysterious until the short transform fault offsetting the northern set of east-west anomalies was discovered. Since transform faults mark the direction of motion between the plates on either side of the ridge, it is impossible for the plate boundary that formed the east-west lineations to be between the same pair of plates as that which formed the north-south anomalies offset by east-west transform faults. Three plates must therefore have been involved, moving in the manner shown in Figure 12, where the Pacific plate is taken to be fixed for convenience only.

One additional feature of Figure 10 is of interest because it shows how transform faults can be formed by a change in the direction of relative motion between plates, rather than by the rationalization of an initial break, as was the case between Australia and Antarctica. At about anomaly 24, both the anomalies and the transform faults show a change in direction from a NNW–SSE to a N–S direction, in the case of the magnetic lineations, and from an ENE–WNW to E–W, in the case of the transform faults. At the same time new transform faults—the Sedna, Sila, and Aja faults—were formed in the Gulf of Alaska in the

(A) Ridge model

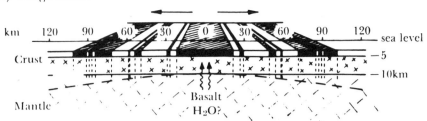

(B) Anomaly map

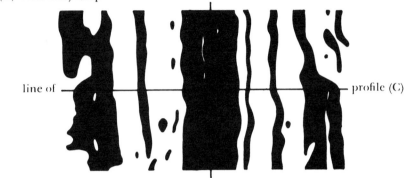

(C) Observed profile

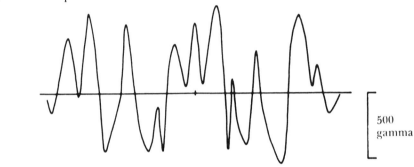

(D) Simulation

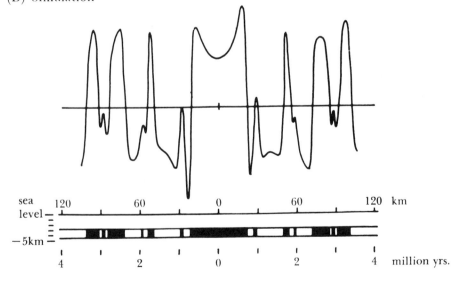

Figure 9. *A*. The generation of magnetic anomalies by a spreading ridge. Shaded material is normally magnetized, unshaded is reversely magnetized. *B*. Part of a map of magnetic anomalies over the Juan da Fuca Ridge in the northeast Pacific. The black regions are positive anomalies, the white are negative. *C*. Magnetic anomaly profile along line shown in *B*. *D*. A profile computed from the reversal time scale.

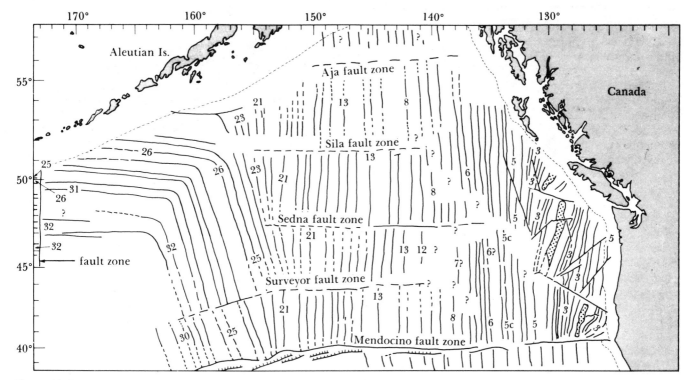

Figure 10. Magnetic lineations in the northeast Pacific. Individual anomalies are identified by numbers, as described in the text (adapted from Atwater, *Bull. Geol. Soc. Amer.* 81:3513, 1970).

northeast corner of the Pacific basin. These new faults offset what was earlier a straight ridge axis. However, since the direction of spreading changed at about anomaly 24, the ridge either had to spread obliquely or had to break up into ridge segments separated by transform faults, as in Figure 5. Obliquely spreading ridges are rare, and in the Gulf of Alaska new transform faults were formed instead. Transform faults of similar origin have more recently been discovered in the North Atlantic and Indian Oceans, and appear to be relatively common.

Trenches

Trenches are much more complicated structures than ridges, and their evolution is much less well understood because they consume sea floor and leave little evidence of the vanished material. The evolution of the continents is closely connected to that of trenches, because vulcanism in the island arcs behind trenches produces rocks similar to those out of which the continents are built. The whole process of generation of continents is more complex than that of the generation of sea floor, and the concepts of plate tectonics have had much less

effect on the study of the deformation of continental regions.

The importance of trenches and island arcs was recognized by Vening Meinesz (*15*) and by Gutenberg and Richter (*16*) more than forty years ago, but a clear understanding of their structure became possible only when accurate locations of earthquakes and good determinations of their mechanisms became available A particularly striking observation was made by Sykes (*17*), who carefully located earthquakes near the Tonga and Kermadec trenches in the southwest Pacific. Figure 13 shows the locations of the deepest earthquakes in this region, with all shocks shallower than 500 km omitted. The striking feature of this plot is the hook at the northern end of the deep earthquake region, which is a reflection of a corresponding hook in the trench, marked by the shaded 6 km contour. This feature was very difficult to understand when it was discovered. It is, however, much more obvious how such a structure can be formed if the Pacific plate is moving beneath the island arc, marked by the volcanoes.

Figure 7 shows diagrammatically what is happening, with the trench on the left representing the Tonga-Fiji-Kermadec trench. Since earthquakes are believed to be restricted to the lithosphere, including the tongues sinking into the mantle beneath trenches, the deep earthquakes in Figure 13 must

mark the shape of the sinking lithosphere. A section through this region in Figure 14 shows the continuous narrow band of earthquakes that connects the deepest with the shallow shocks beneath the trench and represents part of the evidence for our present views about the deep structure of island arcs. The explanation of the hook in Figure 13 is now clear, since the shape of the slab depends on the shape of the trench where it started to sink and does not change as it descends. The motion between the sinking plate and the other plate is the cause of most major earthquakes, and their mechanisms and resulting surface deformation have been used to study the details of the deformation between and within the plates as they bend and sink.

One reason why some oceanographers found the concepts of sea-floor spreading and plate tectonics difficult to accept was that trenches showed very little evidence at the surface of being major consumers of oceanic plate. They often contained undeformed, flat-bedded sediment. Moreover wherever deformation and faulting can be seen on the sea floor, it more resembles the extensional faulting found on ridges than the faulting expected when one plate is overriding another. The explanation of these apparent inconsistencies is gradually becoming clearer as the complications of the deformation are unraveled by very detailed studies of earthquakes.

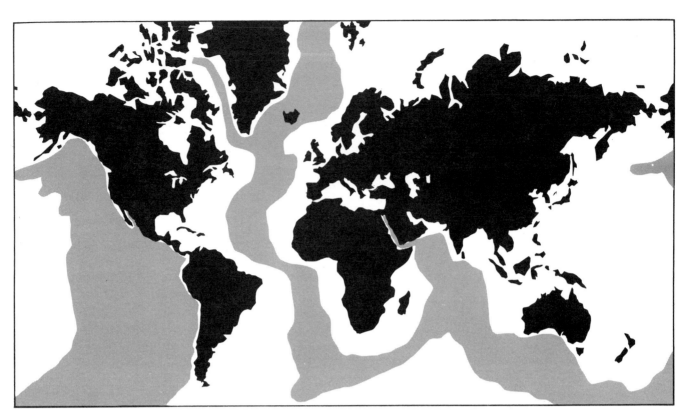

Figure 11. Areas of sea floor created in the last 75 million years (shown in gray) cover a large part of the floor of the ocean basins. In contrast, little continental area has been formed during this time, and considerable portions of many continents (black) are older than 1,000 million years.

Some extension faulting occurs within the plate that is bent, because of the stresses induced by bending. Slow production of sea floor may also be taking place above the sinking slab. Also, oceanographers have now observed the thrusting of the arc over the sea floor, in agreement with the earthquake mechanisms. There are therefore no observations at variance with the plate tectonic interpretation of trenches, though many details and complications remain to be understood, and many observations are not explained by plate tectonics. In particular, it is important to discover whether the sediments on top of the sinking plate are scraped off and added to the front of the island arc, or whether they sink with the rest of the plate and perhaps melt to contribute to the material erupted by volcanoes.

Convection

The success of the geometric ideas of plate tectonics has been remarkable and has surprised everyone who has contributed to their development.

One important consequence of the success of the geometric—or kinematic—theory is the realization that little is at present known about the dynamics of plate motion. One by one the observations that were once believed to be a consequence of convection within the mantle are seen to be a consequence of the creation, motion, and destruction of plates. Measurements of the flow of heat from the earth's interior are high near ridges, because the hot plate that is being formed cools as it moves away by losing heat through the sea floor. Ridges are elevated above the surrounding sea floor because they consist of rock that is hotter and therefore less dense than the older, colder plate; the lighter material is elevated in the same way as icebergs protrude above the sea surface. The place where plates sink into the mantle is marked by a trench because the plate cannot bend sharply and therefore starts to bend before it reaches the island arc. Though they appear to have little surface expression, convective motions within the mantle below the plates must exist to move the material from the island arcs back to the ridges, and hence form a closed circulation.

At present it appears that thermal convection of some type is the only mechanism that can generate enough energy to sustain the energy released

Figure 12. *A*. Plate boundaries and motions about 40 million years ago (anomaly 15) in the northeast Pacific. The Pacific plate is taken to be fixed and the large arrows on the American and Farallon plates show the direction of motion between each plate and the Pacific plate. The arrow on the trench between the Farallon and American plate shows the direction of motion between them. *B*. The arrangement of ridges in the eastern Pacific 75 million years ago (anomaly 32), which formed the Great Magnetic Bight, now in the Gulf of Alaska.

A

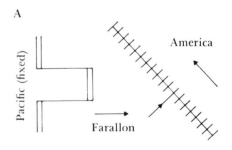

B

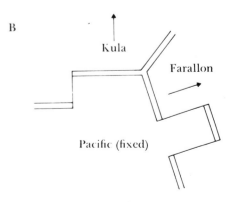

by earthquakes, and therefore several studies are being made of the form such convection can take. The most familiar forms of convection take place when a fluid is heated from below, for instance in a kettle or porridge pan, or when making jam. Convection in the earth must be maintained by radioactive heating from within, and not by heating from below. If convection extends throughout the whole mantle, all the heat must come from within, but if convection extends only to 700 km depth, then heating may be partly from below and partly from within. Convection must extend at least to 700 km because the sinking plates in which deep earthquakes occur reach this depth. Heating from within produces a very different type of convection from that which results from heating from below.

Figure 15 shows the flow lines and contours joining points at equal temperature of convective motions that result from the two cases. When the heat is applied below, it can be moved to the upper surface by transporting a thin layer of hot fluid, called a boundary layer, from the bottom of the box to the top. Most of the fluid is at constant temperature and is not involved in the heat transport. No such flow is possible if heat is generated within the fluid, because all heat must be lost from the upper surface. Therefore, all parts of the fluid must at some time during the flow come close to the upper surface to lose their heat. They are then returned to the bottom of the box and gradually warm up as they rise toward the surface. The cold regions of Figure 15B already resemble the shape of the plates in Figure 7; the major difference is the huge horizontal extent of the plates on earth, which may be the result of the great strength of cold rocks. The fluid in Figure 15 does not become strong when cold, and this property, like the internal heating, should strongly modify the convective flow.

Though the simple models in Figure 15 are very remote from the real earth, they are an attempt to understand the general form of possible motions. They also illustrate an important feature that any dynamic theory of the earth's tectonics must include: the plates and the sinking slabs beneath island arcs are simply the boundary layer of the convective

flow. The main flow cannot be understood without the boundary layer, and vice versa. Since the two flows are strongly coupled mechanically and thermally, they can only be understood dynamically when considered together.

The current absence of a detailed dynamic theory in no way affects the use of the kinematic theory of plate tectonics to describe the evolution of the earth's surface. Although most geophysical phenomena relevant to the theory now seem to have been

discovered, its geological and petrological implications have only recently generated interest, and almost no work has yet been done on the circulation of oceans and atmospheres in the geological past using plate tectonic reconstructions. Such a study might determine whether ice ages are simply the consequence of the changing shape of ocean basins. Another interesting field which can now be studied is the effect continental motions have on the distribution of animals and plants and on the evolution of marine organisms. There-

Figure 13. The distribution of deep earthquakes at the northern end of the Tongan Island Arc mirrors the hook at the northern end of the trench (adapted from Sykes, *J. Geophys. Res.* 71:2981, 1966).

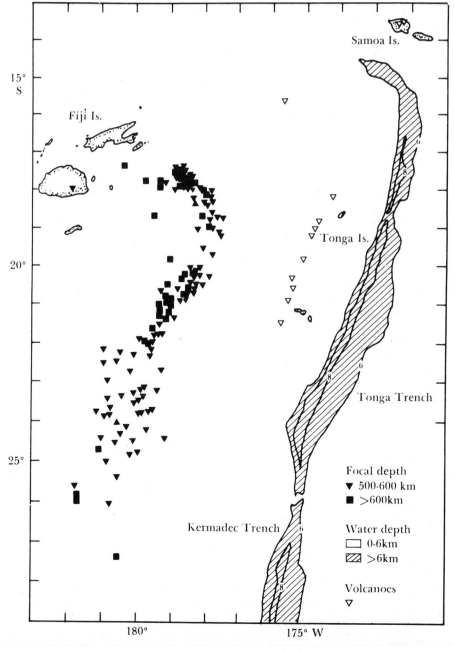

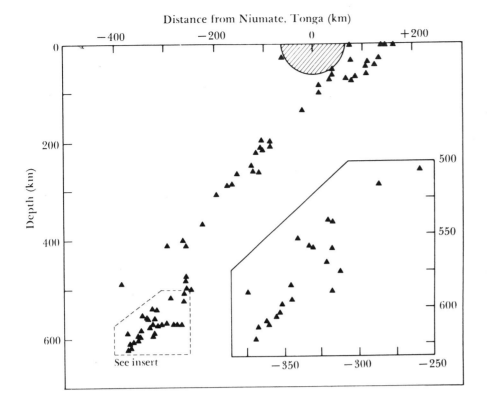

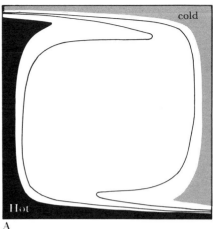

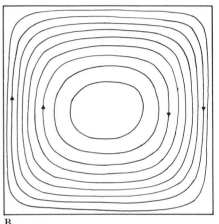

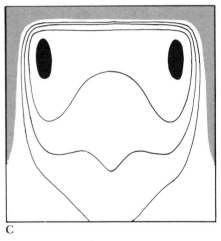

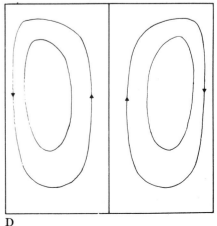

Figure 14. Tonga-Fiji earthquakes. Section at right angles to the northern end of the Tongan Island Arc, showing the locations of intermediate and deep focus earthquakes which lie within the sinking slab (adapted from Sykes, *J. Geophys. Res.* 71:2981, 1966).

Figure 15. Temperature (*A*) and flow lines for convection in a fluid heated from below (*B*). Heat is transferred from bottom of top by hot rising plumes (black) and cold sinking plumes (light gray) . Temperature (*C*) and flow lines for convection in a fluid heated from within (*D*). The cold, sinking plumes remain (gray) but the rising ones are absent. The whole fluid rises toward the surface, as is shown by the flow lines in *D*.

fore, although the kinematic theory of plate tectonics is probably reasonably complete, its important consequences to other subjects have been little studied, and the dynamic theory is scarcely understood at all.

References

1. Wilson, J. T. 1965. A new class of faults and their bearing on continental drift. *Nature* 207:343.

2. Bullard, E. C., J. E. Everett, and A. G. Smith. 1965. The fit of the continents around the Atlantic. *Phil. Trans. R. Soc. London* A258:41.

3. McKenzie, D. P. and R. L. Parker. 1967. The North Pacific: An example of tectonics on a sphere. *Nature* 216:1276.

4. Morgan, W. J. 1968. Rises, trenches, great faults and crustal blocks, *J. Geophys. Res.* 73:1959.

5. Isacks, B. L., J. Oliver, and L. R. Sykes. 1968. Seismology and the new global tectonics. *J. Geophys. Res.* 73:5855.

6. Barazangi, M., and J. Dorman, 1969. World seismicity maps compiled from ESSA Coast and Geodetic Survey epicentral data. *Bull. Seism. Soc. Am.* 59:369.

7. Hess, H. H. 1962. History of the ocean basins. In *Petrologic Studies* (Buddington Memorial Volume), Geol. Soc. Amer., p. 599.

8. Vine, F. J., and D. H. Matthews. 1963.

Magnetic anomalies over ocean ridges. *Nature* 199:947.

9. Vine, F. J. 1968. Magnetic anomalies associated with mid-ocean ridges. In *The History of the Earth's Crust*, R. A. Phinney, Ed. Princeton University Press.

10. Vine, F. J. 1966. Spreading of the ocean floor: New evidence. *Science* 154:1405.

11. Pitman, W. C., and J. R. Heirtzler. 1966. Magnetic anomalies over the Pacific–Antarctic ridge. *Science* 154:1164.

12. McKenzie, D. P. 1969. Speculations on the consequences and causes of plate motions. *Geophys. J. R. Astr. Soc.* 18:1.

13. Atwater, T. 1970. Implications of plate tectonics for the Cenozoic tectonic evolution of western North America. *Bull. Geol. Soc. Amer.* 81:3513.

14. McKenzie, D. P., and W. J. Morgan. 1969. The evolution of triple junctions. *Nature* 224:125.

15. Vening Meinesz, F. A. The earth's crust deformation in the East Indies. *Koninkl. Ned. Akad. van Wetensch.* 43:3.

16. Gutenberg, B., and C. F. Richter. 1938. Depth and geographical distribution of deep focus earthquakes. *Bull. Geol. Soc. Amer.* 49:249.

17. Sykes, L. R. 1966. The seismicity and deep structure of island arcs. *J. Geophys. Res.* 71:2981.

William A. Nierenberg

The Deep Sea Drilling Project after Ten Years

Deep-sea drilling has provided exciting new evidence and taught us much about the earth

Just over ten years ago—August 1967—*Glomar Challenger* went on her acceptance trials preparatory to her first scientific cruise. The project was a daring one—to build, design, and operate a drilling and coring vessel at water depths of 15,000 ft and more in mid-ocean. The vessel had to hold position within a few hundred feet for days at a time, unaffected by reasonable winds and currents. This had never been done and, with one notable exception, has still not been duplicated despite the rapidly developing application of ocean technology to offshore drilling.

The scientific developments of which Deep Sea Drilling Project formed a major, and some would say culminating, part represent one of the great achievements of the twentieth century and have so been recognized over the entire world. The Russians were instantly admiring and very soon bid to join with the U.S. institutions that had sponsored the project—along with a direct annual contribution of $1 million. Megascale geology has been transformed from a vast collec-

William A. Nierenberg is Director of the Scripps Institution of Oceanography. His current scientific interests are in physical oceanography but also include political and social affairs involving the oceans and the atmosphere. His earlier research was in nuclear physics at the University of California, Berkeley. He is a member of the National Science Board, former Chairman of the National Advisory Committee on Oceans and Atmosphere, Advisor-at-Large to the State Department, and Assistant Secretary General to NATO for Scientific Affairs. He is also a member of the National Academy of Sciences, the American Academy of Arts and Sciences, and the American Philosophical Society. Address: Scripps Institution of Oceanography, La Jolla, CA 92093.

tion of apparently unconnected phenomena to a global synthesis of most of what had been observed, all under the rubric of plate tectonics and continental drift. To understand what was accomplished and how, we must go back about fifteen years and review two separate affairs—one that turned out to be political and the other purely scientific.

The first was Project Mohole. It is not my purpose to review the history of that ill-fated project, but one of the major difficulties we faced in the development and launching of our project was the shadow of the Mohole, which bore a mechanical resemblance to our proposal. The Mohole Project, you will remember, had a very straightforward objective—the lowering of a drill string through the water column and drilling and coring more than 6 km through the ocean bottom to reach the Mohorovičić discontinuity. This discontinuity is, as it affects the velocity of seismic waves, the principal, if indirect, source of information on the earth's interior. The importance of the Moho is that it is continuous around the earth and presumably marks an important change of property of the constituents of the upper mantle of the earth. The reason for siting the Mohole in the oceanic crust is that, whereas the Moho averages 35 km in depth below the continents, it is only 6 km or so below the bottom of the sea—making an attempt to reach it seem more achievable.

Project Mohole started off auspiciously enough with successful engineering trials off Baja California, led by Willard Bascom. However, its development was marked by indecision and organizational infighting, and its projected cost escalated to an esti-

mated $125 million, which blew it up on the floor of the U.S. Senate, accompanied by charges of political scandal.

From the beginning, many marine geologists felt that the Mohole Project was one of misplaced priority and that a financially less ambitious project that would sample sediments of the world's ocean basins would be far more rewarding scientifically. Of course, the point of this story is that they were right.

Therefore, as it happened, one or two years before the demise of the Mohole, a few of the country's oceanographic laboratory directors decided to form an informal association to pursue this new objective. There were several attempts at association, and one that succeeded—JOIDES, which stands for Joint Oceanographic Institution Deep Earth Sampling. The original four institutions were the Lamont Geophysical Observatory of Columbia University (now the Lamont-Doherty), the Institute of Marine Science of the University of Miami (now the Rosenstiel Institute of Marine Sciences), the Woods Hole Oceanographic Institute, and the Scripps Institution of Oceanography of the University of California.

Owing to the trauma of Mohole, the four institutions decided to proceed cautiously to test the technical feasibility, but perhaps even more the political feasibility, of the association. These first trials were very successful. With Miami as the grantee of the National Science Foundation and Lamont as the operating institution, a drilling program was successfully completed on the Blake Plateau—a bottom feature off the southeast coast of the United States. J. Lamar Worzel

Figure 1. The *Glomar Challenger*'s four-story laboratory has equipment for about twenty scientists and technicians. The 400-ft vessel, with twelve 800-horsepower diesel engines, has a ship's crew of forty-five. The tower of the permanent drilling rig is 196 ft above the sea. (Photograph by the author.)

was principal investigator, and a ship of opportunity—*Caldrill*—was leased for one month to drill and core in depths of up to 3,400 ft of water. Pleased with the success of this trial venture, the group decided to go forward and nominated the Scripps Institution as the operating institution, and therefore, by inference, myself as the principal investigator.

Progress in the earth sciences

Certain parallel developments in the earth sciences made the Deep Sea Drilling Project initially very interesting. One is a scientific prehistory to this story. The full force of the modern development of plate tectonics and the achievements of *Glomar Challenger* would have been greatly diminished if the earth scientists had paid closer attention to certain observations. They might well have made great discoveries much earlier.

The first of these observations was published by Alfred Wegener, a Swiss meteorologist, in 1910. (It is very little known, but an American, F. B. Taylor, preceded him by two years.) He noted that the east coast of South America and the west coast of Africa fitted together almost perfectly. He proposed that the two continents had originally been one and so produced the concept that is now known as continental drift. It was not accepted seriously, at least by geologists who worked in the Northern Hemisphere.

The second was based on the erosion of continents. Geologists are able to make very good estimates of the deposition of erosional sediments into the world's oceans. The rate is such that the basins would be filled in tens of millions of years. Even allowing for adjustments like isostasy, which accounts for areas of less dense crustal material rising above areas of more dense material, to increase this number, this is far short of the several billion years generally accepted as the age of the earth. The gross inconsistency remained for many years.

A third observation was based on the residual magnetism of rocks. Many rocks will have a residual magnetism that is fixed by the direction of the earth's field at the time the rock cools or is deposited and takes its magnetic character—a time that is often known from independent geological evidence. When the direction of the residual magnetism was measured at widely spaced points for the same epoch it was believed that their intersection fixed the position of the magnetic pole at that time. This led to the concept of polar wandering.

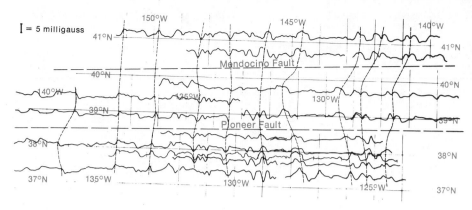

I = 5 milligauss

Figure 2. By measuring the total magnetic field intensity along east-west tracks starting from the west coast of the United States, Mason and Raff were able to identify major east-west escarpments, or transform faults.

The fact that the magnetic poles are supposed to be intimately associated with the earth's rotational axis gave rise to revolutionary concepts like the earth's north pole being located at the Mediterranean basin in the relatively recent geologic past. This is a violent concept, but it had to remain with the other unsorted observations.

We will return to these observations later and see how they fit together and join the body of knowledge extracted from plate tectonics, but here I will make a second digression.

The R & D efforts in antisubmarine warfare of World War II spilled over into the immediate postwar period and were brilliantly exploited by such men of genius as Harry Hess, Maurice Ewing, Victor Vaquier, and Russell Raitt. The first results were from precision bathymetry, noting the physiography of the ocean bottom. The most startling and also relevant feature for our discussion that emerged was the mid-ocean ridge. These continuous mountain ridges of enormous length gird the earth beneath the sea.

Second, exploitation of seismic exploration techniques that had already been widely used by geologists for prospecting on land revealed beneath the sea an upper layer of what were almost certainly sediments. But the sediments are almost nonexistent at the mid-ocean ridges and generally get progressively thicker toward the continents, where they reach thicknesses of several kilometers. It was Russell Raitt who demonstrated the existence of the thin sediments, which alone indicated that something was amiss if there is supposed to be a continuous and reasonably uniform deposition of sediments.

Before I describe the third set of results, I shall digress within the digression to point out that before the advent of *Glomar Challenger*, the occupation of the marine geologist was very grim indeed. Unlike his counterpart on land, his contacts with rocks were few and far between. Knowledge was largely based on the statistical and sparse sampling afforded by dredging and shallow coring from oceanographic vessels. The ocean—15,000 ft thick—is an almost complete barrier between the geologist and his geology. As a result, he is particularly dependent on remote-sensing technology, and he ingeniously and doggedly employs any physical variable for remote sensing, no matter how limited the information or great the effort.

The researches that were perhaps the most revealing in this pre-Deep Sea Drilling period were those that depended on the measurement of the magnetic field. The first measurements were made using ship-towed fluxgate magnetometers adapted from World War II antisubmarine warfare. Fortunately, the proton nuclear magnetometer was developed at about this time. This was a more re-

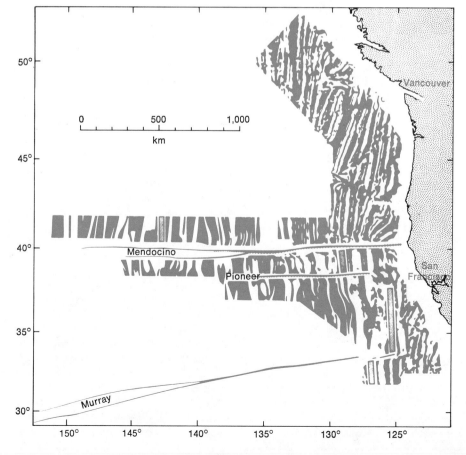

Figure 3. Mason and Raff took the data from Figure 2 and plotted the regions of magnetic field less than the average in black and the regions above average in white. The correspondence of the magnetic shifts on both sides of the ridge led to the theory that new ocean bottoms were formed at the ridges.

Figure 4. The continuous upwelling of magma at the mid-ocean ridge forces the crust to spread laterally away from the center. *N* is the "normal" direction of magnetization; *R* is its opposite.

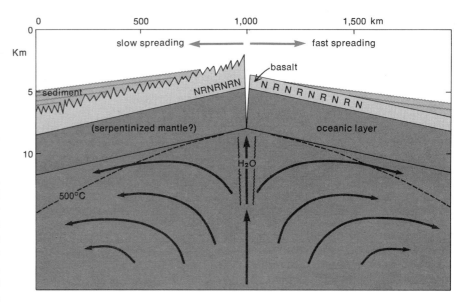

liable and stable sensor very well adapted to shipboard application.

The first breakthrough was achieved by Ronald Mason and Arthur Raff at the Scripps Institution of Oceanography. They made a series of line surveys of the magnetic field of the west coast of the United States; the traverses were naturally east-west, perpendicular to the coast. The composite results, shown in Figure 2, are very revealing. The first feature noted is a vertical (north-south) correlation. The major magnetic features are in north-south alignment, but they are interrupted by east-west displacements. The lines of displacements are identifiable with major east-west escarpments—surface expressions of what geologists now call transform faults. The most prominent of these enormous features is the Mendocino escarpment, which can be traced for 6,000 miles and has a maximum relief of over 10,000 ft.

The Mason-Raff diagram (Fig. 3) is more revealing. It is based on the same data as Figure 2 but smoothed out by representing values of the magnetic field that are less than average in black and greater than average in white, which reveals an extraordinary regularity. The diagram simplifies to a series of vertical black and white stripes that are periodically displaced along the transform faults.

It was these data (incomplete, as it turned out) that led the Canadian earth scientist L. Wilson Morley to his inspired deduction that mid-ocean ridges are loci of creation of new ocean bottoms. The result of upward motion in the upper mantle of the earth at the ridges, the new crust spreads laterally away from the ridge center (Fig. 4). As the plastic material mounts, its temperature drops, eventually passing through the Curie temperature where the material can assume a permanent magnetization, which will be in the direction of the earth's field at that time. However, it is known that the earth's magnetism reverses aperiodically

with a typical interval between reversals of 500,000 years, which clearly would explain the magnetic stripes. Further, the stripes should be symmetric on both sides of the mid-ocean ridge. There is, however, no ridge off the coast of California, and therefore no symmetry visible. Morley brilliantly surmised that this was an exceptional case—that, in fact, the mid-ocean ridge off California was under California and at the location of the San Andreas Fault, to be exact. He guessed that the sea-floor spreading—as it came to be called—was overridden in California by continental drift. Morely's deduction has been spectacularly borne out by further analysis. Figure 5 shows the magnetic trace across a mid-ocean ridge; the symmetry is quite clear. From measurements of igneous se-

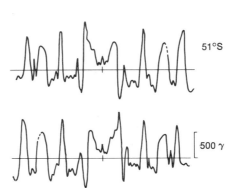

Figure 5. In this actual magnetic trace across the East Pacific rise south of Baja California, the second trace is the left-right reverse of the first. The symmetry confirms Morley's brilliant deduction that the continental drift of California had covered the mid-ocean ridge, part of which is the San Andreas Fault.

quences on land, something was known about the magnetic reversals of the last several million years, and we see that the first few Mason-Raff stripes in Figure 6 correlate very well with the reversal time scale. The correlation of the data gives a spreading rate of between 3 and 10 cm/yr depending on the ridge.

The exciting developments described above were taking place while four of us, T. van Andel, William Reidel, J. D. Frautschy, and I, were busy putting together a proposal to the National Science Foundation for a Deep Sea Drilling Program.

We end our nested digression, leaving aside related researches of the time such as gravity, electricity, and heat flow results, to return to the Project, for, brilliant as Morley's deductions were, particularly when complemented by the work of H. Hess, D. Matthews, F. Vine, and Tuzo Wilson, the earth scientists could never completely accept these hypotheses without actual samples of the ocean's sedimentary basins.

Glomar Challenger

Work on the proposal began almost immediately after July 1, 1965, after the JOIDES decision. After several rewrites and consultations with JOIDES, a final version was accepted and a letter contract was offered the University of California on June 24, 1966. Just one week later, on July 1, W. W. Rand, a geologist from Berkeley and a pioneer in offshore drilling

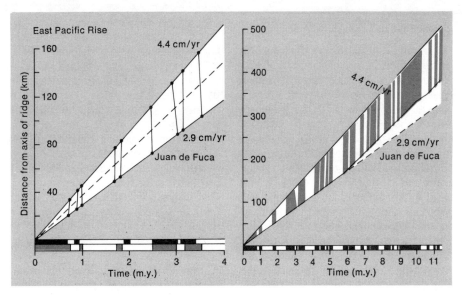

Figure 6. The oceanic magnetic anomalies in distance (Fig. 3) versus time show a spreading rate of between 3 and 10 cm/yr.

technology, was appointed Project Manager. We were determined to build on available technology as much as possible and eschew all unnecessary R & D, and thus we decided to subcontract the vessel and its operations by bid solicitation conforming to our designs, specifications, and program plan. This was very difficult because, owing to some of the missing technology such as the positioning subsystem, we might overspecify, or, because of timidity, we might set our goals too low.

On November 14, 1967, a subcontract was let to the Global Marine Corporation of Los Angeles to build and operate a drilling research vessel. The keel for the vessel—*Glomar Challenger,* a name chosen by us to honor the original *Challenger*—had already been laid on October 18, in Orange, Texas. It was at this point that the immensity of the responsibility took tangible reality. The builder's trials were scheduled for mid-June 1968 (they actually took place June 19–22) and the acceptance trials were in late July and August.

From then on, *Glomar Challenger* was to operate a 55-day leg, 24-day mode, with 5-day in-port changeover, for 18 months at a day rate of $10,000 (today about $20,000). At that cost it was imperative that all available ship time be used for science. Several experienced technical people at Scripps were charged with designing and overseeing the installation of a four-story laboratory in the vessel besides designing a safe core-storage system aboardship and for transporting the

cores to permanent storage onshore. A chief scientist—Melvin Peterson—was named in December 1967 to coordinate all the inputs, arrange for the science team for each leg, and generally oversee all the scientific aspects of the project.

The miracle was that by the deadline of August 11, 1968—the start of the first scientific leg of *Glomar Challenger*—the vessel's laboratories were completely equipped with the scientific and archival equipment needed for the approximately twenty scientists and technicians who made up the team for each cruise. The laboratory is housed in a conventional three-story building with its own elevator. It was designed that way by Global Marine so that, if the renewal options after the 18-month lease in our initial contract were not exercised, the building could be bodily removed and the ship converted to an offshore drilling vessel.

This problem emphasizes the difficulty in initiating scientific projects that require substantial unit capital investment. The NSF budgets on a yearly basis and had about $7 million available that year for the project. However, because of the amortization of the vessel and related costs, a minimum lease of 18 months was needed. The government funds were adequate to cover subcontract cancellation costs, but the University of California would have had the responsibility of covering the remainder. Fortunately, the project was a success and the question never arose.

Returning to the ship, we have a vessel that was unique for its time. There was nothing extraordinary about its dimensions—400 ft in length and 10,500 gross tons—with a ship's crew of 45. However, the vessel has a permanent drilling rig and platform marked by its tower 196 ft above the sea. The tower carries a 60,000 lb traveling block operating to and from the drilling platform. The drill pipe of special steel is fed in triple lengths of 90 ft to the platform by an automatic racker. There it was joined continuously to make up the drill string by one of two 9-man roughneck, or drilling crews. This is an arduous, sometimes monotonous, but never totally safe task: it takes about 170 triple lengths of drill pipe just to get to the sediment, at an average of 12 hours of effort, plus another 12 hours for returning the string to deck.

While a string can be easily jettisoned for cause such as weather, the current cost of a full string with its bottom assembly and drilling tool is close to $500,000, and its abandonment is not to be done lightly. At the beginning of the project, because of the unknowns, we conservatively budgeted for two lost strings a year. In ten years' time, however, we have lost only two full strings—one because of an error in procedure on deck, the other drilling on the Falkland Plateau, east of the Straits of Magellan, where the normally bad weather finally caught up with us. The wind and the currents exceeded the capacity of the ship's automatic positioning equipment, which was unique for its time and which accounted in large part for its great technical success.

In operation, *Glomar Challenger* proceeds to a presurveyed site proposed by the JOIDES planning committee. After a brief corroborative survey, the vessel lowers to the bottom a battery-operated sound source that emits a two-millisecond, one-kilowatt pulse every two seconds. This sound pulse is received by three widely spaced hydrophones attached to the hull, and a computer is used to calculate the position of the vessel from the times of arrivals of the pulse, taking into account the roll and pitch of the vessel as read from a gyrosensor. This information is used to control some of the 11 (now 12) 800-horsepower diesels in the engine room that drive the twin screws and the forward transverse tunnel fans that position the vessel and keep it within 100 or 200 ft of a fixed location in seas with current as high as 2 knots and winds to 40 knots.

From the very beginning, satellite navigation was employed to locate the hole absolutely. Since the vessel can be on-station for 5 to 6 days while drilling, and sometimes longer, the large number of fixes is used to locate the absolute position on the earth with a precision of the order of 100 ft. This is important if the hole has to be reentered—a basic problem of the early years of the project, because very often a drilling program had to be aborted because the drill bit wore out. At the beginning, 30 different bits were on board to meet the variable (and unknown) drilling conditions; the bits varied from the relatively inexpensive ($3,500) tungsten carbide fixed buttons to the costly ($8,500) diamond bits—for once the less expensive variety was the best.

The required reentry capability was eventually developed, with first success on Christmas Day, 1970, in 13,500 ft of water in the Caribbean. This allowed the chief scientist to call for indefinite drillings to the weight limit of the derrick—24,000 ft of drill string. The most that has been suspended is 22,000 ft in 20,500 ft of water. The deepest penetration into the bottom is 5,800 ft and the deepest penetration in the basaltic layers below the sediments is 1,900 ft.

The vessel has sailed 250,000 nautical miles and has drilled in all the world's oceans including the Antarctic, the Mediterranean, and the Black Sea. To July 1, 1977, 429 sites were occupied and drilled: 625,000 ft of ocean bottom have been drilled and 159,000 ft of core recovered and archived. The operation record has been spectacular: the vessel has had only 100 down days in ten years, not counting the planned changeover and overhaul periods. The point of all this, of course, is the collection of cores. These are recovered in 30-ft sections in plastic barrels, and cut into 6-ft lengths and split lengthwise for both ease in handling and storage and sampling and archiving.

Scientific findings

I can only highlight here the scientific accomplishments of *Glomar Challenger,* and even sketching them is not completely possible, for literally every leg returns with at least one major discovery. *Glomar Challenger* is responsible for two kinds of science—that performed on the spot by the shipboard scientists to interpret and guide the immediate drilling and extract the most pertinent and obvious conclusions, and the analysis work that is carried out in the indefinite future on the cores and deductions from them. The results are available in bound volumes of about 1000 pages each, called the Preliminary Report. To date, 38 volumes have been published.

With the Preliminary Report as a reference, scientists all over the world ask for core samples for analysis and research. There is no way to evaluate this work numerically in terms of numbers of papers—it literally pervades the literature of all the earth sciences, including meteorology and cosmic ray physics, for example. Over 85,000 core samples have been distributed to 1,100 investigators, including 230 in foreign countries.

Interestingly enough, the first major discovery was not related to sea-floor spreading. The first leg started immediately from the Texas shipyard, and the first holes drilled were in the Gulf of Mexico on the way out to the Atlantic. Ewing and Worzel of Lamont were the co-chief scientists on the leg, and the second site was one of the Sigsbee Knolls. These prominent features, in 12,000 ft of water, which are part of a large number stretching in a belt from the Texas–Louisiana border to Yucatán, resemble the classical salt dome reservoir of oil in all detail. In spite of this Ewing felt that by drilling in the center cap region, the danger of finding oil would be minimal, but a core was retrieved from about 600 ft below the sea floor that contained unmistakable traces of oil. The string was hurriedly pulled and the hole cemented in. Subsequent analysis of the cores proved the existence of oil, and a small amount of petroleum has since been extracted from them.

It can be optimistically surmised that each of these many knolls is a major oil field that will be tapped when the industry is capable of economic operation at these depths, provided that the petroleum is not in some difficult-to-manage state such as clathrates, a solidified form under the pressure and temperature conditions of these depths. It could be that this first major discovery could also be economically the most important. In subsequent cruises, large beds of iron sediments were discovered, as well as important indications of copper, chromium, vanadium, and other metals. Someday when the economics demand it, there will be a return to and an evaluation of these deposits.

As for sea-floor spreading, the most pervasive evidence found in its favor was the fact that, despite the intensive sampling over all the ocean basins, no rock older than 160 million years (lower part of the Jurassic) has been found. In view of the several billion year age of the earth, this means that the present ocean bottom is but a recent geological feature. This fits the picture of two moving magnetic tapes emerging at the mid-ocean ridge to convey their identical magnetic imprint to the continents on each side, where they descend into the mantle, leaving their accumulated sediment to rebuild the eroding continents from underneath. The age is just right, too, corresponding as it does to a velocity of a few centimeters a year across half an ocean basin.

However, much more detailed and specific evidence came out of the many cruises, some of which can be cited here. On Leg Two (with M. Peterson and T. Edgar as co-chief scientists) and on Leg Three (Arthur Maxwell and Richard von Herzen, co-chief scientists), a traverse across the Mid-Atlantic Ridge showed clearly that the bottom sediments aged progressively as the distance from the ridge center increased. Further, the ages of the sediments in depth correlate from one hole to the next. When the basalt layer was penetrated, its general character was what was to be expected.

Of all the results bearing on sea-floor spreading, my favorite, and I believe the most revealing, is related to the Pacific plate (Fig. 7). This large piece of the earth's crust extends from the San Andreas Fault westward and southward to well south of the equator. This is the plate that gives rise to the question: Is California sliding into

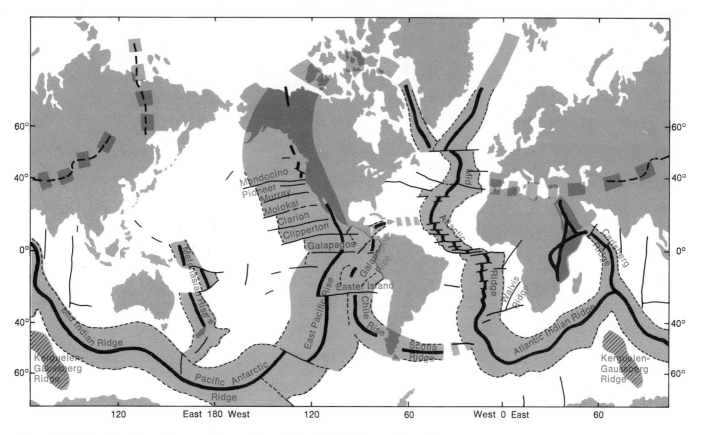

Figure 7. A delineation of the mid-ocean ridge and an outline of the major plates show that the continental drift of North America has over- ridden the East Pacific ridge, positioned along the line of the San Andreas Fault.

the ocean? A suite of holes was cored from south to north across the equator, and the results gave a remarkable confirmation of sea-floor spreading, as well as a major feature of physical oceanography and basic concepts in biological oceanography and paleontology, all in one.

Physical oceanography comes in via the Cromwell current. This remarkable current is subsurface at the equator, flowing from west to east—just opposite to the prevailing winds. It is sheetlike in form, several hundred miles wide, and with peak velocity of 2.5 knots at a depth of 100 m. Given the extent of the equatorial Pacific—6,000 miles—it is an awesome physical phenomenon. Despite its clear connection with the earth's rotation, and its definite parameters, a satisfactory explanation is still forthcoming.

Nevertheless, given the general characteristics, it can be shown that at the very center of the current, exactly at the equator, there is a slow upward movement—an upwelling similar to what takes place at the

continental margins. Just as in that case, the equatorial upwelling brings nutrients to the surface and represents a biologically productive region in an otherwise relatively barren ocean. This rising colder water sharply marks the surface by a measurable dip in temperature. This is a classic phenomenon in biological oceanography, and it is accompanied by a constant rain of the skeletons of microorganisms, a fraction of which reach the bottom and mark the position of the equator as if with a chalk line. However, the two things are occurring simultaneously: the ocean bottom is moving northwestward with the Eastern Pacific plate at a rate of 10 cm per year, and the species of microorganisms are changing and evolving in geologic time. If a series of holes are cored in a line across the equator, there should be evidence of this equatorial line displaced parallel to itself in a northwest sense, with the surface micropaleontology showing a gradual aging, corresponding to the picture I have drawn, and a variation in paleontology with depth that is consistent. This is exactly what was observed. The extra dividend is the

demonstration that the Cromwell current has existed for at least 40 million years.

This kind of result has given impulse to the developing science of paleooceanography and paleoclimatology. The type and location of sediments and kinds of fossils can be interpreted in terms of atmosphere and oceanic climate conditions in the past. One of the more interesting is the demonstration that the Antarctic glaciation has lasted more than 20 million years, or more than four times the age that had been previously accepted.

The interpretation of these climate changes is made especially interesting by the fact that the movement of the continents, which affects the ocean currents and, therefore, sediment and fossil records, also will affect the climate accordingly, and the climate changes and the movement of continents can be correlated either directly or indirectly. The separation of the continents can be very dramatic in the initial breaking stages of the assumed single land mass. When Africa and South America first separated,

Figure 8. This schematic plot of the spreading ridge of the East Pacific rise shows the displacements due to the transform faults. The epicenters of earthquakes, marked with squares, fall almost entirely on the faults.

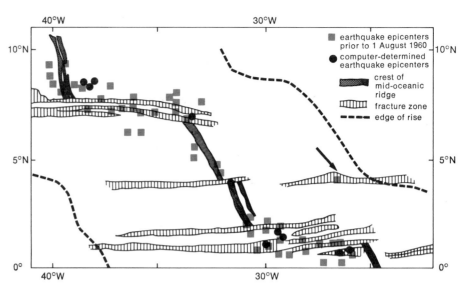

the narrow saline sea that was formed was deprived of free circulation with the rest of the world's oceans for about 20 million years. This was shown in the cores of the region near Africa, which also demonstrated that an amount of about 10 percent of the currently dissolved salt had been deposited in that region.

the most startling result of this kind was the discovery that the Mediterranean was sealed about 12 million years ago and reopened again about 5 million years ago. While the earlier basins were shallower than the present, the Mediterranean must have had the appearance of a gigantic Death Valley and a vastly different climate.

This extraordinary wealth of results is continually accumulating, and yet certain of the original objectives have not been attained. The first was the hope that a continuous paleontologic record would be obtained in at least one drilling operation. One of the virtues of oceanic sedimentary geology is the less violent erosional environment and thus the relatively undisturbed record. Unfortunately, the thick sediments are periodically disturbed by massive gravity displacements, called turbidity currents, that put gaps in the records. Thus the paleontologists' expectations for a complete suite were foiled.

More peculiar is the fact that the magnetic anomaly phenomenon that gave birth to the sea-floor spreading concept—and is unquestioningly tied to it with about 50 stripes having been identified—has not been explained. The physical origin of the anomaly remains as much a mystery as ever. Great efforts are taken to preserve the in situ orientation of the cores, but the variations in magnetism from hole to hole are not consistent with the model. This remains a problem for the future.

We can be consoled by the fact, however, that the prehistory mysteries are completely explained by the plate tectonic hypothesis. The existence of the earth-girding mid-ocean ridges is

easily explained by the uplift of the forces generating the new crust. The accumulation of sediments near the continents no longer presents a problem: the sediments are continually "scraped" from the plate as it takes its downward plunge at the continental margin. The accumulation is presumably impacted into the continent and rebuilds and maintains it in some crude isostatic equilibrium. In the ideal model, the sedimentary thickness would vary from zero at the ridge to a maximum at the continental border, and this is about what was observed. (The heat flow—which I have not discussed—also varies widely in relation to proximity to the ridge, from a maximum near the ridge to a smaller value going away.)

More fun lies in explaining away the extreme polar wandering. Any rigid motion on the surface of a sphere can be interpreted as an instantaneous rotation. Instead of the interpretation that the earth's magnetizing field changed direction in time, it is equally possible that large pieces of the earth's surface rotated in time, carrying the "frozen-in" magnetization with it. Thus the intersection of these two directions need have no special interpretation. In fact, it is not only possible, it is now probable in the light of the new theories.

Before leaving this subject, it is important to display one other important and practical contribution of the hypothesis—to earthquake theory. Figure 8 is a line sketch of a ridge in the Pacific, broken and displaced laterally by two transform faults.

Plotted on the diagram are the loci of a large number of recorded earthquakes, and as you can see, they fall on the ridges—which is not surprising. But they also fall on the transform faults between the displaced parts of the ridges, and essentially nowhere else. This distribution makes sense under the cover of the sea-floor spreading hypothesis, because the fault region between the ridges represents opposing motion, whereas outside this region the two sides move together.

Thus we see the cause of the earthquakes along the San Andreas Fault. The corresponding ridge is exposed in the Juan de Fuca Straits and in the southern part of the Gulf of California and below that on the East Pacific Rise off South America. The origin of Baja California is almost self-evident. We are confident enough, on the basis of this theory, to predict that within 40 million years San Francisco will move northward beneath the sea and Berkeley, across the bay, will move south of San Diego. In any event, this insight helps explain the origin of 90 percent of earthquakes and their beltlike concentration and has given impetus to earthquake research, in the hopes of developing some prediction capability.

Returning to the history of the project, the first nine legs (18 months) were so successful that the project entered a second phase of continuous drilling which lasted 30 months. Meanwhile, the project began to coopt other oceanographic institutions. More or less successively, at somewhat irregular intervals, the following U.S. institutions joined: the departments of oceanography of the University of Washington, Oregon State University, Texas A & M, Rhode Island University, and the Institute of Geophysics of the University of Hawaii.

More significant was the signing up of five foreign countries—USSR, West Germany, France, Great Britain, and Japan—each with a direct contribution of $1 million a year and a much larger indirect one. It has become the most natural functioning international scientific operation yet. The drilling to date and the approved program through August 1979 has somewhat naturally been divided into five phases, the last two of which have been labeled IPOD—the International Phase of Ocean Drilling.

Continental margins

To date, little has been done near the continental margins, primarily because of the deep concern about possible disastrous blowouts. Because the project is not equipped with suitable blowout-prevention equipment, the best talent in the world is employed to review each site to be certain that no hydrocarbon reservoirs exist. Up to now, there have been no failures, which is not too surprising since the reverse operation—the search for oil—does very little better.

The question of what happens at the continental margins becomes more interesting as more data accumulate. The margins have many interesting features that need explaining. One is the trench—the doubling of the ocean depth—that is seen in many places: the Marianas, Tonga-Kermadec, Mindanao, Puerto Rico, Peru, Chile, and so on. Where the margin is active, complex processes take place—a kind of vertical chemical factory on a cosmic scale, where the amorphous sediments are cooked at high pressures and temperatures, while rising isostatically to the surface. Aside from explaining the renewal of continents, or at least the margins, research in this area should give us clues to the discovery of deeper ore-bearing strata than we conceive of now.

To secure safety when drilling in this region, a new platform is called for. Aside from the requirement for a cased hole, if oil-bearing strata are penetrated, it also is necessary to construct a riser reaching to the surface. More time will have to be spent on preparing for drilling and less on the actual coring. One preliminary design calls for flotation rings to support the more than 15,000 ft of riser that would be required. It has been noted that *Glomar Explorer* (the vessel of the Russian submarine-lifting fame) has the weight-lifting capacity to carry the entire riser string and otherwise serve as an excellent drilling platform. If the momentum of these researches on the continental margins continues, there is a high probability that *Glomar Explorer* will become part of the drilling.

The ultimate phase in the explanation of the lithosphere will be the analytic continuation into the continents. The Russian scientific community had, on their own, started a continental Mohole project at about the same time as ours, and it collapsed at about the same time. They regrouped, reorganized, and recommenced a slow, methodical program involving a succession of progressively deeper holes. One operation is in the Kola Peninsula, the other in the Kuril Valley, about 40 km west of Baku. The first geophysical object of the northern hole is the Conrad discontinuity—shallower than the Mohorovičić and not worldwide, but still important. However, the Russian work is not specifically designed to connect up directly with the ocean drilling. In the United States as well, a country-wide program of scientific drilling is being developed, but here, also, it is in no continuous relation to the deep-sea drilling.

The ultimate phase will consist of a transect of cored holes starting from the ocean side of a trench, continuing on to the continent and through the interactive region on the land side. Deep-sea drilling and coring will be of higher priority in this research than it was for the ocean basins. The quite notable successes of pre-drilling marine geology were primarily due to the fact that the ocean sediments are relatively less disturbed and more horizontal and planar than those on land. Hence, the seismic profiling data were far more easily interpreted and, as it turned out, largely correct. A technology recently introduced by the oil exploration communities and adopted by a number of oceanographic institutions is the use of many individual seismic sensors in an array to improve signal quality and geometrical resolution. On land, as many as 128 sensor channels have been employed.

Extracting the information in something like real time requires extensive software development. In addition, physically encompassing this cumbersome towed array (up to 5 km in length) and concomitant equipment, such as compressors, computers, and data storage, onboard ship would be very expensive in capital and very expensive in operation. Nevertheless, the effort would be well worth it to help understand the complicated structures that will be found at the margins.

It does not seem reasonable, however, that, given what we know today, this multichannel seismic research will be adequate in delineating the geological and geophysical features. Drilling and coring seems the only possible non-random method of sampling the diagenetic processes—those complex transmutations of the oceanic sediments into continental rocks.

Many important problems in the earth sciences have emerged, and it is still an exciting time for geologists. Paradoxically, despite the wealth of questions that has been answered, the principal observation that broke the code—the Mason-Raff diagram of the magnetic anomalies—remains unexplained as to physical origin. The records of the horizontal orientation of the basaltic cores are more carefully observed, but no correlation with the observed magnetic anomalies has been shown.

The fundamental problem is to figure out the source of the earth's magnetism and a good explanation of the aperiodic field reversals. Plausible models have been constructed, but they remain just that—plausible. Parallel researches are under way that may help. Very ingenious towed vehicles that remain close to the ocean bottom are accumulating information on the magnetic field—fine structure, for example. Some new ideas, nevertheless, are needed.

The hypothesis has been put forward that dramatic changes take place while the field goes through zero, and, if correct, evidence of the change should be recorded in the ocean sediments. There is fragmentary evidence from the remains of marine microorganisms that accelerated evolutionary changes in species occur at the magnetic reversal epochs. Two possible explanations both depend on the redistribution of the cosmic-ray flux. With a vanishing magnetic field, the cosmic-ray flux will be greatly increased over much of the earth. According to the first explanation, an increased rate of mutations is presumed to accelerate the evolutionary process. According to the second and more plausible explanation, the change in cosmic-ray flux is supposed to change the climate, and it is the climate change that induces the break in the evolutionary process. A change in climate should be legible in the sedimentary record independently from, for example, isotope ratios.

The initial advisory group JOIDES counted four institution members, and it has since expanded to fourteen. Under pressure from NSF, an American corporation, the Joint Oceanographic Institutions, or JOI, was formed, consisting of the nine United States members. Its ultimate purpose is to undertake projects like the Deep Sea Drilling Project, which the international character of JOIDES may make difficult. JOIDES, in effect, becomes a division of JOI, Inc.

The success of the Deep Sea Drilling Project has obscured the fact that bigness was required by the instrument itself—the drilling platform—as is the case with the large nuclear accelerators. Unfortunately, sometimes bigness has been sought for its own sake, partly to solve political and institutional problems, rather than technical ones. It is a matter of some concern that this technically successful project is not also held to blame for being the phototype of projects whose sole characteristic is bigness.

"I thought continental drift was much slower."

Richard K. Bambach
Christopher R. Scotese
Alfred M. Ziegler

Before Pangea: The Geographies of the Paleozoic World

Pre-Pangean configurations of continents and oceans can be reconstructed according to our knowledge of geologic processes still operating in the modern world

Although the earth is known to be 4,600 million years old, most of our geologic information is from rocks formed in the last 570 million years, the time during which organisms have left a good fossil record. This span is subdivided into the traditional geologic time scale (Fig. 1). During this time, life colonized the land, the vast majority of deposits of fossil fuels accumulated, and all of the present mountain ranges of the earth formed. While this was going on, continental drift—a result of plate-tectonic processes—caused the geography of the earth to change constantly.

The paleogeographic history of the last 240 million years, the Mesozoic and Cenozoic eras, is well understood, because this part of the geologic record is relatively well preserved and the pattern of dated magnetic reversal stripes on the ocean floor allows us to reposition the continents as they were during this interval. Reconstructions of continental positions by

Richard K. Bambach is Associate Professor of Paleontology at Virginia Polytechnic Institute and State University. During 1978–79 he was Visiting Associate Professor at the University of Chicago. His research interests cover a wide range of topics in paleoecology. Christopher R. Scotese is a Research Assistant in the Department of Geophysical Sciences at the University of Chicago. He is a specialist in the use of computer graphics for depicting paleogeography and is currently studying the paleomagnetism of Paleozoic rocks. Alfred M. Ziegler is Professor of Geology at the University of Chicago. He has published in the areas of community paleoecology, stratigraphy, and paleogeography. All three authors have been working on an atlas of paleogeographic maps. Address: Richard K. Bambach, Department of Geological Sciences, Virginia Polytechnic Institute and State University, Blacksburg, Virginia 24061.

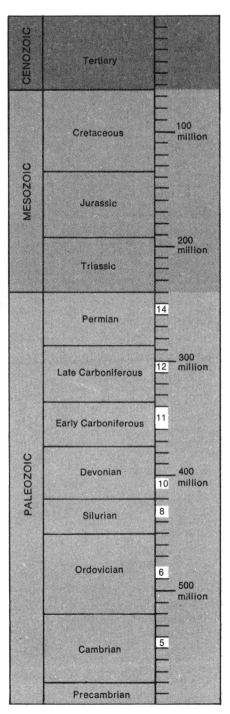

Figure 1. Geologic time. (White blocks represent time spanned by the indicated map.)

Dietz and Holden (1970a,b), Van der Voo and French (1974), and Smith and Briden (1977) show that about 240 million years ago the present continental blocks were grouped together into a "supercontinent" called Pangea. The most recent quarter billion years of geologic time has seen the breakup of Pangea, the formation of the "new" Atlantic and Indian Ocean basins, and the collision of some of the fragments of Pangea to form the Afro-Eurasian landmass, the nucleus of a new supercontinent. These events have been documented by data collected by projects such as the Deep Sea Drilling Project (Nierenberg 1978).

But because no pre-Pangean geographic relationships are preserved, the positions and paleogeography of the continents in the preceding 330 million years, the Paleozoic Era, is much more difficult to determine. The pioneering effort at mapping the positions of continental blocks in the Paleozoic was a set of four maps by Smith, Briden, and Drewry (1973), based solely on paleomagnetic information. The first reconstructions utilizing paleoclimatic and tectonic information combined with paleomagnetic data were presented in a study of the Silurian by Ziegler, Hansen, and others (1977) and another covering times from the start of the Paleozoic to the Cenozoic prepared in the Soviet Union (Zonenshayn and Gorodnitskiy 1977a,b). These maps identified most of the separate paleocontinents of the Paleozoic and proposed a logical, consistent sequence for Paleozoic plate motions.

The emerging pattern of geographic change during the Paleozoic allows us to return to a uniformitarian view of

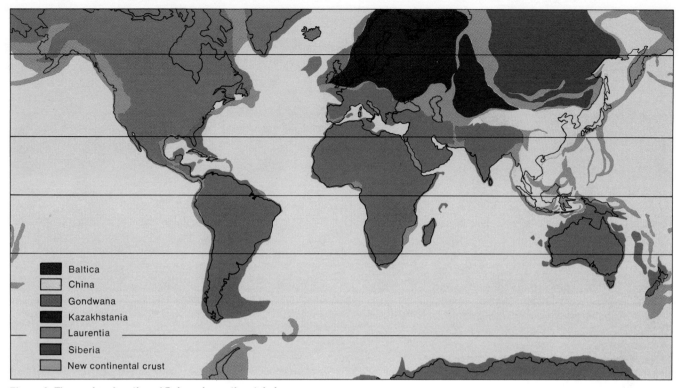

■	Baltica
□	China
▨	Gondwana
■	Kazakhstania
▨	Laurentia
▨	Siberia
▨	New continental crust

Figure 2. The modern location of Paleozoic continental pieces

earth history. No longer do we need to invoke improbable concepts of borderlands, land bridges, and worldwide shifts in climatic equilibrium to explain the seemingly odd locations in the modern world of ancient salt deposits, coral reefs, and evidence of glaciation or the widely scattered occurrence of similar fossil flora and fauna. Rather, we can understand how these features are the result of movements of the continents themselves, by geologic processes still operating today.

Making paleogeographic reconstructions

Reconstructions of the changing Paleozoic world are made by identifying areas that acted as separate continents, positioning these paleocontinents in their correct orientations, compiling data indicative of geographic and climatologic features, and interpreting the distribution of environmental conditions on each paleocontinent. The interpretation of Paleozoic geography requires synthesis of data derived from many fields of geology. Paleomagnetism has served as the cornerstone of this work (McElhinney 1973), but paleoclima-

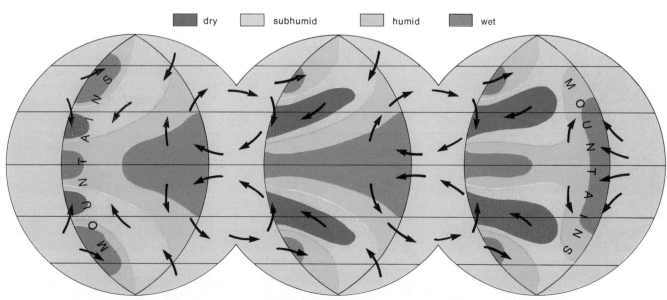

Figure 3. Idealized model of climatic conditions as a function of latitude and geographic configuration

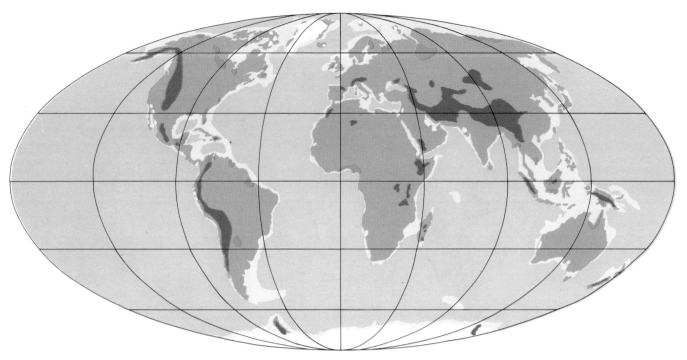

Figure 4. Generalized geography of the modern world

tic, paleobiogeographic, and tectonic data are also vital in compiling detailed reconstructions. Our methods and details of the results have been presented in a series of recent papers (Ziegler, Hansen, et al. 1977; Ziegler et al. 1977, 1979; Scotese et al. 1979), in which the basic data for the following reconstructions were presented.

Geography changes constantly. If too great a period is covered by a single reconstruction, the geographic conditions will be represented by too broad an average to give a clear picture of actual conditions at any particular time. The 12 geologic periods shown in Figure 1 are too long to be summarized by general reconstructions. These periods, however, are divisible into 75 stages, a stage corresponding to the smallest subdivision of time-equivalent rocks recognized worldwide. Each stage lasted from 5 to 15 million years, and since plates move at a rate of 2–8 cm/yr, changes in the relative and absolute positions of the continental blocks during such time spans are contained enough to be realistically summarized. A stage is thus a useful basis for a single reconstruction, though shoreline positions and the extent of glaciation shown are still averages of changing conditions within a stage.

The continental blocks of the Paleo-

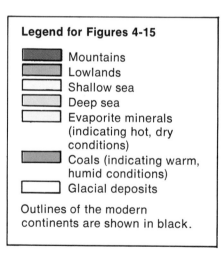

Legend for Figures 4-15

- Mountains
- Lowlands
- Shallow sea
- Deep sea
- Evaporite minerals (indicating hot, dry conditions)
- Coals (indicating warm, humid conditions)
- Glacial deposits

Outlines of the modern continents are shown in black.

zoic that collided to form Pangea were not the same as the continents that formed as Pangea split apart during the Mesozoic or as those which exist today. Identifying ancient continents requires the recognition of features that mark the outlines or margins of these paleocontinents. A major achievement of plate-tectonic theory has been the definition of criteria for recognizing these ancient continental boundaries (Mitchell and Reading 1969; Dewey and Bird 1970; Dickenson 1970; Burk and Drake 1974; Burke et al. 1977). The torn and rifted margins of once-associated continents are represented by certain geologic features, especially belts of basaltic

igneous rocks associated with elongate basins bordered by normal faults. Other features, especially belts of andesitic igneous rock and mountain belts with strongly folded rocks, represent the deformed margins of continental blocks under which oceanic crust was subducted as lithospheric plates moved together. Where such belts cross a continent, such as the Ural Mountains in Eurasia, two former continents appear to have collided and been sutured together.

We recognize six major paleocontinents during various parts of the Paleozoic: Gondwana, Laurentia, Baltica, Siberia, Kazakhstania, and China. Figure 2 shows the parts of the modern continents that belonged to each of these paleocontinents and areas of the present continental crust that have accreted to the margins of the continents during Paleozoic, Mesozoic, and Cenozoic times.

Gondwana was a supercontinent composed of what are now South America, Florida, Africa, Antarctica, Australia, India, Tibet, Iran, Saudi Arabia, Turkey, and southern Europe. Laurentia comprised most of modern North America and Greenland, with the addition of Scotland and the Chukotski Peninsula of the eastern USSR. The missing parts of eastern North America were either part of Gondwana or associated with

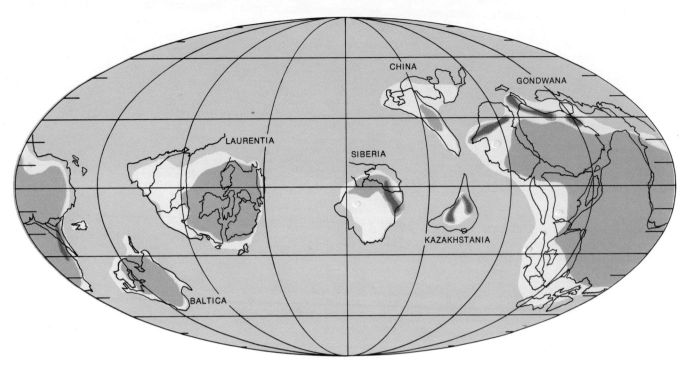

Figure 5. Late Cambrian (550–540 million years ago)

small microcontinents similar to modern New Zealand. Baltica was composed of Russia west of the Ural Mountains, Scandinavia, Poland, and northern Germany. Siberia from the Urals east to the Verkhoyanski Mountains was a separate continent in the Paleozoic. Its southern margin crossed Asia north of present Kazakhstan and south of Mongolia. Kazakhstania was a triangular continent centered on what is now Kazakhstan, with one part extending up between the Urals and southwestern Siberia and another part extending east between the Altai on the north and the Tien Shan Mountains on the south. China is a complex area that may have been subdivided into more than one block at times in the Paleozoic, but there are general similarities which imply that the pieces were not widely separated. For now, we treat all of southeast Asia, including China, Indochina, and part of Thailand and the Malay Peninsula, as a single continent.

Positioning the continents is at the heart of any reconstruction of world paleogeography. Assigning correct positions to paleocontinents involves both the latitude-longitude location and the orientation of the continents relative to the appropriate paleonorth direction. We believe, for instance, that in the Silurian period Siberia was centered at about lat. 30° north and

was rotated 180° from its modern orientation, so that its present Arctic coast faced south.

Paleomagnetic information is the basis for all continental positioning. Paleomagnetism provides direct quantitative evidence for both latitude and the north-south orientation of the paleocontinents. Although the magnetic poles may stray from the rotational poles, the earth's magnetic field has a north-south polarity and maintains an alignment that over geologic time on the average parallels the rotational axis. The lines of magnetic force also vary in their inclination to the earth's surface as a function of latitude, being vertical at the magnetic poles and parallel to the surface at the equator. The remnant magnetism of adequately preserved rock samples indicates the azimuth direction to the north (or south) magnetic pole and the latitude of the rock at the time the magnetism was imposed. Thus, from oriented samples we can determine the latitude and orientation of continental blocks for times in the past. Paleoclimatic indicators serve as independent checks on latitude and as guides to latitude when paleomagnetic information is not available.

Longitudes are determined by integrating biogeographic relationships and plate-motion constraints. Al-

though it is not possible to assign absolute longitude (relative to the prime meridian) in the Paleozoic, the whole interval is bracketed by times when relative longitude can be determined within narrow limits. At the time of our earliest reconstruction, in the Late Cambrian, all the continents except Baltica and China straddled the equator (see Fig. 5) and, with their surrounding oceans, occupied much of the space available in the equatorial belt. Since their relative order in sequence around the globe is clear from biogeographic evidence, space constraints alone fix their longitudinal positions within rather narrow limits.

At the end of the Paleozoic (Fig. 14), most of the continental blocks were grouped together in the supercontinent of Pangea. The spatial fit of the continents at this time tells us their relative longitudes quite precisely. Thus, with the endpoints of the Paleozoic well defined, we can follow a regular pattern of plate motion as the plates, bearing the separate continents which ringed the equator in the Cambrian, shifted and brought the continental blocks together to form the pole-to-pole mass of Pangea by the Permian.

The final task in preparing paleogeographic reconstructions is interpreting the geographic features and

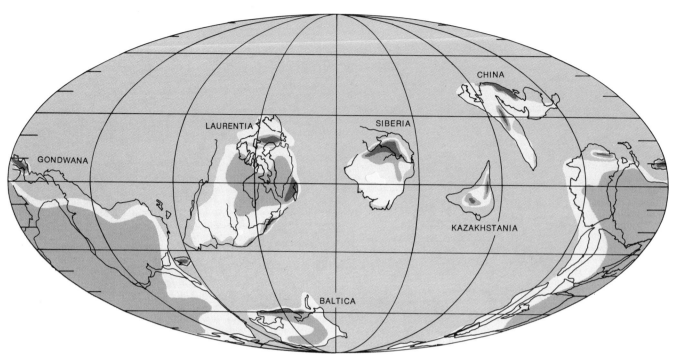

Figure 6. Middle Ordovician (490–475 million years ago)

distribution of environmental conditions on the continental blocks. In our reconstructions we identify regions that were highlands and mountains, lowlands and floodplains, coastal areas, shallow and deep marine platforms, and submarine-slope and deep-sea environments. Environmental conditions are interpreted from the processes known to have formed particular types of rocks.

Sands and coarser-grained detrital sediments are deposited in high-energy environments such as stream channels and along beaches, while fine-grained sediments such as clays and muds accumulate in low-energy environments such as floodplains, lagoons, and deep offshore marine environments. Limestones typify warm, shallow marine conditions at some distance from a source of land-

derived detrital sediments. Sands and muds, for example, are present along both the Atlantic and Gulf coasts of modern North America where rivers bring them to the shore, but in the Florida Keys and the Bahamas—areas far removed from sources of detrital sediments—limestones are accumulating from the skeletal secretions of abundant marine life. Fossils help differentiate terrestrial,

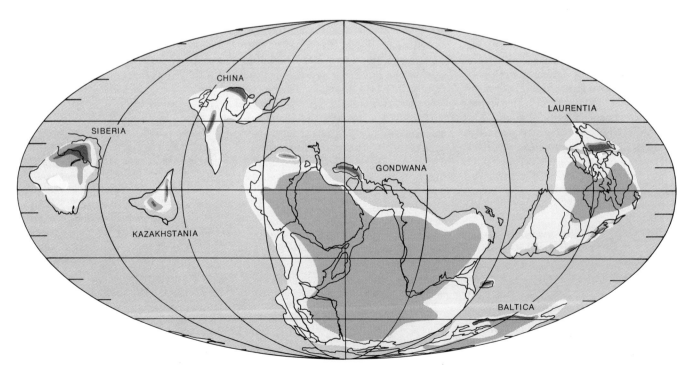

Figure 7. Middle Ordovician, view of earth rotated 180° from the view in Figure 6

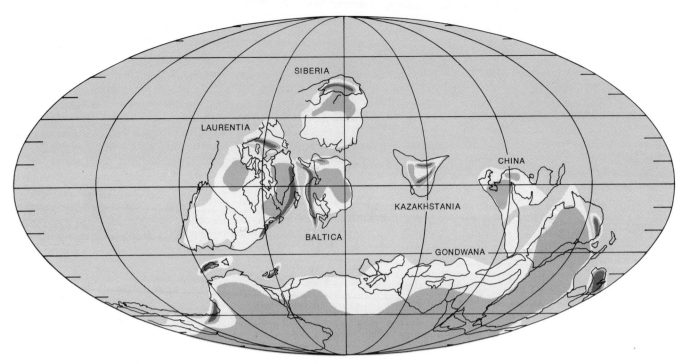

Figure 8. Middle Silurian (435–430 million years ago)

freshwater, and marine deposits, as do suites of sedimentary structures unique to particular depositional settings.

Some sediments reflect climatic conditions. Detrital sediments are indicative of humid environments. Coals formed in freshwater swamps where temperature and rainfall were both adequate for abundant plant growth. Hot, dry shorelines and protected basins along coasts in desert regions are the sites for deposition of salts and other evaporite minerals from brines left by evaporating seawater or playa lakes. In cold environments, glaciers leave distinctive deposits, called *tillites,* with features imposed by ice flow and the peculiar conditions of freezing and thawing at the melting edges of the ice. In areas where nutrients from deep water are supplied in abundance to surface waters, such as zones of coastal upwelling, the sediments may include cherts, phosphorites, or deposits rich in organic material, the result of high biologic productivity.

Geographic features can also be inferred from the type and structure of rock bodies. Delta complexes and

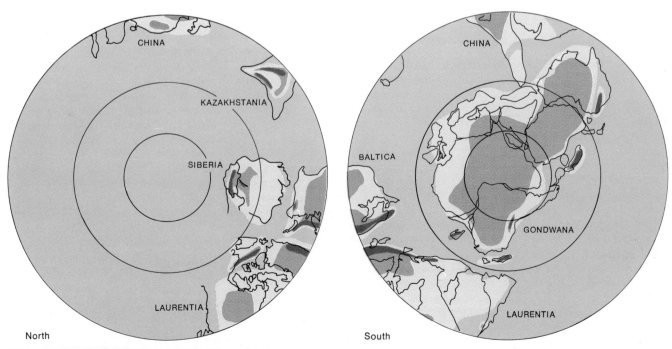

Figure 9. Middle Silurian, North and South Polar projections

deep-sea fans have typical internal features and distinctive three-dimensional forms. Belts of folded and thrust-faulted strata associated with metamorphic rocks and igneous intrusions represent areas that were mountain ranges at the time the deformation and metamorphism occurred, even if they are eroded to lowlands today. Andesitic igneous rocks were erupted along continental margins where subduction of oceanic crust occurred, as is happening in western South America and in the Aleutians today. Basalts were erupted and normal fault basins developed where rifting, associated with seafloor spreading, started, as in the modern African rift valleys or in Iceland.

The distribution of environments on the modern earth is related to the systematic arrangement of topographic features associated with plate-tectonic processes and to the climatic belts produced by atmospheric and oceanic circulation. These patterns are not random now, and they weren't in the past. The circulation of the atmosphere and oceans is an unchanging geophysical system produced by heat-transfer activities on a rotating earth as it is warmed by the sun. Despite the changes in geography over geologic time, some of which have produced local anomalies, the general climatic pattern of the earth is fixed over time. The present earth serves very well as the basis for a general model of global climate structure.

The modern world as a model

The sun is a stable main-sequence star and probably has not altered its intensity of radiation very much in the last billion years. The earth has always been spherical and has orbited at the same distance from the sun. Therefore the heat budget of the earth, with excess heating in the equatorial region, heat transfer by atmospheric and oceanic circulation toward the poles, and cooling in the polar regions, can be regarded as fixed. Although the earth's rotation has been slowed somewhat by tidal friction, it has not decreased by more than about 15% in the last half billion years. This means that the Coriolis effect, which deflects motions to the right in the Northern Hemisphere and to the left in the Southern

Hemisphere, has remained nearly constant.

The influence of the Coriolis effect on the airflow generated by heating and cooling of the earth's surface creates a latitudinally zoned pattern of atmospheric circulation (Strahler and Strahler 1978). The equatorial belt is characterized by hot, humid conditions with irregular surface winds, because the heated air is primarily expanding and rising rather than moving in a particular direction. Dry air, cooling and contracting in the upper atmosphere, sinks toward the surface of the earth at about lats. 30° north and south of the equator (the horse latitudes), causing dry climates with irregular surface winds. The sinking air flows out from this belt both toward the equator and toward the poles. The surface flow toward the equator is deflected by the Coriolis effect into strong prevailing easterly winds (the trade winds). The surface flow of air poleward is also deflected by the Coriolis effect and becomes the prevailing westerly winds of the temperate belts at lats. 40°–50° north and south of the equator.

In the polar regions cooling causes the air to contract and sink. This cold polar air flows as surface winds toward the equator, and these winds are deflected by the Coriolis effect into easterly winds in high latitudes. The polar front, where the equatorially trending polar air intersects the poleward flow of temperate air, is a region of high precipitation. The 23.5% tilt of the earth's rotational axis generates the seasonal fluctuations in climate as the earth orbits the sun. These fluctuations are most pronounced in the high temperate latitudes, where the polar front shifts back and forth, causing seasonal temperature changes from below freezing to above.

The interaction of this regular pattern of atmospheric circulation with the land-sea distribution (geography) of the earth produces a climatic regime that is also regular but does not consist of simple latitudinal belts. Except for locations very near the equator or at extremely high latitudes, most continents have quite different climates on opposite coasts at the same latitude (Fig. 3). The tropical humid zone widens toward the eastern sides of continents, where moisture is brought from the ocean by

prevailing easterly winds, but it is narrow on the western sides, confined to the narrow zone of intense heating where surface air is rising and losing moisture. The arid belts rise in latitude across the continents from west to east. They extend closer to the equator in the belts of easterly winds on the rain-starved western sides of continents and extend poleward in the midlatitude regions of prevailing westerly winds on their similarly rain-starved eastern sides. As with the wet eastern sides of continents in the tropics, the wet belts in temperate latitudes are much broader on the windward, west-facing margins of continents.

In the modern world these features are seen in the extensive arid belt which rises from low latitudes of the Sahara in eastern Africa to the high-latitude Gobi desert of western Asia. The humid regions of southeast Asia (Indochina, Burma, south China) are at the same latitude as the Sahara in the trade-wind belt (of easterlies), while France and Austria are at the same latitude as the Gobi Desert in the belt of prevailing westerlies. In the Southern Hemisphere, the Atacama Desert extends equatorward on the west coast of South America into latitudes occupied by the Amazon rain forest to the east, and the south Chile rainy zone is at the same latitude as the dry Argentinian pampas. The Andes create intense contrasts even across the narrow South American continent because of the rain shadow caused by their height.

The distribution of features in the modern world serves as a model for understanding climate patterns in the past. Because so many features of sedimentary rocks reflect climatic influence, we can map major climatic features on paleogeographic reconstructions. And the fact that climatic systems are predictable from their modern distribution means, as we have said, that we can use paleoclimatic data to cross-check paleomagnetic determinations of latitude. Paleoclimatic features also provide latitudinal information about regions and times for which paleomagnetic data are not available.

The Paleozoic world

Figures 5, 6, 7, 8, 10, 11, 12, and 14 are reconstructions of world paleogeography at seven times during the

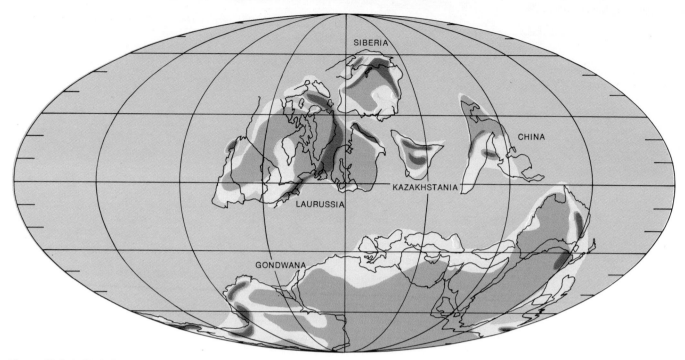

Figure 10. Late Early Devonian (410–405 million years ago)

Paleozoic. These reconstructions show the evolution of the Paleozoic world from the Late Cambrian through the Late Permian. We have used a Mollweide projection, which is an equal-area projection that avoids the excessive area expansion in higher latitudes of the familiar Mercator projection. Although there is angular distortion near the edge of the map at high latitudes, the Mollweide projection also shows the entire earth's surface, including the polar regions, which a Mercator projection cannot do.

The pageant of geographic change through the Paleozoic is profound. None of the Paleozoic geographies is similar to that of our modern earth. In the modern world (Fig. 4) the continents are grouped into three extensive north-south masses—the Americas, Europe/Africa, and eastern Asia through Indonesia to Australia—which partly isolate three equatorially centered oceans—the Pacific, Atlantic, and Indian oceans. The modern continents are mostly emergent, with only narrow continental shelves flooded by shallow seas. There are close connections between almost all the continents surrounding the North Pole, and the small Arctic Ocean is virtually landlocked by the belt of high-latitude land extending from Greenland across northern North America and Siberia to Scandinavia.

Antarctica covers the South Pole. Extensive high mountain belts are characteristic, most prominently the Alps-Caucasus-Himalayas system, extending across the Eurasian supercontinent, and the Andes–Rocky Mountains system, extending from Tierra del Fuego to Alaska in the Americas. Areas forming climatically sensitive sediments today are shown on the reconstruction for correlation with modern climate distribution and for comparison with the distribution of ancient deposits.

The ancient Late Cambrian world (Fig. 5) contrasts sharply with the world of today. In the Late Cambrian the continents of the Paleozoic were isolated from each other and dispersed around the globe in low tropical latitudes. The ocean basins were extensively interconnected and the polar regions were occupied by broad open oceans. There was no land above lats. 60° north or south. Shallow seas had transgressed onto the low-lying continental platforms earlier in the Cambrian period and covered large areas of Laurentia, Baltica, Siberia, Kazakhstania, and China in the middle Late Cambrian. The major highlands were in northeastern Gondwena (Australia and Antarctica today), eastern Siberia, and central Kazakhstania. Erosion had reduced the topography of Laurentia and Baltica to low levels.

During the Ordovician (Fig. 6) and Silurian (Fig. 8) Gondwana moved southward from its Late Cambrian position on the equator halfway around the globe from Siberia to a position straddling the South Pole. This change can be followed by comparing the latitude of Gondwana in Figures 5, 7, and 9. The oldest record of glaciation in the Paleozoic era is of Late Ordovician age in what is now the Sahara Desert: tillites were deposited at the time that this part of Gondwana was actually crossing the South Pole.

During the movement of the plate containing Gondwana, Baltica also shifted position along a parallel path from the Cambrian to the Silurian. The two paleocontinents may have been located on one lithospheric plate. Siberia also shifted from equatorial to north temperate latitudes during the early Paleozoic.

These movements were marked by mountain building along the eastern margin of Laurentia (the Taconic orogeny) in the middle and late Ordovician and later mountain building in western Baltica during the Silurian (the Caledonian orogeny), as the ocean between Laurentia and Baltica closed. Shallow seas were widespread on the continents throughout most of the Ordovician and Silurian. Climatic zonation is detectable in the distri-

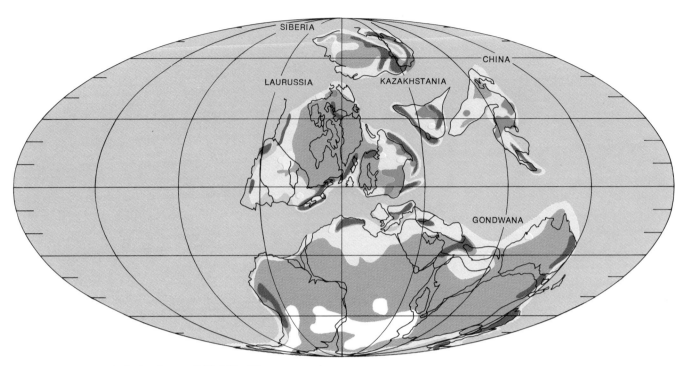

Figure 11. Middle Early Carboniferous (360–340 million years ago)

bution of evaporite deposits, which are concentrated between lats. 15° and 30° north and south, as in the modern world.

Frigid polar climates are not detectable in the Cambrian or most of the Ordovician. This may be because the open polar oceans of these times were always warmed by oceanic circulation, or it may simply be that the lack of polar land prevented preservation of a geologic record of polar climates. As mentioned above, the first land areas to enter polar latitudes, the North African portion of Gondwana, was glaciated in the Late Ordovician. Marine faunas of relatively low diversity—and therefore probably from temperate climates—are known from the higher south latitude areas of Gondwana in the Ordovician and Silurian, and a distinctive low-diversity fauna is found in the Silurian along the margin of Siberia, which first moved as far north as lat. 40° north.

Major reorganization of world geography is apparent by the Silurian. The shift southward of Gondwana from its Cambrian position had opened the former North Polar ocean basin until it not only circled the world at high northern latitudes but extended in an unbroken expanse southward across the equator to high southern latitudes. The former South Polar ocean basin of the Late Cam-

brian had been displaced to middle southern latitudes by the shifts of Gondwana, Baltica, and Siberia and had become a partly enclosed basin between Baltica, Kazakhstania, and Gondwana. This was the start of the development of the Tethys Sea, a region characterized by distinctive marine faunas throughout the Late Paleozoic and Mesozoic.

By the Early Devonian (Fig. 10), Laurentia and Baltica had collided to form a larger continent, Laurussia. The collision began in the Late Silurian with the Caledonian orogeny in northwestern Baltica. Mountain building continued into the Devonian as the Acadian orogeny in eastern Laurentia. These uplands along the suture between the formerly separate paleocontinents were located in the equatorial belt, and large volumes of detrital sediments were eroded from them. Nonmarine fluvial sediments covered large parts of eastern North America (the "Catskill Delta") and northern Europe; these sediments were deeply weathered and are stained red from iron oxides—hence the name Old Red Sandstone in Great Britain. Although land plants had begun to evolve in the Silurian, it is in these tropical nonmarine Devonian deposits that abundant larger fossil land plants first appear.

The Early and Late Carboniferous

reconstructions (Figs. 11 and 12) show that Gondwana continued to move across the South Pole and entered the same hemisphere as Laurussia, closing the ocean between them. Their collision in the Late Carboniferous resulted in the Hercynian orogeny, which extended across central Europe, and the Alleghenian orogeny in eastern North America. Baltica had begun colliding with Laurentia in the Silurian and Devonian at a location relative to Laurentia far to the south of the position it occupied later. Repositioning took place in the Carboniferous, during the collision between Laurussia and Gondwana, when what had been Baltica was displaced northward along a series of faults extending from coastal New England, across Newfoundland, and through Scotland along the zone of weakness at the original suture.

From the Silurian through the Carboniferous, Siberia moved to high latitudes and Kazakhstania and China moved westward. By the Late Carboniferous, Kazakhstania and Siberia had collided and all the paleocontinents were clustered tightly as Pangea began to take shape (Fig. 13). Mountain belts extended along the suture between Laurussia and Gondwana (the southern Appalachian and Hercynian belts) and along the reactivated suture where

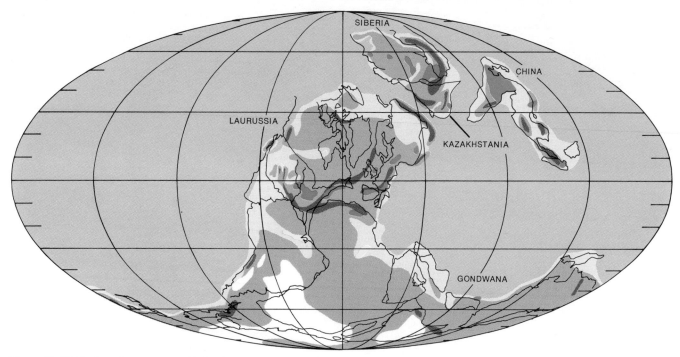

Figure 12. Middle Late Carboniferous (310–300 million years ago)

Baltica sheared northward relative to Laurentia (the northern Appalachian-Caledonide belt). Mountain systems marking subduction zones also developed during the Carboniferous on the eastern side of Baltica, as the ocean between Baltica and Kazakhstania-Siberia was closing, and in the Mongolian portion of Siberia, as the ocean between Siberia and China also closed.

The distribution of climatically indicative deposits in the Late Carboniferous (Fig. 12) reconstruction is particularly interesting. The great coal reserves of eastern North America, western Europe, and the Donetz Basin of the USSR lie in the equatorial zone. The coal swamps developed on marshy delta platforms built by rivers bringing detrital sediments from the adjacent mountain ranges.

The large volume of both detrital sediments and plant remains testifies to the high rainfall in this tropical belt. Plant fossils in the coals do not show strong seasonal growth rings, which implies that they grew under constantly warm rather than temperate seasonal conditions. The belt of tropical coals is narrow in the west and broadens to the east, as predicted by the climate models (see Fig. 3).

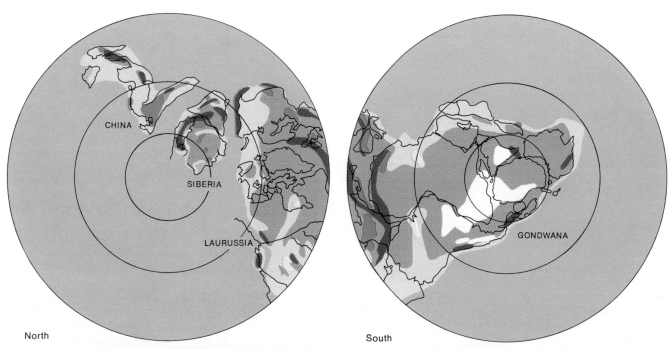

North

South

Figure 13. Middle Late Carboniferous, North and South Polar projections

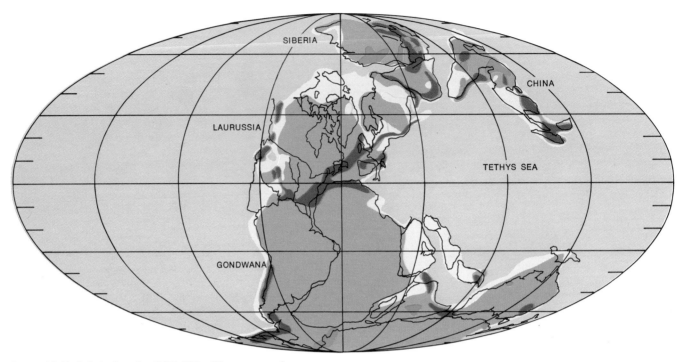

Figure 14. Early Late Permian (260–250 million years ago)

Evaporite deposits, indicative of low rainfall, occur in far western Laurussia at low latitudes and farther east between lats. 15° and 30° north and in a belt in Gondwana between lats. 20° and 30° south, again as expected from the climatic model. In the north temperate belt (lats. 40°–60° north) extensive detrital sediments and coals in Siberia and China indicate abundant rainfall. The seasonal nature of

the climate of this belt is indicated by well-developed growth rings in plant fossils from Siberia. Seasonal growth rings, imposed by interruption of growth during cold winter months, are also found in plant fossils of both Carboniferous and Permian age from the south temperate latitudes of Gondwana.

Glaciation is recorded by widespread

tillites in southern Gondwana. Ice sheets were present above lat. 60° south from the Early Carboniferous into the Early Permian. At their most extensive they flowed equatorward as far as the middle temperate latitudes, just as the Pleistocene glaciers flowed south from the Arctic regions of North America and Europe less than 100,000 years ago. The fact that the Carboniferous glacial deposits are

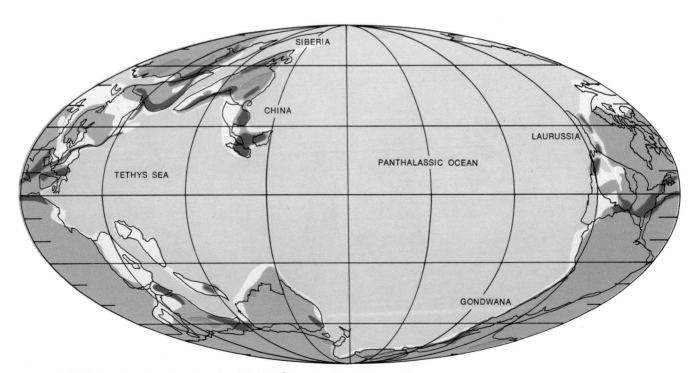

Figure 15. Early Late Permian, view of earth rotated 180° from the view in Figure 14

now found on continents scattered across half the earth's surface (from South America through southern Africa to Antarctica and India) was one of the strongest intimations of continental drift before the theory of plate tectonics.

Pangea was nearly assembled by the Late Permian (Fig. 14). It is worth noting, however, that a Pangea containing all continental blocks probably never formed completely. In eastern Gondwana during the Permian, rifting had begun that pulled Tibet, Iran, and Turkey away from the Tethyan margin of Gondwana as a separate, isolated block or blocks before China collided with the Mongolian region of the former Siberian block in the Triassic.

The effects of continental collision reach far beyond the simple suturing together of two formerly separate continental blocks. Collision deforms rocks through the entire thickness of the lithosphere along the colliding margins of the plates. The resulting folding and thrust faulting cause the rocks to "pile up" on themselves, thus thickening the continental crust in the zone of collision. Because lighter continental crust is buoyed up on the denser mantle, these belts of thickened, deformed continental crust form uplifted mountain belts.

And because the same mass of crustal material as formerly existed in undeformed, thinner crust is packed into these thickened belts, the area covered by continental crust is decreased during collision, just as the area covered by a rug is less if it is crumpled up against a wall rather than spread flat on a floor. The decrease in continental crustal area during collision is taken up by an increase in the area of the ocean basins, with a concomitant increase in their volume. This can cause a general lowering of sea level in relation to the continents, as shallow seas flooding continental platforms drain into the enlarging ocean basins.

The series of continental collisions through the late Paleozoic (Devonian–Permian) formed extensive mountain belts and decreased the total area of continental crust. As a result of this process and possibly also because of changes in the volume of mid-ocean ridges at the spreading centers, sea level was effectively lowered and most of the shallow seas that had flooded the low continental platforms through much of the Paleozoic were drained. The larger areas of exposed land contributed to increased climatic severity.

Permian terrestrial sediments indicative of arid and semiarid conditions were widespread. The mountains in the tightly sutured region between Laurussia and Gondwana were high enough to disrupt the subtropical easterly winds and create an intense rain shadow even in tropical latitudes, much as the Andes do in western South America today. Evaporites of Permian age extend to what was the equator in western Laurussia. Desert conditions also extended north of lat. 40° in eastern Laurussia on the side of the continent downwind from the westerly winds, just as deserts extend northward in eastern Asia today.

The clustering of the continents into Pangea had an extraordinary geographic consequence. An enormous single interconnected ocean developed (Fig. 15). This "world ocean," sometimes called Panthalassa, not only spanned the globe from pole to pole but extended for 300° of longitude at the equator, twice the distance from the Philippine Islands to South America across the modern Pacific. Circulation in this giant ocean had to have a major impact on Permian climates. For example, the equatorial currents driven by the trade winds flowed uninterrupted around five-sixths of the circumference of the earth. The east-facing (Tethyan) coast of Pangea, against which these currents impinged, must have been extremely warm. Ancient "Gulf Streams" would have circulated these warm waters into higher latitudes, causing an especially strong climatic asymmetry between eastern and western Pangea. The Permian was indeed a time of geographic extremes.

Paleogeography and geologic history

The old dictum "the present is the key to the past" has a rather specific meaning for geologic history. The processes of geology, including plate tectonics, have probably operated much as we observe them today over long spans of geologic time, and this is certainly the case if we accept reasonable variation in their rates. But although geographic features have always reflected the operation of processes and systems we observe today, the order of geographic change is a unique historical sequence. Thus the present *is* the key to understanding past processes, but it *is not* the key to describing past configurations.

Because plate-tectonic processes operate constantly, there has been no stable or even average geography of the earth during the past half billion years. Paleogeography changed continuously, passing from one extreme configuration through a series of intermediates to a different extreme, as illustrated by these maps, and then changing still further. The geographies of the Cambrian, Permian, and the present are simply single steps in this dynamic pattern.

The Cambrian, with its isolated, equatorially distributed continents and two polar but interconnected oceans, was totally unlike the Permian, with its single concentration of continents stretching from pole to pole and its immense, equatorially centered Panthalassic Ocean. The widespread shallow seas that flooded continental platforms in the Early Paleozoic contrast with the large proportion of exposed land in the Late Paleozoic. These two extremely different periods also differed totally from the modern world, which is the product of the breakup of Pangea into nearly interconnected north–south continental belts and large, semi-isolated ocean basins.

The modern world, of course, is just another transient stage in geologic history. Sound—though broadly outlined—predictions on the future course of geographic change indicate that the Atlantic Ocean, which has been opening for the past 150 million years as the Americas have been moving westward, will grow larger. The Atlantic coast of North America will probably develop into an Andean type of mountain system in the not-too-distant geologic future. There is very little subduction in the Atlantic today, but it is likely to begin in the next 50 million years, especially in the relatively old western North Atlantic.

On the other side of our continent, Baja California and the part of Southern California west of the San

Andreas Fault will continue on their present course of northwestward motion as part of the Pacific plate. California will not "fall into the sea," but the part west of the San Andreas fault may become a New Zealand-like small continental island moving away from mainland North America in the next 50 to 100 million years. Across the Pacific Ocean, Australia, which has been moving northward away from Antarctica for the last 40 million years, will move north past Indochina and may collide with China, Japan, or far eastern Russia. The world's largest ocean then will be the interconnected Indian-Antarctic-South Pacific.

The historical development of our earth has followed a nonrepetitive path through time. The world of the Permian was a world alien to the one in which we live; the world of the Cambrian was equally alien, to the Permian as well as the present; and the world 100 million years from now will be alien to our present one. It is within this framework of changing geographies, with markedly different extreme configurations and long intervening transitions, that we must cast our ideas of geologic history.

References

Burk, C. A., and C. L. Drake, eds. 1974. *The Geology of Continental Margins.* New York: Springer-Verlag.

Burke, K., J. F. Dewey, and W. S. F. Kidd. 1977. World distribution of sutures: The sites of former oceans. *Tectonophysics* 40: 69–99.

Dewey, J. F., and J. M. Bird. 1970. Mountain belts and the new global tectonics. *J. Geophys. Res.* 75:2625–47.

Dickenson, W. R. 1970. Relation of andesites, granites, and derivative sandstones to arc-trench tectonics. *Revs. of Geophysics* 8: 813–60.

Dietz, R. S., and J. C. Holden. 1970a. The breakup of Pangea. *Sci. Am.* 223(4):30–41.

————. 1970b. Reconstruction of Pangea: Breakup and dispersion of continents, Permian to present. *J. Geophys. Res.* 75: 4939–56.

McElhinny, M. W. 1973. *Paleomagnetism and Plate Tectonics.* Cambridge Univ. Press.

Mitchell, A. H., and H. T. Reading. 1969. Continental margins, geosynclines, and ocean floor spreading. *J. Geol.* 77:629–46.

Nierenberg, W. A. 1978. The deep sea drilling project after ten years. *Am. Sci.* 66:20–29.

Scotese, C. R., R. K. Bambach, C. Barton, R. Van der Voo, and A. H. Ziegler. 1979. Paleozoic base maps. *J. Geol.* 87:217–68.

Smith, A. G., and J. C. Briden. 1977. *Mesozoic and Cenozoic Paleocontinental Maps.* Cambridge Univ. Press.

Smith, A. G., J. C. Briden, and G. E. Drewry.

1973. Phanerozoic world maps. In *Organisms and Continents through Time,* ed N. F. Hughes, pp. 1–42. Paleontological Association Special Papers in Paleontology, no. 12.

Strahler, A. N., and A. H. Strahler. 1978. *Modern Physical Geography.* Wiley.

Van der Voo, R., and R. B. French. 1974. Apparent polar wandering for the Atlantic-bordering continents: Late Carboniferous to Eocene. *Earth Science Reviews* 10:99–119.

Ziegler, A. M., K. S. Hansen, M. E. Johnson, M. A. Kelly, C. R. Scotese, and R. Van der Voo. 1977. Silurian continental distributions, paleogeography, climatology, and biogeography. *Tectonophysics* 40:13–51.

Ziegler, A. M., C. R. Scotese, W. S. McKerrow, M. E. Johnson, and R. K. Bambach. 1977. Paleozoic biogeography of continents bordering the Iapetus (pre-Caledonian) and Rheic (pre-Hercynian) oceans. In *Paleontology and Plate Tectonics,* ed. R. M. West, pp. 1–22. Milwaukee Public Museum, Special Publications in Biology and Geology, no. 2.

————. 1979. Paleozoic paleogeography. *Ann. Revs. Earth and Planet. Sci.* 7:473–502.

Zonenshayn, L. P., and A. M. Gorodnitskiy. 1977a. Paleozoic and Mesozoic reconstructions of the continents and oceans, article 1: Early and Middle Paleozoic reconstructions. *Geotectonics* 11:83–94.

————. 1977b. Paleozoic and Mesozoic reconstructions of the continents and oceans, article 2: Late Paleozoic and Mesozoic reconstructions. *Geotectonics* 11:159–72.

Bruce D. Marsh

Island-Arc Volcanism

Volcanic arcs ringing the Pacific owe their origin to magma generated in a mysterious fashion when tectonic plates are consumed by the inner earth

Figure 1. Looking east from Mt. Moffett (1,200 m) on Adak Island, one sees Adagdak volcano (621 m), on the same island, and Great Sitkin volcano (1,722 m). The narrowness of the volcanic front is apparent, but the summits of the three volcanic centers are not aligned here, for this is a knee in the volcanic front, as can be seen in Fig. 3. The Aleuts call Great Sitkin "Great Emptier of the Bowels," because of its frequent volcanic activity. (Photograph by the author.)

Bruce David Marsh, Associate Professor of Geophysics and Igneous Petrology at The Johns Hopkins University, received his Ph.D. in 1974 from the University of California, Berkeley. His main research activity is concerned with the physics and chemistry of the generation and evolution of magma, and the large-scale dynamics of the inner earth. For the last six years he has spent much time studying volcanic centers in the Aleutian Islands, Alaska. He is also interested in the geomorphology of the winter shorelines of the Great Lakes. His work is sponsored by an NSF grant to The Johns Hopkins University. Address: Department of Earth and Planetary Sciences, The Johns Hopkins University, Baltimore, MD 21218.

At this very moment, somewhere beneath the Pacific, magma is ascending toward the earth's surface. It may erupt spectacularly as fountaining red-hot lava visible for miles in the night; it may explode dangerously; or it may quietly become encumbered at depth and cool very slowly, over millions of years, its heat eventually producing a field of geysers. The origin of magma and the way it moves to the earth's surface are among the oldest mysteries of nature; they worried even Aristotle. The mere existence of volcanoes has so far defied simple, universal explanation.

The greatest number of unsubmerged volcanoes—and the most impressive—are the large, steep-sided, conical monuments of lava surrounding the Pacific and some other oceanic areas. Mt. Fuji, Mt. Ranier, Mt. Shasta, Mt. Katmai, Krakatoa, and Mt. Mayon are but a few of the more well known of these commonly picturesque volcanoes. Cropping up in long, narrow, bow-shaped chains (Fig.

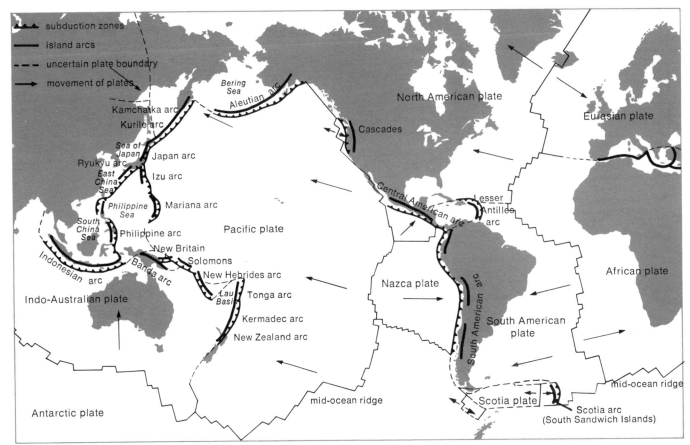

Figure 2. Island arcs and volcanic fronts grow parallel to the trenches or deeps where oceanic plates are consumed by the inner earth. Plates are originally formed at the mid-ocean ridges, but a smaller amount of spreading also occurs in the floors of some (not all) of the marginal seas behind island arcs, such as the South China Sea and the Sea of Japan.

1)—loosely referred to as island arcs—these volcanoes parallel the oceanic deeps or trenches that mark the line of descent of one plate beneath another. Volcanoes in island arcs have a characteristic chemical composition different from those of volcanoes associated with other parts of the tectonic cycle. (For a review of plate tectonics, see D. P. McKenzie's July 1972 article, "Plate Tectonics and Sea-Floor Spreading," and William A. Nierenberg's article of January 1978, "The Deep Sea Drilling Project after Ten Years," both in this journal.)

The intimate relationship between island-arc volcanism and sinking (subducting) plates is easily seen, even on casual inspection. We thus have many more clues to the mechanism of magma formation and ascension in this environment than we do for volcanoes in other areas. A fairly detailed explanatory model is now emerging, although opinions differ on certain aspects of it. Before explaining the model, I will describe the general features for which any model must account.

One vivid instance of an observable correspondence between the shape of an island arc and the structure of the oceanic trench it parallels is the Aleutian Islands of Alaska. The Aleutian trench marks the northern edge of the Pacific plate. Because the Aleutian trench is curved relative to the direction of movement of the Pacific plate, subduction ceases west of about long. 176° E, beyond which the plates move tangentially to one another (Fig. 2). Where subduction ceases, so too does Aleutian volcanism.

The long, thin volcanic front of an island arc is commonly segmented. Each segment consists of two to a dozen volcanoes arranged in a straight line, which is offset from neighboring segments. Although this segmented pattern was originally suggested by work in Japan and Central America (1), it is particularly well established in the Aleutian arc (Fig. 3). Surprisingly enough, the pattern of segmentation in the Aleutians seems to reflect the geometry or structure of the subducting plate and not that of the plate upon which the islands are built. Each break thus far located in the subducting plate correlates precisely with a break in the volcanic front (2).

Some of these breaks occur along fracture zones that formed millions of years ago, during plate formation along an irregular oceanic ridge. Other breaks are new and form in response to the incompressible curved plate or spherical cap being forced to subduct along a suture in the earth's surface, the curvature of which is different from that of the earth itself. The reason the plate splits into staves in subducting is easily appreciated by trying to fold a paper beanie cap over a straight edge. If the suture zone has greater curvature than the earth, the plate may buckle at depth rather than splaying into staves. The angle of subduction of each stave in the plate may be different, and this attitude seems to influence the position of the volcanic front.

The volcanic centers of an island arc seem to grow according to a consistent pattern. The first appearance of a young island arc, such as Tonga, is a string of widely but regularly spaced volcanoes barely peaking above the ocean. As these volcanoes become

source regions supplying magma to the individual volcanic centers are in communication with each other. That is, there may be a single, locally (i.e. 100–300 km) continuous source beneath the entire arc.

Once the arc becomes well established, which takes about 3–4 million years, a new group of volcanoes may

The segmentation of the front, the equal spacing of the volcanic centers, and the secondary front, with its lavas containing proportionally greater quantities of potash, are the first-order characteristics of island-arc magmatism that any model must explain in answering the basic questions: What is the source of magma? How is it produced? How does it get

Figure 3. If the summits of historically active and potentially active volcanoes are marked as points on a map, and lines are drawn to connect as many points as possible, the segmented character of the front becomes evident, as in this portion of the Aleutian chain. Each known break in the subducting Pacific plate correlates with a break in the Aleutian front. Bogoslof Island, north of the main front at long. 168° W, represents the developing secondary front.

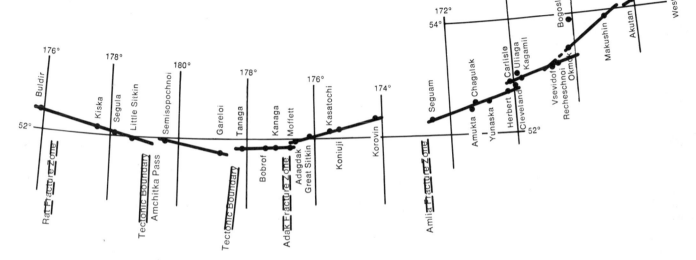

extinct, they are replaced by new ones appearing in the immediate neighborhood of the initial volcanoes. These volcanoes in turn become extinct, and new ones continually appear, until large islands or volcanic centers exist. Thus during the growth of the arc the areas of volcanism remain relatively fixed spatially. Volcanoes come and go, but they usually appear within a few kilometers of one another. It seems once the magma makes a path to the surface, whatever its means of travel, it tends to use the same path for the life of the arc.

The distance between neighboring volcanic centers in many areas is amazingly regular. In the Aleutian Islands, the Alaskan Peninsula, and the Cascades, for example, the most frequent spacing is about 70 km between centers (3). The volume of lava emitted per unit time seems to be roughly proportional to the distance between neighboring volcanic centers. Whatever the source of the magma, this rough proportionality and the regularity of spacing implies that the

appear here and there in a line parallel to and behind the existing front. The volcanoes of this secondary front are contemporaneously active with their nearest neighbors on the front, but the secondary front never becomes as fully developed as the original front. The volume of lava produced per volcanic center is only about 15–20% of that of the first front.

Clear examples of secondary-front development are in the northern Kurile Islands and in northern Honshu, Japan, where the Nasu volcanic front on the east is backed by the Tyokai secondary front, about 80 km to the west (4). In the Aleutian Islands, the secondary front began developing only recently with the appearance of Amak Island, some 7,000 years ago, and Bogoslof Island, which the Aleuts watched emerge from the sea in 1796 (see Fig. 3). There seems to be little difference between the compositions of the lavas emitted from each front, except that those of the secondary front are consistently richer in potash.

to the surface? The two general clues we have to the answers are the life history of a lithospheric plate and the chemical composition and volume of island-arc lavas.

History of a plate

Submarine mountain belts (mid-ocean ridges) mark complex breaks in the crust, beneath which hot viscous rock ascends from great depth, partially melts, then spreads laterally and cools, welding itself to the sides of the break (Fig. 4). Melting is caused by the great reduction in pressure during ascent, coupled with the fact that the rock ascends much faster than it can possibly cool. This is how the plates on either side of the ridge grow. Plates generally move a few centimeters per year, about as fast as one's fingernails grow, but speeds can vary by nearly a factor of ten.

During ascent and partial melting, tholeiitic basalt is extracted from peridotite, the mantle or source rock (see Table 1), at a depth of about 25

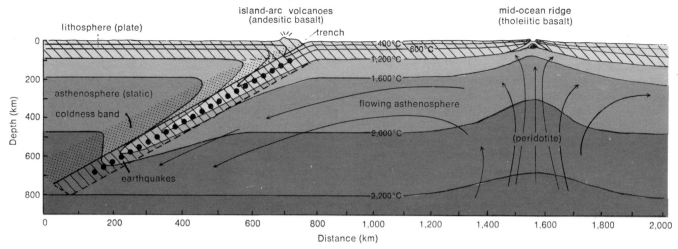

Figure 4. In a model of plate subduction in which the asthenosphere does not flow, a thick band of cool asthenosphere develops along the upper edge of the plate—which is moving at 8 cm/yr—as the plate absorbs heat from its surroundings. When the plate reaches the same temperature as the adjacent asthenosphere, it is resorbed. There is hardly any frictional heating at the upper edge of the plate. Notice how cold the upper edge of the plate is at a depth of 100–150 km. (After Toksöz, Minear, and Julian, ref. 5.)

km and a temperature of about 1,250°C. This magma erupts onto the seafloor and intrudes as dikes and plutons at depth to form a thin (5–8 km) layer, the oceanic crust, which everywhere caps the oceanic part of plates. Below the oceanic crust the plate consists of barren or depleted peridotite. It is depleted because it has already given up a substantial part of itself (15–25%) to form tholeiitic basalt.

As the plate creeps away from the ridge, continued cooling from the top promotes further solidification at its base, causing the lower boundary to grow downward, until after 50 million years it reaches a total steady-state thickness of about 80–100 km. During cooling and contraction, the oceanic crust fractures, allowing seawater to circulate, at least for a short time, deeply into the crust and to chemically alter a small amount of the basalt from its initial pristine igneous state. Just prior to subduction, then, the lithospheric plate consists of about 5–8 km of oceanic crust covered by a thin veneer (about 500 m) of pelagic sedimentary material and underlain by about 75 km of depleted peridotite.

When the plate plunges back into the mantle, its once-cold upper surface, formerly the seafloor, touches increasingly hotter rock, and heat begins to flow into the plate through both its top and bottom. Since the plate is initially thick and cold, and since heat flow by diffusion through

rock is very slow (about 100 km over 300 million years), the plate may penetrate to great depth before coming to thermal equilibrium with the mantle, at which point it is resorbed (see Fig. 4). The resorption time or penetration depth depends on the initial thickness of the plate just prior to subduction, which is proportional to the distance between the subduction trench and the mid-ocean ridge, and the speed of the plate. A thick, fast-moving plate such as that beneath Tonga may penetrate to a depth of 700 km, while a thin, slow plate such as that beneath the Pacific Northwest is almost immediately resorbed upon subduction. Lack of appreciation of this relationship between plate thickness, subduction rate, and depth of earthquakes has misled many geologists into believing that there is no plate beneath some areas of volcanism.

Owing to the uneven heating of the plate during subduction and the dynamical rigors of subduction, stresses develop within the plate that cause it to fracture in its coolest, most brittle parts, and this is where earthquakes occur. Elastic waves emitted by the earthquakes are recorded around the world, allowing the position of an earthquake to be determined essentially by triangulation. Careful positioning by ray-tracing techniques has shown that the earthquakes originate in a band about 20 km thick and probably about 15–20 km below the upper edge of the plate (see Fig. 4). This earthquake band—called the

Benioff zone, after a man who undertook early work in this field—closely indicates plate position within the earth.

Observations show that the Benioff zone is inclined at various angles beneath different island arcs and that the subducting plate must have penetrated to a depth of about 120 km before the volcanoes appear. Thus, a shallower angle of subduction will cause the volcanic front to form farther from the trench or suture zone. Of course, the Benioff zone is always deeper beneath the secondary front than beneath the original front. It also seems that if the plate subducts at an angle of less than about 25°, as in several areas beneath the Andes (6), no volcanism appears. Subduction rate has no apparent effect on the character of the associated volcanism.

Lava types

It is a commonly held belief that andesites (~60% SiO_2), which have a composition intermediate between basalts (~50% SiO_2) and granites (~70% SiO_2), are volumetrically dominant in island arcs. This does seem to be so for some island arcs, especially those built on continents or subcontinental masses such as Japan and the Lesser Antilles. But in many island arcs, especially those built upon oceanic crust, such as Tonga, Scotia, Izu-Mariana, the western Aleutians, and even Indonesia, the predominant lava is not andesite, but something nearer to basalt. Close inspection shows that every island arc has basaltic lavas of this type, though they may be volumetrically subordinate to andesite. Useful volumetric data are difficult to establish in many areas because erosion rapidly destroys many of these poorly constructed volcanoes in less than about one million years, and the more siliceous lavas commonly erupt explosively and scatter widely.

Island-arc basaltic lavas are distinct from other basalt lavas (Table 1) in that they contain less TiO_2 (<1%), low levels of MgO (<~6%), high levels of Al_2O_3 (>~16%), low levels of base metals such as nickel and chromium (<~50 ppm), and usually, but not always, about 1% K_2O. These characteristics are common to most lavas of island arcs and persist throughout the series regardless of SiO_2 content. To underscore this persistent character of the island-arc lavas, basaltic lavas are called andesitic basalts. Lavas that have more silica, but less silica than would qualify them as andesites, are called basaltic andesites.

Lavas generally contain two size populations of crystals: those of larger size (0.1–1 mm) grown slowly at depth and those of smaller size grown quickly upon the rapid cooling during eruption. The proportion of larger crystals (phenocrysts) varies from almost none in some lavas from Scotia and Tonga to about 40–50% of the volume of some Aleutian lavas. The degree of crystallinity appears proportional to the size, or age, of a volcanic center. Young, small volcanic centers on oceanic crust are apt to have phenocryst-free lavas, while older volcanic fronts hardly ever erupt a phenocryst-free lava. Crystallinity is a measure of the magma's residence time within the upper levels of the earth; in established arcs, therefore, magma ascent is on the average slower. It seems, overall, dynamically impossible for the earth to evict a magma containing more than about 50% phenocrysts, for such lavas are rarely observed.

The minerals in island-arc lavas are neither many nor varied. In andesitic basalts, plagioclase (ranging in composition from $NaAlSi_3O_8$ to $CaAl_2Si_2O_8$), olivine (Mg_2SiO_4 to Fe_2SiO_4), magnetite (Fe_3O_4), and monoclinic pyroxene ($CaMgSi_2O_6$ to

Table 1. Composition by % weight of rocks from various parts of the plate-tectonic cycle*

	Mantle	Oceanic crust	Island-arc lavas				Continental crust
	Peridotite	Tholeiite	Andesitic basalt	Basaltic andesite	Andesite	Dacite	
SiO_2	45.10	49.34	49.93	53.29	58.24	66.03	57.80
TiO_2	0.13	1.49	0.90	0.95	0.69	0.46	0.8
Al_2O_3	3.92	17.04	19.46	18.31	17.33	16.12	15.2
Fe_2O_3	1.00	1.99	3.24	3.83	3.91	1.69	2.3
FeO	7.29	6.82	6.30	4.71	3.25	2.64	5.5
MnO	0.14	0.17	0.18	0.20	0.16	0.13	0.18
MgO	38.81	7.19	5.07	3.92	3.24	0.64	4.00
CaO	2.66	11.72	10.43	8.82	7.10	2.36	8.7
Na_2O	0.27	2.73	3.31	3.29	3.74	4.74	3.2
K_2O	0.02	0.16	1.05	1.11	2.00	4.60	2.2
P_2O_5	0.01	0.16	0.23	0.26	0.19	0.15	0.3
Total	99.35	98.81	100.10	98.69	99.85	99.56	100.18

* Small amounts of CO_2, S, H_2O, and a host of trace metals also occur in all of these rocks.

$CaFeSi_2O_6$) appear, while in more siliceous lavas (>~52% SiO_2) olivine is replaced by orthorhombic pyroxene ($MgSiO_3$–$FeSiO_3$). Hydrous minerals such as hornblende and biotite are exceedingly rare. Yet hornblende is commonly found in domes and plugs, rather reluctant and late-stage extrusions that have undergone a protracted history of cooling. When hydrous phases appear, it is generally in island arcs that sit on continental plates, such as the Cascades. The misconception that hornblende is a common and therefore important mineral in island-arc lavas has led to a great deal of confusion concerning the role of water in island-arc magmatism.

The density of island-arc magma is generally less than that of the mineral crystals which form within it. Consequently, unless the magma is well stirred by convection, crystals settle from the magma and change its composition—the crystals, of course, are of a different composition from the magma itself. Crystal fractionation or differentiation results in the large diversity of lava compositions issued from any volcano. Silica content is the principal measure of degree of fractionation. In andesitic basalts, however, the constituent compound minerals contain an amount of silica similar to that of the magma. Large quantities of these minerals (50% of magma volume) may therefore settle out with little effect on the silica content of the magma.

One distinctive feature of island-arc lavas is the remarkably close correlation between their potash content and the depth to the underlying Benioff zone. Volcanic centers over deeper Benioff zones have lavas richer in potash. A general increase in alkali content with increasing depth to the earthquake zone was delineated by Japanese petrologists, notably H. Kuno, in the 1950s. They demonstrated this trend not only across the volcanic front (e.g. near Tokyo) but also in scattered volcanoes as far away as Manchuria. With the coming of plate tectonics and accurate positioning of the Benioff zones, W. R. Dickinson and T. Hatherton demonstrated a systematic global correlation of potash and depth using data principally from volcanic-front lava. Other components, such as soda, sometimes show a weak local correlation, but no other global correlations have been found.

Magma sources

The inspired work of R. R. Coats gave birth to nearly all of the ideas of magma generation near Benioff zones. After studying Aleutian volcanism and tectonics for many years, Coats recognized in about 1960—before the concept of plate tectonics had surfaced—that the distribution of earthquakes and their compressive stress pattern suggested that oceanic crust was being thrust beneath the Aleutian island arc. Coats's work suggested two possible sources for the

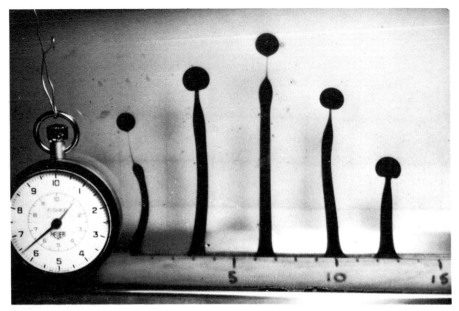

Figure 5. Gravitational instability results when a light fluid is overlain by a heavier one. The lighter fluid forms evenly spaced fingers as it transfers to the surface of the heavier fluid. In this experiment, the fluids are housed in a cylinder completely open at the top; the stage shown in the photograph has been reached after about 100 seconds. Magma may rise in a similar fashion to form the equally spaced volcanic centers of an island arc.

arc magmas. First, the oceanic crust, carrying its veneer of pelagic sediment and water, could eventually melt upon subduction. Second—the alternative Coats favored—an aqueous fluid migrating upward from the subducted crust might induce melting in the overlying wedge of peridotite, because water greatly lowers the melting point of rocks. To this day, these ideas dominate the controversy surrounding Benioff-zone magmatism.

The various arguments for and against each source on chemical grounds have yet to produce universal acceptance of any mechanism of magma production. The strongest argument against melting of peridotite is that it is difficult to match the bulk chemistry of the lavas. To get the peridotite to melt at the expected temperature, a large amount of water must be introduced, but experiments show that the introduction of large amounts of water (~75 mole %) makes the magma much more siliceous than the typical andesitic basalt lavas. It is at least clear from such experiments that water is not a major factor in magma production, because if it were, typical island-arc lava would be much more siliceous than it actually is.

The strongest argument against melting of the subducted crust is that

theoretical studies of the thermal regime show the crust to be much too cold to melt at the proper depth. Melting temperatures can be approached if the oceanic crust contains an unusually large amount of water or if there is intense frictional heating along the upper edge of the plate. But the balancing of these special circumstances seems too delicate to occur globally beneath island arcs. Moreover, the pattern of concentrations of rare-earth elements displayed by arc lavas has so far been difficult to match using a subducted-crust source. Isotopic measurements show that the veneer of pelagic sediment can contribute no more than about 2% (volume) to the composition of magma, implying either that the sediment is scraped off early in the subduction process or that magma is not produced in its vicinity.

Compiling all the other intricate geochemical arguments only produces two equal-sized heaps of circumstantial evidence. The physical evidence presented earlier, while also circumstantial, is thus crucial: if the source can be decided on these grounds, the chemical data can be reconciled. Let us therefore review the physical evidence.

First, we know that lithospheric plates must subduct to a depth of about 120 km for island-arc volcanism

to appear. This suggests that the plate must encounter a certain minimum mantle temperature before magma is produced. It also seems that plates must penetrate the asthenosphere at an angle greater than about 25°. This implies that the asthenosphere is a critical mechanical factor in producing magma. That the asthenosphere is not a critical chemical factor is further suggested by the fact that it is chemically hardly different from the lower portion of the overlying lithosphere. The fact that the appearance of volcanism is independent of subduction rate implies that frictional heating is not a principal factor in producing magma.

The association of a segmented subducting plate with a segmented volcanic front suggests that the magma is generated very near, if not within, the subducting plate. If plate position were only indirectly related to the pattern of volcanism, the correlation would probably be more diffuse. The constancy in the average chemical composition of island-arc lavas over millions of years points to a quickly and easily replenished source, whatever the material. The correlation of potash level in the lava with depth of the underlying subducting plate implies that the chemical process of magma production is pressure dependent: it could reflect the change in pressure in the melt–source rock equilibrium constant. The meaning of the secondary front is unclear, but that front certainly seems to be a result of the maturation or growth of the source region.

What kind of mechanism can account for the extremely regular spacing of the volcanic centers? The simplest explanation stems from a consideration of a fluid gravitational instability (3). When a light fluid lies beneath a heavier one (e.g. oil under water) any perturbation from equilibrium will cause the lower, light fluid to form a series of equally spaced fingers as it penetrates the upper fluid and transfers to the surface (Fig. 5). This instability, which is classically known as a Rayleigh-Taylor (R-T) instability, has been studied under a wide variety of applications, from salt-dome dynamics to the settling of diatoms from the sea (7). (A convenient experiment can be performed by swirling port around in a wine glass, leaving a rim of liquid. Equally spaced "wine tears" will form as the rim becomes unstable under the effects of gravity and surface tension. This is best seen in candlelight.) The spacing (λ) of the series of fingers, or *diapirs,* is controlled by the thickness (h_2) of the lower, buoyant layer (assuming $h_1/h_2 \gg 1$) and the relative viscosities (μ_1/μ_2) of the two fluids—

$$\lambda = \frac{2\pi h_2}{2.15}\left(\frac{\mu_1}{\mu_2}\right)^{1/3}$$

—but it is rather insensitive to the geometry of the lower layer.

Since molten rock is always less dense than its more solid equivalent, a buried layer or ribbon of magma would, upon undergoing an R-T instability, produce a regularly spaced array of volcanic centers on the surface. The spacing of the volcanic centers in many areas is about 65–70 km, but this length is not universal; in South America and the Kurile Islands it is only about 40–50 km. Why the spacing varies is not yet clear.

The width of the source layer can only be guessed at. If the layer were wider than the characteristic spacing there would usually be more than one volcanic front; in fact, the appearance of the weak secondary front late in the history of an arc may imply that the region has grown in width. Judging from the small volume of magma arriving at the surface, it is unlikely that the diapiric fingers of magma are continuous tubelike bodies stretching from the source to the surface. Rather, the magma probably ascends as smaller discrete bodies, which always move along the same paths to the surface. The diapir diameter (d) is related to the properties of the source region as

$$d = h_2\left(\frac{\mu_1}{\mu_2}\right)^{1/4}$$

The initial stage of diapirism forces a path through the upper layer, perhaps by melting, and thus succeeding parcels of magma can easily follow.

All estimates of the initial thickness of the magma ribbon, using either the above equation or the observed flux of lava, show it to be be, geologically speaking, fantastically thin, about 100–500 m. But since the viscosity ratio is uncertain and the flux of lava at the surface is only a lower bound on the actual supply of magma, this estimate is somewhat uncertain. The diapir diameter is similarly found to

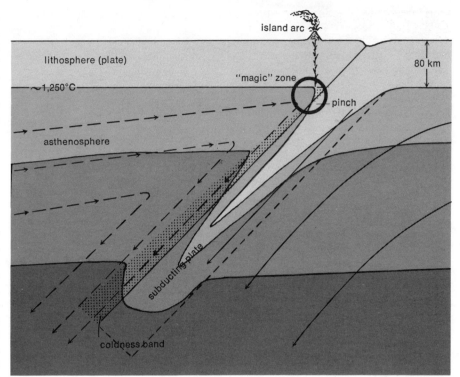

Figure 6. This revised model of plate subduction has the asthenosphere flowing into the corner formed by the opposing plates, as a result of the downward movement of the solid subducting plate, which forces the adjacent fluid to move with it. The black dashed lines indicate the path of the fluid (i.e. stream lines; 8). Since the upper corner of the plate is struck constantly by uncooled fluid, it is a likely source for the magma that forms the overlying island arc. Compare the thickness of the coldness band with that of Fig. 4.

be about 2–5 km, in good agreement with solidified diapirs that have been unearthed near the earth's surface.

Asthenosphere corner flow

In the thermal model depicted in Figure 4, the asthenosphere did not flow in the region of the subducting plate. As the cool plate moved into the stagnant asthenosphere, it removed heat from the adjacent region. Since replacement could not take place fast enough to maintain the high temperature of the asthenosphere as a whole, the temperature pattern developed as shown. Allowing the wedge of asthenosphere above the subducting plate to flow introduces a new dimension of heat and mass transfer in this region (Fig. 6). Where one plate is thrust beneath another, mantle material below the opposing plate is forced to flow in conformity with the motion of the subducting plate. A time-honored principle of fluid mechanics, the "no-slip boundary condition," states that fluid adheres to a solid surface, and right at the interface fluid and solid must move with the same velocity. A simple

observation reveals that dust on a moving fan blade is not swept away: this is because the air at the interface is moving at the same speed as the blade. Since the mantle beneath the plates seemingly behaves everywhere else as a fluid, so too must it here in obeying the no-slip condition.

The asthenosphere flowing with the moving inclined plate means that new, hot mantle material is continuously drawn into the corner, toward the subducting plate. A thick cool area of asthenosphere no longer develops next to the upper edge of the plate near the corner. In fact, if the asthenosphere moves toward the plate fast enough, it will maintain essentially the same high temperature until it turns to flow parallel to the plate. Heat will then keep flowing from the asthenosphere as it moves along the subducting plate, and a band of coolness will develop farther down the plate, thickening with distance. How fast it will thicken with depth depends on the speed and thickness of the plate and the thermal transport properties of the material.

The coldness band will thus be thin-

and velocity of the subducting plate. If the plate dips steeply enough, the arc plate may break near the volcanic front. D. Karig has shown that this happened once in the Tonga arc.

Marginal basin lavas are also helpful in understanding arc volcanism because the basalts produced give a magmatic sample of mantle peridotite near the arc volcanic front. The marginal basin basalts chemically belong to the group of ocean-ridge tholeiitic basalts—distinct, as I have said, from island-arc andesitic basalt in their higher MgO, TiO_2, and base metals (12). But perhaps the strongest dissimilarity is the significantly lower radiogenic strontium ([87]Sr) content in the basin basalts. Heavy isotopes are useful in fingerprinting the source regions of lavas because their fractionation by natural processes is very difficult. The enrichment of [87]Sr in island-arc lavas may point to the subducted oceanic crust as the magma sources. The [87]Sr content of the oceanic crust becomes enriched between the ridge and the subduction trench through interaction with seawater (13).

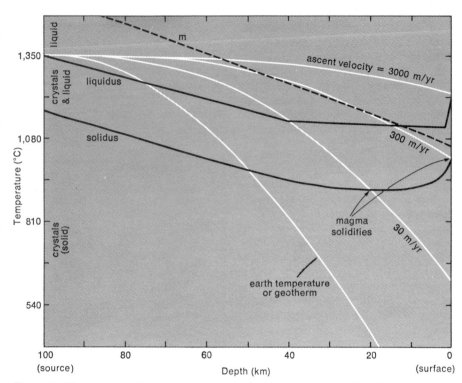

Figure 7. The rate at which an ascending spherical body of magma with a radius of 0.8 km will cool depends upon the speed of its ascent. Three white curves in the graph represents different ascent velocities. Between the curves labeled *solidus* and *liquidus,* the magma is a mixture of liquid and crystals. At the junction of any cooling curve with the solidus, the magma becomes solid and hence ceases to ascend. The curve labeled M marks the melting point of the surrounding earth (estimated from the works of P. J. Wyllie). All successful ascents lie between the two cooling curves on the right; their relation to the M curve suggests that the magma may actually melt its way along over some of its ascent path.

Magma ascent

We have now construed a mechanism for the production of island-arc magma. How magma moves from the magic zone to the surface, however, is a mystery. It may travel as a propagating magma-filled elastic crack or as a hot, viscous globe slowly forcing its way along. Each of these transport mechanisms is known to occur: rapid injection of magma forms dikes, and slow viscous intrusion into the uppermost crust by large bodies of granitelike magma forms plutons. Between these extremes lies an infinite variety of combinations of these mechanisms.

Whatever the means of transport, clearly the magma must traverse the lithosphere without solidifying. This, then, is the major constraint on the mechanism of transport: magma must be supplied to the surface with a maximum of 50% crystallization. We know, of course, that the ascent velocity must initially, at least, be greater than the subduction velocity (~3 cm/year, or 10^{-7} cm/sec), or magma generated in the magic zone would be carried downward by the flow of the asthenosphere. A simple (kinematic) model of the heat trans-

fer from magma traversing the lithosphere might be useful in more closely estimating a mean ascent velocity, which can in turn be used to investigate a dynamic model of ascent.

Imagine a body of magma of unknown size and shape moving toward the surface. If it moves sufficiently fast it will cool only by adiabatic decompression (~60°C/100 km). On the other hand, if the magma moves very slowly, it will always maintain thermal equilibrium with the wall rock and solidify at great depth. Obviously, between these extremes lies a set of cooling curves that give the temperature of the magma at any depth as a function of its ascent velocity (Fig. 7). A likely ascent velocity can be estimated by selecting the curve that best fits the thermal history of the magma as deduced from the degree of crystallinity of the lavas.

The rate of cooling is proportional to the product of the surface area of the magma and the temperature difference between it and the surrounding wall rock. Since the wall rock becomes

progressively colder with approach to the surface, the cooling rate increases strongly during ascent. Cooling is also affected by the ascent rate. If the body is stationary, cooling is solely by diffusion of heat into the wall rock; as the wall rock heats up, cooling becomes increasingly slower. If the magma is moving, it is continually brought into contact with unheated wall rock; and heat need only diffuse across the magma–wall rock interface to be convected away. Movement with respect to the wall rock plays the same role as the all too familiar wind-chill factor.

Within the body of magma itself, only the mean temperature is of interest, because the magma is surely well mixed, owing to its low viscosity (~10^3 poise), the horizontal temperature gradients, and the shear stress at its margins. It is clear that the rate-controlling factor is the ability of the wall rock to carry heat away from the magma. In the actual calculation, the magma is given an ascent velocity, the wall-rock temperature or geotherm is prescribed, and a set of cooling curves is calculated (14).

nest near the corner where the uncooled mantle material first encounters the plate. The flux of heat into the subducting plate will be inversely proportional to the thickness of the coldness band, so the highest flux will also be near the corner. If the coldness band is pinched enough, hot asthenosphere will strike the plate and cause melting of the oceanic crust.

The improvement that this model represents over other explanations of localized melting can be appreciated by a simple consideration. It has been noted by many authors (5) that the plate could be made to melt if friction (viscous dissipation) with a shear stress of 1 to 2 kilobars were applied at the upper edge of the plate. A shear stress of that magnitude, however, is well beyond the strength of the mantle here. If heat is instead supplied by the flow of hot asthenosphere around the corner and against the plate, the fluid can supply this heat with a loss in temperature of less than 25°C for each kilobar of shear stress needed to cause melting with no flow. This convective heat source which causes the pinch would only be effective near the corner, for once the flow becomes parallel to the plate, it cools increasingly with distance from the corner.

The numerical results of Andrews and Sleep (9), although intended for another purpose, show the coldness band to be pinched as thin as 4 or 5 km, the resolution limit of their grid. My own analytical results show it to be much thinner. Recent unpublished work by M. N. Toksöz and A. Hsui shows that if the flow is constrained to move as a large eddy trapped in the corner, coldness once again accumulates and no pinch is formed. This may imply that to maintain the pinch new uncooled mantle material (fluid) must continually be supplied to the corner area.

Moreover, in passing through a constriction such as the corner, the very viscous asthenosphere (its viscosity is about 10^{20} poise; pancake syrup has a viscosity of about 10 poise) may actually become more heated. Even when the fluid is still about 100 km from the corner, it is heating up at the rate of about one degree per year, and it will take at least a few million years for a parcel of fluid to pass into and out of the corner area. Obviously much of this heat is lost to conduc-

tion, but analysis does show that as the fluid mantle approaches the corner, it may be hotter than usual. This heating will enhance the pinch effect. The flux of heat to the plate from flow of the asthenosphere coupled with some frictional heating seems ample to cause the plate to melt near the corner in the pinch zone.

Just as heat diffuses into the plate, water and other volatiles may diffuse from the plate into the flowing asthenosphere. And just as a band of coldness formed next to the plate, so a volatile band will form in the asthenosphere. But since the diffusivity of water through rock is at least one thousand times smaller than that of heat, the chemical boundary layer will be about one thousand times smaller than the thermal boundary layer, or coldness band. It will probably become no thicker than about 10 m; and near the pinch it will be infinitely thin. Thus volatiles from the plate have no effect on the asthenosphere near the corner where magma is produced.

Magma, then, is probably produced somewhere in the small triangular region nearest the corner, which I will call the magic zone in honor of the myriad of mutually exclusive and supernatural mechanisms of magma production suggested to go on in this region of the earth. The very appealing aspects of the coldness-band pinch are that it is localized; it occurs, according to my calculations, at the right depth; and it necessitates mechanical involvement of the asthenosphere.

The volume of magma erupted and intruded in island arcs may be typically about 5 km^3 per km of arc per million years; estimates vary between 1 and 10 (10). The upper kilometer alone of oceanic crust supplies the magic zone with about 60 km^3 m.y.$^{-1}$ km^{-1} by subduction. Peridotite, the material of the asthenosphere, is supplied by the induced flow at the rate of 300 km^3 m.y.$^{-1}$ km^{-1} (included is all material within 50 km of the corner). Volumetrically, about 8% of the upper 1 km of oceanic crust or only about 1.6% of the asthenospheric peridotite within the magic zone must be melted to supply magma continually to the arc. Either source seems ample.

But the critical quantities to supply

are those elements which are strongly enriched in the arc magma relative to the source materials (11). Typical arc lava contains 0.5–1.5% (weight) potash, while oceanic crust contains about 0.25% and peridotite about 0.025%. To supply the potash to an arc magma with 1% potash, then, requires about 30% of the 1 km of crust and about 70% of the entire 300 km^3 of peridotite near the corner. For an arc magma containing 2% potash, such as in some parts of Indonesia, these percentages must be doubled. Similar calculations for other elements in arc magmas lead to the same conclusion: if peridotite is the source, magma production must be extremely efficient, whereas if the oceanic crust is the source, efficiency can be much lower.

Further light may be shed on some aspects of the corner-flow model by examining a feature associated with island arcs: the marginal seas or basins commonly found behind them (see Fig. 2). In many of these seas—e.g. the South China Sea and the Sea of Japan—secondary seafloor spreading occurs, though not in all; it is not found, for example, in the Bering Sea. Where it does occur, plate formation appears to be a more sluggish process than that associated with the great oceanic ridge system, but ridgelike tholeiitic basalt is, nevertheless, produced. Marginal-basin spreading evidently results from the differential motion of the two plates relative to the subduction zone, a result of plate movements on a larger scale across the earth's surface.

At a glance it seems odd that the plates do not separate along their line of contact or at the volcanic front where the arc plate has been weakened by the transmission of magma. (Plates have a very low tensile strength.) That the plate instead breaks a few hundred km behind the front may be a result of the flow induced in the wedge of asthenosphere between the plates. The subducting plate causes the asthenosphere immediately beneath the arc plate to flow toward the corner, producing traction on the arc plate, which binds the plates tightly together. The traction decreases with distance from the corner, and perhaps once it drops below some threshold (i.e. at some distance), the plate breaks. The magnitude of the traction or drag depends directly on the angle of dip

These cooling curves show that a sphere of magma with a radius of 1 km must ascend at least 1 km per year if it is to reach the surface without solidifying. As expected, the cooling curves inherit their shape from that of the geotherm, and therefore they are concave to the liquidus and solidus of the magma. This has important consequences. If the magma initially contains less than about 10% crystals, as it may very well in pulling slowly away from the source, its temperature will be above its liquidus and thus preclude crystallization. With no crystals present in the magma, crystal settling (differentiation) can hardly take place to alter the magma's composition. Over part of the ascent, principally that nearer the surface, the magma may be hot enough to fuse the wall rock and generate another magma.

These general implications are in fact supported by a few shreds of evidence. High-pressure minerals reflecting growth at great depth are absent in island-arc lavas: all the minerals indicate growth at depths no greater than about 10 km. Although foreign inclusions are rare in island-arc lavas, when a piece of mantle-wall rock is found, only the mineral with the highest melting temperature, olivine, is left.

Although these calculations have been carried out for a magma undersaturated with water, similar studies for a water-saturated magma (15) show that even when the magma ascends adiabatically, it will solidify at great depth. The inclusion of water makes the initial magma temperature (i.e. liquidus temperature) at depth less than its liquidus temperature at the surface. To erupt, the magma thus must gain heat during ascent, and this can only be partly accomplished through the heat gained by exsolution (i.e. "boiling") of water from the magma. Allowing for heat loss to the wall rock, the results show that the magma may rise only about 30% of the distance to the surface before solidifying. So even if a water-rich magma were generated in the magic zone, it is highly unlikely that it could ever reach the surface.

If the magma ascends by elastic-crack propagation, the rate of cooling will be greater, owing to the large ratio of surface area to volume. For a magma-filled crack having an aspect ratio of R (length/width) with the same volume as a sphere of magma, the crack must travel $R^{4/3}$ times faster than the sphere to arrive at the surface at the same temperature as the sphere (14). Using a rather conservative value for R—say, 1,000—implies that a body of magma equal to a 1-km–radius sphere must ascend at least 30 km per day to reach the surface without solidifying. If the area of ascent has been heated by the earlier ascent of other bodies, then everything can ascend proportionally slower and still reach the surface.

These ascent rates are minimum estimates, for the magma may ascend in bodies of many different shapes and at uneven rates. To know how the speed varies during ascent, a dynamic model of ascent that depends on the rheological and elastic properties of the mantle must be coupled with the heat-transfer model.

I have mentioned the two extreme models: propagation of a magma-filled elastic crack, and a hot, viscous globe of magma ascending by softening the material about it and causing it to flow around the body. These extremes tend to transport magma in very different ways geometrically. An elastic crack propagates upward and laterally, essentially dispersing the magma relative to its source. A viscous ascent, on the other hand, concentrates magma, producing a body with a large positive buoyancy that can overcome the drag on its walls.

We have seen that the volcanic centers in island arcs repeatedly issue lava from very localized areas that remain fixed over millions of years. This concentration of volcanic activity at discrete points may imply that the mode of magma transport is closer to the second model than the first. In other regions, such as Hawaii, which is not an island arc, great, long fissure eruptions take place and the volcanic centers lie on large regional fractures in the plate. The mode of transport in these regions is surely closely linked with elastic-crack propagation.

But there is a problem with the viscous model: a simple calculation shows that if ascent is governed by Stokes's law, then the great viscosity of the lithosphere (about 10^{25} poise, if it is viscous at all) ensures that the ascent velocity will be about ten thousand times smaller than that

necessary to prevent solidification. A successful ascent could be made only by unrealistically large bodies of magma. But the assumption upon which Stokes's law rests—an infinite Newtonian fluid of constant viscosity—is surely not satisfied here. Near the hot magma, the wall-rock viscosity is apt to be much lower than usual (the viscosity of silicates is extremely sensitive to temperature around the melting point), which will reduce the drag on the magma and thereby hasten ascent. Calculations show that if the magma melts about 50% of the adjacent wall rock, it can ascend at a satisfactory speed (~1 km/yr).

But the magma will still solidify long before reaching the surface if it is moving through lithosphere of normal temperature, because of the tremendous amount of heat (equivalent to about 3,000°C) lost in fusion. The obvious conclusion is that the first few bodies of magma do not in fact reach the surface, but they leave behind a hot passageway that insulates succeeding magmas, allowing them to ascend slowly without solidifying. Under these circumstances a typical ascent velocity may be about 10 to 100 meters per year, and the trip from the magic zone to the surface may take 1,000 to 10,000 years, a short time geologically.

Benioff-zone magmatism

It is impossible to sift through the many facets of island-arc magmatism without gaining affection for some mechanism of magma production. I believe that, of the two principal Coats mechanisms of magma production beneath island arcs, the upward migration of water from plate dehydration to cause melting in the overlying peridotite fares less well than direct melting of the subducted oceanic crust. Peridotite melting induced by water gives a magma richer in silica than the andesitic basalt of island arcs; a magma saturated with water at great depth will solidify long before reaching the surface regardless of its means of ascent; slowly subducting plates dehydrate (are thermally resorbed) at very shallow depths, yet the position of the volcanic front is independent of convergence rate; the lavas do not generally contain a hydrous mineral; and they are enriched in ^{87}Sr.

The flow induced by the subducting

plate in the asthenosphere below the arc plate supplies a large amount of heat to the plate near the corner. This, coupled with viscous heating, probably causes melting of the subducted oceanic crust. The mismatch between the abundances of rare earth elements in arc lavas and that derived from a hypothetical subducted crust source more than likely reflects an inadequate knowledge of the mineralogy of the crust and the thermodynamic properties of the distribution of the rare earth elements among these minerals and magma.

It seems, then, that melting of the oceanic crust precipitates island-arc volcanism. The magma, once generated, may sometimes interact chemically with the overlying peridotite, and this may explain some of the present chemical inconsistencies. But, on the whole, this interaction cannot be too important, for the delicate monotony of the major element and isotope composition of the lavas can be produced only by a steady, systematic process such as subduction itself.

A general model, then, is depicted in Figure 8. The magma produced from partial melting of the oceanic crust collects as a ribbonlike body near the upper edge of the plate. Owing to its local continuity and gravitational instability, the ribbon sends up magma in fingers that produce regularly spaced (~70 km in many areas) volcanic centers at the surface. The final configuration of the volcanic front is heavily influenced by the structure of the upper edge of the subducting plate, which controls the position of the ribbon (Fig. 9).

A few million years after establishment of the front, volcanoes often appear behind the front, forming a weak secondary front. The distance separating the two fronts is generally somewhat less than the spacing of the volcanic centers along the front. In fact, in many areas it is less so exactly by a multiplicative factor equal to cosine of the angle of dip of the subducting plate. If the ribbon of magma grows far enough (i.e. >~70 km) downdip, a secondary gravitational instability will occur, producing at the surface a volcanic center separated by only 50 km from the front (assuming that the plate dips at 45°, 70 × cos 45 ≃ 50 km).

The spatial configuration of the two

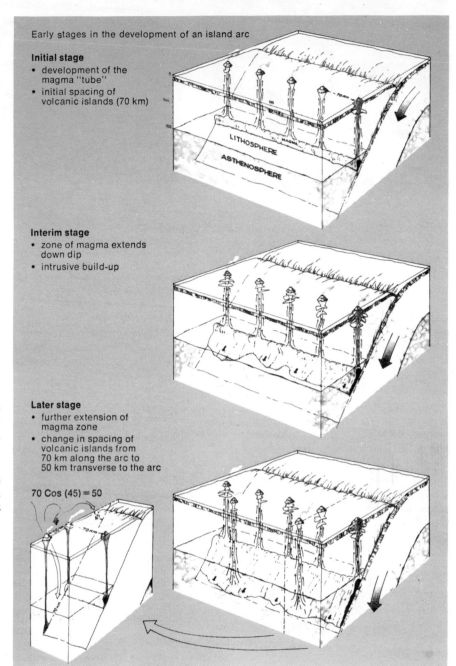

Early stages in the development of an island arc

Initial stage
- development of the magma "tube"
- initial spacing of volcanic islands (70 km)

LITHOSPHERE
ASTHENOSPHERE

Interim stage
- zone of magma extends down dip
- intrusive build-up

Later stage
- further extension of magma zone
- change in spacing of volcanic islands from 70 km along the arc to 50 km transverse to the arc

70 Cos (45) = 50

Figure 8. If the model shown in Fig. 6 works, a thin ribbon of magma will form near the corner at the upper edge of the downgoing plate. Diapiric conduits for the magma form in accordance with the Rayleigh-Taylor model of gravitational instability, creating the evenly spaced volcanic centers or islands of an island arc. Over 3–4 million years, the ribbon of magma is drawn out by the downward movement of the subducting plate. When it has become as wide as the distance between the adjacent centers in the original front, the Rayleigh-Taylor model takes hold in a new dimension, and new diapirs will form behind the earlier ones. Since the band of magma is not parallel to the surface of the earth but rather at about a 45° angle to it, the new front will be separated from the original front by less than the spacing of the centers in the first front. Development of the secondary front tends to be patchy, suggesting that the lower edge of the magma band is at the borderline of the pinch in the coldness band. (Drawing by Peter Van Dusen.)

fronts and their volcanic centers implies a dipping, ribbonlike layer of magma, the width of which is probably about 100 km. The growth downward of the initial ribbon may be caused simply by the motion of the subducting plate, for in the time until the secondary front appears (~3 million yrs), the initial melting spot will be dragged downward about 100 km (~3 cm/yr × 3 million yrs). But since the depth of initial melting re-

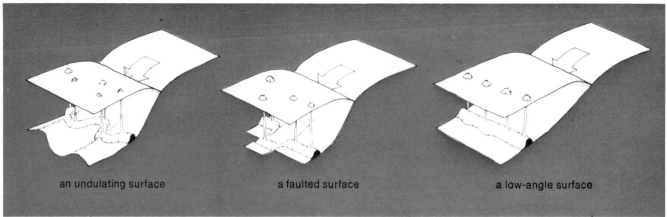

an undulating surface   a faulted surface   a low-angle surface

Figure 9. Variations in the surface or angle of subduction of a plate will cause different arrangements of the islands in an arc. A plate with a simple, smooth surface subducting at about a 45° angle, shown in Fig. 8, produces a straight line of volcanoes fairly close to the trench. An uneven surface will disrupt the straight line, and subduction at a smaller angle will cause the front to form farther from the trench. (Drawing by Peter Van Dusen.)

mains fixed relative to the corner, the whole ribbon of magma grows wider in being pulled downward by the subducting plate. The weak and patchy development of the secondary front implies that the pinch in the coldness band, and hence melting, cannot be sustained for more than about 100 km downdip. As originally suggested by Dickinson (15) and later shown theoretically (3), the correlation of lava potash content with increasing depth probably arises from the effect of pressure on the equilibrium constant describing the distribution of potash between magma and subducted oceanic crust.

Island-arc volcanism, simply viewed, is but a link in the distillation of the mantle to produce continents. Peridotite ($\sim$45% SiO_2) gives rise to basaltic oceanic crust ($\sim$50% SiO_2), which in turn yields the island-arc andesitic suite (andesitic basalt to dacite, $\sim$55% SiO_2) of continental-type rocks. Worldwide, the crust around island-arc volcanic centers receives an input of energy, which, over the life of the arc, is higher than at any other continental location. The crust here simply sits and collects magma for millions of years. With a steady stream of energy-carrying parcels of magma penetrating a geologically tiny space of crust, it is perhaps not surprising that ore deposits, geysers, huge explosions, etc., occur in these areas. The probability of any single magma body forming an ore deposit is vanishingly small, but a large number of bodies greatly increases the chances.

Benioff-zone magmatism provides a working model of earth evolution. Magmatism in other areas may be far more difficult to understand, for with their mysteries come fewer clues.

References

1. M. J. Carr, R. E. Stoiber, and C. L. Drake; 1973; Discontinuities in the deep seismic zones under the Japanese arcs; *Geol. Soc. Am. Bull.* 84:2917–30. R. E. Stoiber and M. J. Carr; 1973; Quaternary volcanic and tectonic segmentation of Central America; *Bull. Volcanologique* 37:304–25.

2. J. D. Van Wormer, J. Davies, and L. Gedney; 1974; Seismicity and plate tectonics in south central Alaska; *Bull. Seismol. Soc. Am.* 1467–75. J. C. Holden and J. Kienle; 1977; Geometry of a subducted plate and atiarc segmentation of Aleutian volcanic chain (abstr.); *Trans. Am. Geophys. Union* 58:168. W. Spence; 1977; The Aleutian arc: Tectonic blocks, episodic subduction, strain diffusion, and magma migration; *J. Geophys. Research* 82:213–30.

3. B. D. Marsh and I. S. E. Carmichael; 1974; Benioff zone magmatism; *J. Geophys. Research* 81:975–84.

4. K. Yagi, Y. Kawano, and K. Aoki; 1963; Types of Quaternary volcanic activity in northwestern Japan; *Bull. Volcanologique* 26:223–35.

5. M. N. Toksöz, J. W. Minear, and B. R. Julian; 1971; Temperature field and geophysical effects of a downgoing slab; *J. Geophys. Research* 76:1113–38. E. R. Oxburgh and D. L. Turcotte; 1970; Thermal structure of island arcs; *Geol. Soc. Am. Bull.* 81:1665–88. D. L. Turcotte and G. Schubert; 1973; Frictional heating of the descending lithosphere; *J. Geophys. Research* 78:5876–86. B. D. Marsh; 1976; Mechanics of Benioff zone magmatism; in *Am. Geophys. Union Monograph 19*, pp. 337–52.

6. B. L. Isacks and M. Barazangi; 1977; Geometry of Benioff zones; Lateral segmentation and downwards bending of subducted lithosphere; in *Island Arcs, Deep Sea Trenches, and Back-Arc Basins*, ed. M. Talwani and W. C. Pitman III, pp. 99–114; Am. Geophys. Union.

7. F. Selig; 1965; A theoretical prediction of salt-dome patterns; *Geophysics* 30:633. J. A. Whitehead, Jr., and D. S. Luther; 1975; Dynamics of laboratory diapir and plume models; *J. Geophys. Research* 80:705–17.

8. M. N. Toksöz and P. Bird; 1977; Formation and evolution of marginal basins and continental plateaus; in *Island Arcs, Deep Sea Trenches, and Back-Arc Basins*, ed. M. Talwani and W. C. Pitman III, p. 379; Am. Geophys. Union. D. P. McKenzie; 1969; Speculations on the consequences of plate motions; *Geophys. J. Roy. Astro. Soc.* 18:1–32.

9. D. J. Andrews and N. H. Sleep; 1973; Numerical modeling of tectonic flow behind island arcs; *Geophys. J. Roy. Astro. Soc.* 38:237–51.

10. A. Sugimura and S. Uyeda; 1973; Island arcs, Japan and its environs, p. 101; New York: Elsevier Scientific. A. R. McBirney, J. F. Sutter, H. R. Naslund, K. G. Sutton, and C. M. White; 1974; Episodic volcanism in central Oregon Cascade Range; *Geology* 2:585–9. R. E. Stoiber and M. J. Carr; 1973; Quaternary volcanic and tectonic segmentation of Central America; *Bull. Volcanologique* 37:304–25.

11. R. Armstrong; 1971; Isotopic and chemical constraints on models of magma genesis in island arcs; *Earth & Planet. Sci. Letters* 12:137–42.

12. S. R. Hart, W. E. Glassley, and D. E. Karig; 1972; Basalts and seafloor spreading behind the Mariana island arc; *Earth & Planet. Sci. Letters* 15:12.

13. C. J. Hawkesworth, R. K. O'Nions, R. J. Pankhurst, P. J. Hamilton, and N. M. Evensen; 1977; A geochemical study of island-arc and back-arc tholeiites from the Scotia Sea; *Earth & Planet. Sci. Letters* 36:253–62.

14. B. D. Marsh; 1978; On the cooling of ascending andesitic magma; *Phil. Trans. Roy. Soc., London* (A) 288:611–25. B. D. Marsh and L. H. Kantha; 1978; On the heat and mass transfer from an ascending magma; *Earth & Planet. Sci. Letters* 39: 435–43.

15. D. M. Harris; 1977; Ascent and crystallization of albite and granitic melts saturated with H_2O; *J. Geol.* 85:451–9.

"Frankly, I don't see how we can keep it burning
through eternity."

G. Brent Dalrymple
Eli A. Silver
Everett D. Jackson

Origin of the Hawaiian Islands

*Recent studies indicate that the Hawaiian volcanic chain
is a result of relative motion between the Pacific plate
and a melting spot in the Earth's mantle*

The idyllic, palm-fringed outposts in the middle of the Pacific Ocean known as the Hawaiian Islands lie at the southeastern extremity of the Hawaiian Archipelago, a broadly linear

G. Brent Dalrymple, Eli A. Silver, and Everett D. Jackson are Research Geologists for the United States Department of the Interior, Geological Survey. Dr. Dalrymple joined the Geological Survey's Branch of Theoretical Geophysics after receiving his Ph.D. in geology from the University of California at Berkeley in 1963. Now in the Branch of Isotope Geology, his primary research interests include the development and improvement of isotopic dating techniques and their application to geomagnetic field reversals, the evolution of volcanoes, and various aspects of the Pleistocene history of the western United States. He has been a Visiting Professor and Research Associate at Stanford University and a Principal Investigator for Apollo Lunar Samples. His bibliography includes more than 50 research papers and the textbook Potassium-Argon Dating *(W. H. Freeman and Co., 1969).*
Dr. Silver received his Ph.D. in oceanography from Scripps Institution of Oceanography in 1969 and joined the Geological Survey's marine geology program in 1970. He has concentrated much of his research on the effects that moving lithospheric plates have on the structure and development of continental margins. His major regions of study have been the continental margins off the western United States and northern Venezuela. Dr. Silver's interest in the Hawaiian-Emperor chain has been in helping to develop a useful tool for measuring the history of plate motions and a better understanding of the driving mechanism of plate tectonics.
Dr. Jackson received his A.B. in 1950 and his Ph.D. in 1960 from UCLA. He is a member of Phi Beta Kappa and Sigma Xi and a fellow of The Geological Society of America, The Mineralogical Society of America, and a number of other scientific organizations. Well-known for his work on layered intrusions and their ore deposits and for studies of the nature of the upper mantle of the earth, Dr. Jackson has also been heavily involved in the lunar program, particularly in astronaut training and the petrology of lunar rocks. He is the author of more than 80 publications in these fields. Dr. Jackson joined the Geological Survey in 1951 where he now works in the Branch of Field Geochemistry and Petrology. The authors wish to thank D. W. Scholl, L. J. P. Muffler, and D. W. Swanson for their helpful reviews of this paper. Address: U. S. Department of the Interior, Geological Survey, 345 Middlefield Rd., Menlo Park, CA 94025.

chain of more than fifty huge undersea volcanoes that extends 3,500 kilometers across the central Pacific sea floor (Fig. 1). Individual volcanoes rise as much as 9,000 meters above the sea floor and reach a diameter of nearly 120 km at their base. In their prime, the volcanoes stand nearly 5,000 m above the sea. These great volcanic edifices are among Earth's largest mountains, although they are now almost entirely submerged beneath the sea (Fig. 2). All these volcanoes cap the Hawaiian Ridge, a pronounced topographic high on the ocean floor, bordered by a moat as much as 700 m deep on both sides.

The volcanoes of the Hawaiian Islands become progressively younger to the southeast, culminating in the presently active volcanoes of Kilauea and Mauna Loa, two of the five that comprise the island of Hawaii. This age progression was first recognized by J. D. Dana (1849, 1890), who noted that the volcanoes in the northwestern part of the archipelago were progressively more eroded than those in the southeast. Dana's conclusions were confirmed by McDougall (1964), who obtained potassium-argon ages on the major islands of the Hawaiian group and showed that the volcanoes decrease systematically in age from Kauai (5.6 million years) to Hawaii (< 0.7 million years).

Near latitude 32° N, longitude 172° E, in the vicinity of Yuryaku Seamount, the Hawaiian Ridge makes an abrupt bend to become the Emperor Seamounts (Fig. 1), a chain of about thirty additional submerged volcanoes that continues northward another 2,500 km, ending near the juncture of the Kurile and Aleutian trenches. The Emperor Seamounts, like the Hawaiian Ridge, form a

linear chain, and the individual shield-shaped volcanoes cap a gentle rise that is bordered by a moat. With few exceptions, the volcanoes of the Emperor chain resemble those of the Hawaiian chain in form and size, and the two great linear features are thought to have a related origin.

This elbow-shaped lineament of more than eighty huge volcanic mountains is one of the most remarkable geologic features on Earth. What processes, operating over what period of time, could account for it? The answers to these questions are incomplete, but it now seems probable that the origin of the Hawaiian-Emperor chain of volcanoes is related to movements of the Pacific plate, the great crustal slab that underlies the Pacific Ocean (see D. P. McKenzie, *American Scientist*, July 1972).

In 1963, J. Tuzo Wilson proposed that the Hawaiian volcanoes were formed as the lithosphere (the term applied to the Earth's crust and the rigid part of the upper mantle) beneath the Pacific Ocean moved slowly northwestward over a fixed "hot spot" in the deeper part of the mantle. This heat source successively supplied magma to form the volcanoes (Wilson 1963). In 1972, W. J. Morgan extended Wilson's idea to include the Emperor Seamounts, which, he argued, were a continuation of the Hawaiian Ridge. He suggested that

Figure 1. Bathymetry of the Hawaiian Ridge and Emperor Seamount chain (after Chase et al. 1970). These two chains are composed of more than 80 huge shield volcanoes, all of which are now thought to have a common origin. The contours indicate the depth, in fathoms, below sea level; the contour interval is 600 fathoms.

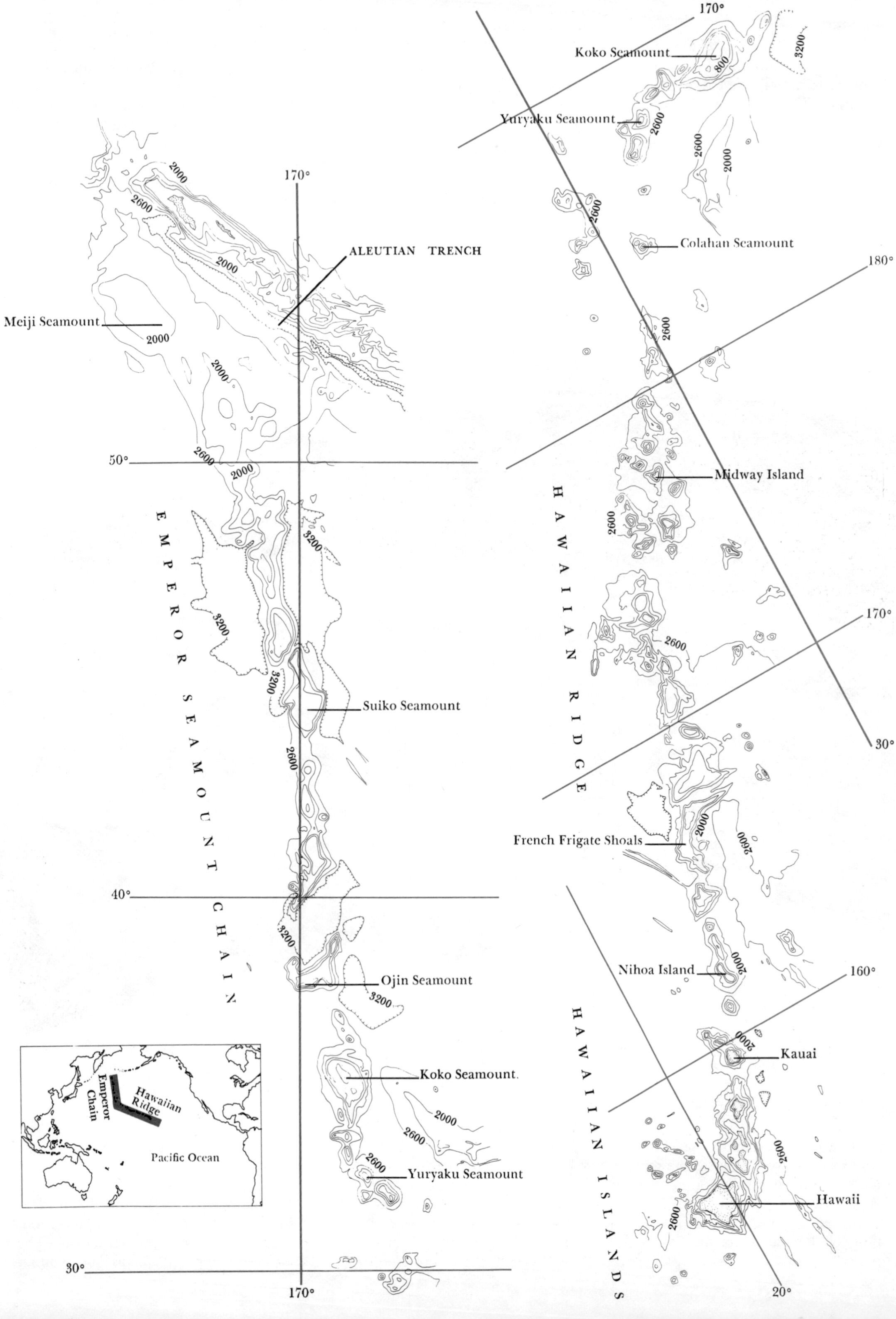
Koko Seamount
Yuryaku Seamount
Colahan Seamount
Midway Island
HAWAIIAN RIDGE
French Frigate Shoals
Nihoa Island
HAWAIIAN ISLANDS
Kauai
Hawaii
ALEUTIAN TRENCH
Meiji Seamount
EMPEROR SEAMOUNT CHAIN
Suiko Seamount
Ojin Seamount
Koko Seamount
Yuryaku Seamount
Emperor Chain
Hawaiian Ridge
Pacific Ocean
170°
180°
160°
50°
40°
30°
20°

the difference in trend of the two chains resulted from a change in the direction of motion of the Pacific plate millions of years ago (Morgan 1972a, 1972b). Variations of these hypotheses have been proposed, but all begin with the idea that the Hawaiian volcanoes form a chain as a result of relative motion between the Pacific plate and some sort of melting spot in the mantle.

These hypotheses, like other broad generalizations about the Earth's behavior and history, evoke a flood of specific questions without simple answers. Are the Emperor Seamounts

dence related to these questions, discusses several alternative models for the origin of the Hawaiian-Emperor chain, and points out the directions of current research designed to test these models.

The volcanoes

Hawaiian volcanoes are excellent examples of a type known as shield volcanoes. Above sea level, they are typically broad and smooth with gently sloping flanks that commonly rise less than 50 m per km. Below sea level, the flanks slope more steeply (Moore and Fiske 1969) but are still

out far beneath the sea. The surface expressions of these rift zones are open fractures, collapse pits, lines of spatter and cinder cones, and low lava domes. At depth, the rift zones consist of closely spaced, parallel dikes (Fig. 5).

Fiske and Jackson (1972) have noted that the rift zones of isolated Hawaiian volcanoes, such as Kauai, tend to lie at an angle of about 20° to the trend of the Hawaiian Ridge and may be controlled by fundamental planes of weakness in the lithosphere. Where volcanoes are closely clustered, however, rift zones tend to be

Figure 2. Cloud cover partly obscures the Hawaiian Islands as they were seen by the Apollo 9 astronauts. This view of the island chain is a mosaic of NASA photographs AS-9-3505–3508. (Photos courtesy of the National Aeronautics and Space Administration; mosaic by P. Y. W. Ho and R. L. Tyner.)

really a continuation of the Hawaiian Archipelago? What causes the melting spot? Is the spot fixed or in motion? What does the Hawaiian-Emperor chain tell us about the rate and direction of Pacific plate motion? This paper summarizes the available evi-

gentle relative to such classic stratovolcanoes as Mounts Rainier or Fuji. While in their constructional stage, shield volcanoes erupt not only from vents near the summit but also from linear rift zones (Figs. 3 and 4) that pass through the summit and extend

oriented parallel to the flanks of neighbor volcanoes.

Hawaiian shield volcanoes erupt lavas of variable chemical composition, usually in a definite sequence. The earliest observed lavas of young, rapidly growing volcanoes and the deepest exposed parts of older ones are invariably composed of a primitive type of basalt called tholeiite; it is therefore generally assumed that the main structure of the volcanoes was formed by copious eruptions of tholeiitic lava. This phase is followed within a few hundred thousand years by a thin capping of alkalic basalt, which is poorer in SiO_2 and richer in K_2O and Na_2O than the tholeiite. During both these volcanic phases, the eruption rate is greater than the erosion rate, and the volcanoes maintain their smooth, shield-shaped forms.

Near the end of the second phase of volcanism, the lavas become more silicic, the eruption rate decreases steadily, and the shields become dissected by deep canyons. Finally, small amounts of nephelinitic basalt, which is even poorer in SiO_2 and richer in alkali metals than the alkalic basalts, may be erupted from scattered vents several million years after formation of the main volcanic edifice. Individual volcanoes may cease activity at any stage before the cycle is complete.

The eruption rates of Hawaiian tholeiitic basalt are the highest known on Earth. Macdonald and Abbott (1970) have estimated that Mauna Loa, which has a volume of about 30,000 cubic km and rises nearly 9,000 m above the sea floor, could have been built in only 1.5 m.y. if it has erupted at its historic rate throughout its lifetime. According to Moore (1970) the combined eruption rate for all volcanoes on the island of Hawaii (derived principally from data at Mauna Loa and Kilauea) averages about 0.05 cubic km per year. Swanson (1972) calculated that the actual magma supply rate to Kilauea has been about 0.1 cubic km per year during the last 20 years, at which rate the volcano would have required less than 0.4 m.y. to reach its present size.

These exceedingly rapid eruption rates have been supported by other data. McDougall (1964) concluded from radioactive age measurements using the potassium-argon technique that most of the subaerial part of each Hawaiian volcano formed in 0.5 m.y. or less. Paleomagnetic data (Doell and Cox 1965) and potassium-argon ages (Dalrymple 1971; Mc-

Dougall and Swanson 1972) show that the most deeply exposed part of the oldest of the five shields that make up the island of Hawaii formed less than 0.7 m.y. ago, and probably within the last 0.4 m.y. Present-day eruption rates may be reasonable approximations of past eruption rates of all shield volcanoes in the Hawaiian chain. However, some recent observations, to be discussed later, suggest that growth rates may be episodic and that present-day eruption rates may be close to maximum values for the chain as a whole.

Details concerning the exact source of the lava are lacking, but several lines of evidence suggest that it is located deep within the mantle. Earthquakes and seismic tremor have been observed as deep as 50 to 60 km directly beneath Kilauea, and Eaton (1967) has suggested that these phenomena result from the transport of lava from at least these depths to the surface. Seismic refraction studies indicate that the oceanic crust in the vicinity of Hawaii is of normal (5 to 6 km) thickness but that it thickens to 10 to 20 km beneath the Hawaiian Ridge. This crustal thickening is thought to have resulted from the transfer of molten material from the mantle to the crust during growth of the volcanoes. In addition to geophysical evidence, some of the Hawaiian lavas contain angular fragments of exotic, coarsely crystalline rocks (xenoliths), whose mineralogy and texture indicate a depth of origin of more than 60 km (Jackson and Wright 1970).

The volcanoes of the Emperor Seamounts have smooth shield shapes that are very similar to those of the Hawaiian segment, but they are quite distinct from the rougher and steeper topography of other seamounts elsewhere in the Pacific Ocean. The Emperor shields are about the same size as the Hawaiian shields, and in places the bathymetry is detailed enough to reveal probable rift zones. The main difference between the Hawaiian and Emperor shield volcanoes is their spacing. There are about 18 volcanoes per 1,000 km along the Hawaiian Ridge, whereas for the Emperor Seamounts the spacing is about 13 per 1,000 km. Detailed examination of the bathymetry at the junction of the Hawaiian and Emperor chains shows that neither chain projects beyond the bend (Fig.

1). The available evidence, therefore, supports the idea that the Emperor Seamounts are shield volcanoes that form an older continuation of the Hawaiian chain.

Structure of the chain

Volcanoes of the Hawaiian-Emperor chain appear to have no obvious relation to the age or structure of the sea floor from which they rise. Magnetic data and sediment ages indicate that the sea floor in this area is Cretaceous in age and ranges from about 80 to 120 m.y. old (Heirtzler 1968; Larson and Chase 1972; see Fig. 6). Potassium-argon ages of basalt samples dredged from seamounts near, but not on, the Hawaiian-Emperor chain are also Cretaceous. The oceanic crust in this region of the Pacific appears in no way unusual, and presumably it originated normally at midoceanic spreading ridges during Cretaceous time. Thus the sea floor is definitely older than any of the dated volcanoes in the Hawaiian-Emperor chain.

The position and orientation of both the Hawaiian Ridge and the Emperor Seamounts is oblique to all nearby sea-floor structures. Major fracture zones cross the Hawaiian and Emperor chains at angles of 30° to 40° and 75° to 80°, respectively (Fig. 6). None of these fracture zones detectably disrupts the linearity of the volcanic chain, even though older magnetic anomalies to the east of Hawaii are offset along these fractures by as much as 800 km. Likewise, projected linear magnetic anomalies cross the Hawaiian-Emperor chain at large angles with no obvious change in trend (Fig. 6). Thus it appears that this chain of more than eighty volcanoes was erupted across nearly 6,000 km of older sea floor with little regard for pre-existing sea-floor structures.

Thus far, we have described the alignment of shields in the Hawaiian-Emperor chain as linear, but inspec-

Figure 3 (*overleaf*). "Curtain of fire" on the southwest rift of Kilauea Volcano, island of Hawaii, during the September 1971 eruption. Note new lava flows in foreground. Shield volcanoes grow by frequent eruptions along such rift zones as well as by summit eruptions. (Photo courtesy of D. W. Peterson, Hawaiian Volcano Observatory, U.S. Geological Survey.)

tion of Figure 7 shows that, in fact, the individual volcanoes appear to lie along a series of short, sigmoidal loci of volcanic centers that are subparallel to each other (Jackson et al. 1972). Looking only at the southeastern Hawaiian Islands, it is apparent that the loci of individual shields, from Molokai south, fall along two curved lines (inset Fig. 7). In addition, available information on the ages of these volcanoes indicates that the order of tholeiitic eruption during the last 1.8 m.y.

Waianae, and Kauai shields, which are progressively older. Beyond Kauai, however, no simple extension is possible, and the locus appears to end.

Northwest of Kauai, the volcanoes are either submerged or deeply eroded, and their central vent areas cannot be directly located. In the well-exposed southeastern shields, however, the central vent area of each volcano is marked by a topographic high, a positive gravity anomaly, and a pattern of radiating rift zones with

Figure 4. The east rift zone of Kilauea Volcano, island of Hawaii, in eruption March 1965. Approximately 10 km of the rift zone is active in the photograph; the entire east rift system is more than 100 km long. (Photo courtesy of J. G. Moore, U.S. Geological Survey.)

was West Molokai, East Molokai, Lanai, West Maui, Kahoolawe, Haleakala, Kohala, Hualalai, Mauna Kea, Mauna Loa, and Kilauea. The shields apparently have been advancing simultaneously, but somewhat out of phase, along both the north and south loci. The southern locus can be reasonably extended northwestward through the Koolau,

unique topographic expression. Using these features as guides, we have located the probable central vent areas of all shields in the Hawaiian-Emperor chain and found that these too lie on smoothly curved loci (Fig. 7).

Although the cause of these loci is uncertain, their general *en echelon* arrangement suggests that they may be related to regional extensional strain. The idea is supported by the fact that the rift zone orientations of isolated volcanoes are parallel to the loci. Presumably, both kinds of structures reflect the direction of maximum principal stress in the Pacific plate in this area.

Geochronology of the chain

The Wilson hypothesis and its subsequent variants predict that the extrusive rocks were quickly buried by succeeding ones, and thus we must be satisfied with a less direct approach. As a practical alternative, we have taken the oldest potassium-argon date for tholeiitic volcanism on individual shields as the best available approximation of the age of each volcano. This approach is probably reasonable because we know that Hawaiian shields grow rapidly, particularly when erupting tholeiitic lava. In fact, for volcanoes older than about 30 m.y., the growth time is probably

shield volcanoes of the Hawaiian-Emperor chain are progressively older toward the northwest. Thus the radiometric ages of individual volcanoes are a most critical test of mechanisms proposed to explain the progression. Ideally, we would like to know the time that each volcano in the Hawaiian-Emperor chain first erupted onto the sea floor, but these inceptive

smaller than the error in the available dating techniques, which typically is 3 to 5 percent.

Only fifteen volcanoes in the Hawaiian-Emperor chain have been radiometrically dated; their ages are plotted as a function of distance from Kilauea in Figure 8. Ages of the volcanoes from Kauai to Kilauea fall along a

Figure 5. Interior of a typical Hawaiian rift zone exposed by erosion after the volcano ceased to erupt. The rift zone consists of dense vertical dikes that once acted as pathways for lava moving toward the surface. Note that the spacing between the vertical dikes decreases away from the center of the rift zone, which is located near the right edge of the photograph. The horizontal bands between the dikes are lava flows. (Photo by E. D. Jackson.)

remarkably smooth curve, which indicates that the progression of volcanism on the two most recent loci has been systematically accelerating. Shaw (in press) has noted that an increase in lava volume accompanied this acceleration (Fig. 9); it is apparent that eruption rates increased as volcanism progressed southeastward.

Unfortunately, the age data are insufficient to determine the rate of progression of volcanism along other loci, but the general relation between distance and lava volume does occur elsewhere in the Hawaiian chain.

Few radiometric ages of Emperor seamounts have been determined, and we can say little about the progression of volcanism along the Emperor segment. We are not entirely sure, for example, that the direction of progression actually occurred as hypothesized, i.e. from north to south. A minimum age of 70 m.y. was obtained from fossils in sediments on Meiji Seamount in the northernmost part of the Emperor chain (Scholl et al. 1971). An apparent age of 41 m.y. for Suiko Seamount was measured on highly altered rocks (Ozima et al. 1970) and is almost certainly

Emperor chain are scanty, it seems reasonable to include them in general hypotheses. Before we can explore the relation of melting to plate motion, we must introduce three terms that will be useful in discussing melting processes in the mantle and transfer of magmatic liquids to the surface: *volcanic conduit*, *mantle source volume*, and *melting spot*.

We use the term volcanic conduit to describe a system of feeders in the lithosphere through which magmatic liquids rise to feed an individual shield volcano. An individual conduit is not, therefore, a single tube but a complex system of pockets, tubes, and fissures leading upward to a near-surface rift system.

Jackson and Wright (1970) used the distribution of xenoliths on Koolau volcano to suggest that lava rose through a cylindrically shaped conduit with a diameter of about 20 km. Koyanagi and Endo (1971) have shown that present-day swarms of tremor and earthquakes are confined to a conduit about 25 km in diameter at depths of 20 to 40 km beneath Kilauea. The axis of the conduit extends almost vertically downward beneath the summit of the volcano. A 20- to 25-km diameter is consistent with the fact that individual shields maintain their physiographic and compositional identity in time, although they are spaced as close as 40 to 50 km apart (Dana 1890; Powers 1917; Macdonald 1949; Wright 1971).

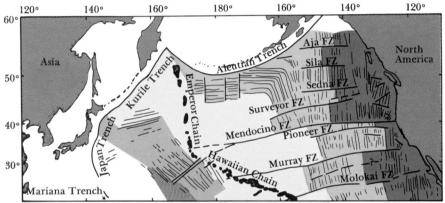

Figure 6. Magnetic lineation and structural features of the North Pacific (after Hayes and Pitman, 1970, by permission of the Geological Society of America). FZ = fracture zone; lines perpendicular to fracture zones represent magnetic anomalies on the ocean floor. Colored areas show ages of the ocean floor (after Larson and Chase 1972). Red areas are <40 m.y. old; blue 40–80 m.y.; yellow 80–120 m.y.; green 120–150 m.y.; and orange >150 m.y.

The limited data suggest that volcanism along the chain occurs in discrete pulses, each pulse forming volcanoes along either a single locus or a set of adjacent loci. During each pulse, volcanism progresses at an accelerating rate along the chain, accompanied by an increase in eruption rates, until the pulse ends and a new one begins. Moreover, the age-distance relations we now have indicate that one pulse may begin before the previous one has ended.

Although several workers, including ourselves, have used linear extrapolations to predict the age of the Hawaiian-Emperor bend, this approach must be used with caution for the reasons given above. As we shall see later, the relation between the volcanic loci and the ages of individual volcanoes is an important factor in deciding between alternative hypotheses concerning the origin of the chain.

a minimum value. The only reliable rock age is 46 m.y. for Koko Seamount, which lies about 300 km north of the bend (Clague and Dalrymple 1973). Considerable effort is now being directed toward obtaining datable samples from the Emperor shields, but it will probably be several years before our knowledge of the Emperor volcanoes is sufficient to allow more than speculation about their exact nature, rate and direction of progression, and origin.

The melting spot

Most recent investigators agree that the Hawaiian chain probably owes its origin to relative motion between the Pacific plate and a zone of melting in the Earth's mantle (Wilson 1963; Jackson and Wright 1970; McDougall 1971; Morgan 1972a, 1972b; Jackson et al. 1972; Shaw, in press). Although data from the

The mantle source volume, by our definition, is the region in the asthenosphere from which basalt is melted from its parent rock. The source volume can be no smaller than the diameters of the conduits through which lava rises to feed individual shields. Shaw (in press) calculates the cross-chain width of the source volume to be about 75 km, considering that four loci are accommodated across the 300 km width of the chain at Midway. From considerations involving the rate of volcanism, he estimates that the source volume is about 140 km long parallel to the chain, and that it is about 50 km thick. From these data, he calculates a source volume for each shield of about 500,000 cubic km. If the average volume of a shield is taken as about 20,000 cubic km, then the percentage of mantle melted to obtain the basalt is approximately 4 percent.

We use the term melting spot to describe a region of finite size in the mantle within which tholeiitic magma is generated and whose vertical projection on the Earth's surface is an area within which genetically related tholeiitic eruptions have occurred or may occur simultaneously (Jackson et al. 1972). So defined, the melting spot contains more than one volcanic conduit and may contain more than one mantle source volume.

What can we say about the size of the Hawaiian-Emperor melting spot? From the data in Figures 7 and 8 it appears that volcanoes that occur along different loci in a given region may erupt simultaneously. For example, Waianae and Niihau were in their tholeiitic eruptive phases at roughly the same time, and Nihoa is not much older than Kauai. The length of the melting spot along the chain, therefore, is on the order of 200 to 400 km. The width between adjacent multiple loci in the southeastern part of the chain is presently only about 50 km. The multiple loci at Midway, however, give cross-chain distances of about 300 km.

On the basis of present information, the projection of the melting spot on the Earth's surface is, to a first approximation, a circle about 300 km in diameter, whose previous path across the Pacific has been wide enough to include all the volcanoes of the chain (Fig. 7). We do not think the spot is currently centered on Kilauea, but rather northeast of there, because the curvature of the two loci that contain Kilauea and Mauna Loa has carried the volcanic centers far south of the median line of the chain. We predict that a new locus of centers will begin, in the very near geologic future, about 40 to 50 km north or east of Kilauea, perhaps coincidentally on its currently active east rift zone.

In summary, the melting spot within which simultaneous volcanism may occur appears to be roughly circular in plan and to have a diameter of about 300 km. The source volumes for individual shields within the melting spot may be regions about 75 km by 140 km, which lie at depths between 50 and 100 km. The conduits between the source areas and the surface at depths of less than 50 km appear to be roughly cylindrical and about 20 to 25 km in diameter.

Movement of the melting spot

The question of whether the melting spot has remained fixed geographically throughout its history or whether it is in motion relative to the lower mantle is fundamental and extremely important. Using the geometry of linear magnetic anomalies, transform faults, spreading ridges, and ocean trenches, and current knowledge about the way these features form (see D. P. McKenzie, *American Scientist*, July 1972), it is possible to deduce the relative motion, past and present, between adjacent crustal plates.

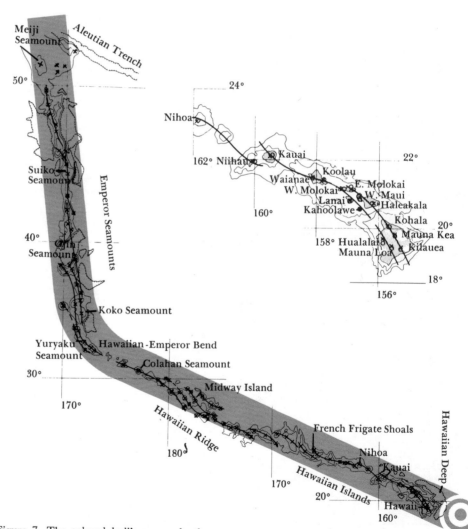

Figure 7. The colored bull's-eye marks the present location of the melting spot in the Hawaiian-Emperor chain, and the colored band the presumed path of the spot across the Pacific during the last 70 m.y. or so. The heavy black lines indicate loci of shield volcanoes. Inset shows the detailed relation between topographic highs (crosses), Bouguer gravity anomaly highs (circles), and loci for the principal Hawaiian islands. (After Jackson et al., 1972, by permission of the Geological Society of America.)

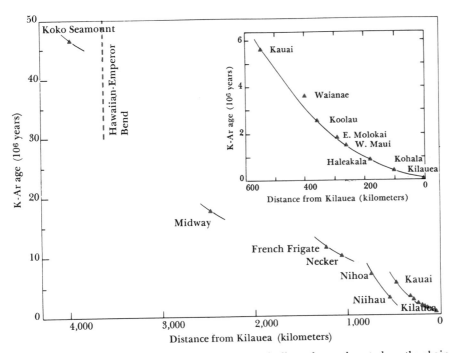

Figure 8. Potassium-argon age of tholeiitic volcanism on the Hawaiian-Emperor chain as a function of distance from Kilauea. Inset shows the detailed relations for the principal Hawaiian islands. Lines connect ages of shield volcanoes that lie on common or adjacent loci (Fig. 7). Although the ages of the volcanoes increase more or less systematically to the northwest along the chain, the age-distance relations are not linear. (Data from McDougall 1964; Dalrymple 1971; Doell and Dalrymple 1973; McDougall and Swanson 1972; Clague and Dalrymple 1973; and unpublished data of the U.S. Geological Survey.)

Morgan (1972a, 1972b) suggested that melting spots in the mantle may provide a fixed reference system for the analysis of plate motions. He proposed that volcanic island chains and aseismic ridges are formed as crustal plates move over these melting spots, that the spots are stationary relative to each other and to the Earth's geographic poles, and that approximately twenty such spots can be identified. In contrast, McDougall (1971) preferred a counterflow system in which the melting spot in the mantle moves at about the same speed but in a direction opposite to that of the crust. Shaw (in press) believes that rates of progression of volcanism along the Hawaiian chain are determined by the physicochemical processes involved in shear melting (described below) and may have a more complex relation to the rate of plate motion.

Morgan concluded that the Hawaiian-Emperor chain is not unique in the Pacific. There are two, and perhaps three, others with similar orientation, length, and geometry (see Fig. 10). Two of these are the Tuamoto-Line Islands and the Austral-Marshall Islands, each with recent volcanic activity at their southeast ends—Easter Island and Macdonald Seamount, respectively. A third possible chain runs northwest from Cobb Seamount, located near the northwest coast of North America, through the Gulf of Alaska to the Aleutian Trench near Kodiak Island.

Morgan has shown that all four volcanic chains could have been formed by movement of the Pacific plate over four fixed melting spots. The Hawaiian, Tuamoto, Austral, and Gulf of Alaska chains all lie on small circles with a common pole at latitude 76° N, longitude 73° W. The Emperor, Line, and Marshall chains lie on small circles with a pole at latitude 23° N, longitude 110° W. The northern end of the chain in the Gulf of Alaska is missing, possibly because it has been consumed in the Aleutian Trench. The peculiar geometry of these island chains does not favor completely independent moving spots; the melting spots must be more or less fixed relative to one another, although they could be moving with respect to some other reference frame. Data on the ages of volcanoes in these other island chains are practically nonexistent, and only

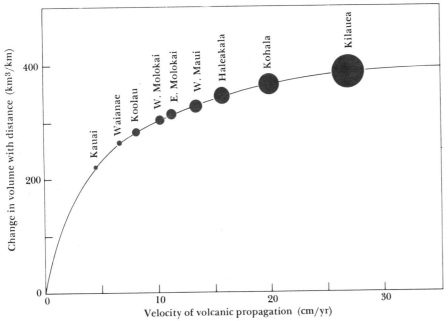

Figure 9. Change in volume of lava with respect to unit distance versus change in distance with respect to time (velocity of volcanic propagation) for the principal Hawaiian islands. The diameter of the circles is proportional to the apparent eruption rates, which have increased during the last 5 to 6 m.y. from about 0.01 cubic km/yr during the formation of Kauai to about 0.1 cubic km/yr for Kilauea Volcano. (Volume estimates from D. A. Swanson; figure after Shaw, in press.)

the Hawaiian chain is sufficiently well studied to verify, at least partly, the predicted age progression.

Grommé and Vine (1972) have made an elegant test of the fixed melting spot hypothesis. They determined the latitude of formation of Midway Island by measurement of the inclination of the remanent magnetism in tholeiitic basalt flows from two holes drilled into the atoll (Ladd et al. 1967). Lava flows, as they cool through the Curie temperatures of their magnetic minerals, record the direction and intensity of the Earth's magnetic field at the time and place of their formation. Because the magnetic and geographic poles are coincident over long periods of time, it is possible to infer ancient latitudes from paleomagnetic data. The results of Grommé and Vine indicate that Midway formed at 15.1 ± 4.1° north latitude. This is not significantly different from the present latitude of Kilauea (19.5° N) but is significantly different from either the present latitude of Midway (28.2° N) or the latitude of about 24° N predicted by McDougall's (1971) moving spot hypothesis.

An indirect test of the fixed melting spot hypothesis has been made by investigations of sediment thickness in the equatorial Pacific Ocean. Present-day sedimentation rates are highest at the equator because of the high productivity of organisms owing to upwelling of nutrient-rich waters. Deep-sea drilling has shown that this equatorial sediment bulge in older sediments is displaced progressively northward from the present equator (Winterer 1973; Clague and Jarrard 1973). If the melting spot is fixed and the Pacific plate has indeed moved northward in time with respect to the Earth's spin axis, one would expect just such a northward shift of the sediment bulge. While these data do not prove that the melting spot is fixed, they are not inconsistent with that idea.

Origin of the melting spot

The hypotheses so far advanced to explain the Hawaiian melting spot have many points in common but generally fall into four categories:

1. The "hot" spot hypothesis
2. The plume hypothesis

3. The propagating fracture hypothesis
4. The shear-melting hypothesis

The "hot" spot hypothesis was originally proposed by Wilson (1963), who envisioned the Pacific oceanic lithosphere as being transported over an area of anomalously hot mantle. The chain of volcanoes that formed by melting from this hot spot drifted northwest like the smoke from a stationary fire (Fig. 11). Jackson and Wright (1970) noted that if this model were correct, the source of materials for melting must be entirely in the

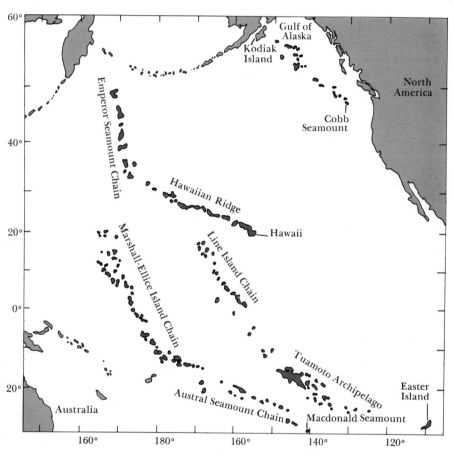

lithosphere, which would be heated as it passed above such a thermal anomaly. They calculated that if the lithosphere were 100 km thick, as much as 50 percent melting of the source volume would be necessary to account for the amount of basalt in the shields. This percentage of melting is larger than most recent estimates but is not so high as to be completely unreasonable.

The plume hypothesis, a more advanced version of the hot spot hypothesis, was proposed by Morgan (1971; 1972a; 1972b) to account for the location of certain volcanic centers on the Earth and to provide a mech-

Figure 10. Seamount and island chains in the Pacific Ocean. Morgan (1972a, 1972b) has proposed that the four chains were formed by motion of the Pacific plate over four melting spots, each now located at the southeast extremities of the chains.

anism to drive crustal plates. Plumes are supposedly the result of convection and originate deep within the Earth's mantle, possibly near the core (see Fig. 12). They arise because of thermal instabilities (excess heat) which cause upward convection of a hot plume of mantle rock with relatively low viscosity, about 150 km in diameter, in much the same way that thermal instabilities in the atmosphere cause so-called thunderhead clouds, which have a similar plume shape. According to Morgan, these plumes convect upward at a rate of about 2 m per year and not only are responsible for volcanoes but also provide the main driving force for plate motions.

Convection in the mantle is not a new idea, but Morgan was the first to suggest that it was concentrated in such a narrow, high-velocity stream. As the plumes impinge upon the base of the lithospheric plate, the rising material is dispersed and returned to the deep mantle over large areas as a sort of gentle "rain" that migrates slowly downward. Morgan proposes that a low-density fraction (presumably higher in silica and the alkali metals and lower in magnesium and the transition metals like iron) from the Hawaiian plume is "plastered" at the plate-mantle interface, where it accumulates in pockets much like oil trapped in a reservoir. Because the material is gravitationally unstable, it rises through vents to the surface where it forms shield volcanoes. As the plate moves away from the plume head, a new pocket forms, and the existing pocket ceases to be supplied with new material. Material in such an isolated pocket is free to fractionate chemically and provide the late-stage alkalic and nephelinitic lavas common to Hawaiian volcanoes.

The propagating fracture hypothesis has been advanced by several workers (Betz and Hess 1942; Jackson and Wright 1970; Green 1971) but probably is best developed by McDougall (1971). He proposed that the Hawaiian chain is formed by propagation of a tensional fracture caused by a thermal high or incipient upwelling that reflects a concentration of heat-producing radioactive elements in the mantle. This fracture invites further diapiric upwelling of mantle material from the asthenosphere into the lithosphere, and the accompany-

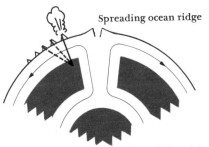

Spreading ocean ridge

Figure 11. Wilson's proposed possible origin of the Hawaiian Island chain. The diagram shows that if lava is generated in the stable core of a convection cell in the mantle, and the surface is carried along by plate motion, then one source can give rise to a chain of extinct volcanoes. (After Wilson, 1963, by permission of the National Research Council of Canada.)

ing pressure release causes partial melting and the formation of tholeiitic lava (Fig. 13).

As the plate and the heat source move away from each other, the pocket of mantle material is isolated in the lithospheric plate, and the cycle of volcanism for that particular volcano is completed in a manner similar to the plume hypothesis. In contrast to the fixed plume model, and in

contrast to the propagating fracture models of Betz and Hess and Jackson and Wright, McDougall postulates that the heat source moves with equal velocity as the Pacific plate but in the opposite direction.

The shear-melting hypothesis was recently advanced by Shaw (in press) as an alternative to the plume and propagating fracture models. This hypothesis is based on the observation that a viscous medium will heat up when sheared. As temperature rises, viscosity decreases, allowing an increase in rate of shearing which, in turn, allows a further increase in temperature. Shaw proposed that shearing of the lithosphere over the asthenosphere results in increased temperature and potential melting. When melting begins, magma rises to the surface to form volcanoes. This carries off excess heat, temperature drops rapidly, viscosity increases, and the system begins a new cycle. The process proposed by Shaw thus contains a thermal feedback mechanism that is capable of initiating cycles of rapid melting and accelerated volcanism which terminate abruptly.

At present there are insufficient data

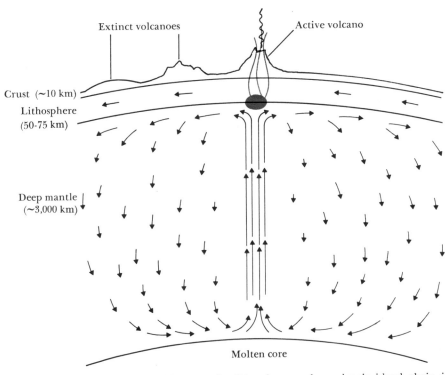

Figure 12. In this schematic diagram of Morgan's proposed plume under Hawaii, hot mantle material streams up in a narrow zone at high velocity (~2 m/yr) from deep in the mantle, causing melting in a region near the lithosphere-asthenosphere boundary. The magma generated pierces the lithosphere, and a volcanic island chain is formed as the Pacific plate moves westward. According to this hypothesis, the plume provides the force for driving the plate. Return flow of mantle material is at low velocity over a large area beneath the Pacific.

Figure 13. Schematic diagram of Mc-Dougall's model for the origin of the Hawaiian chain. Relative velocities of the lithosphere and asthenosphere are indicated by arrows to the left of the diagrams. According to McDougall, a propagating tensional fracture initiates diapiric upwelling in the asthenosphere (*top*), resulting in partial melting and tholeiitic volcanism (*second from top*). Relative motion between the lithosphere and the asthenosphere eventually decapitates the diapir (*third from top*), and the cycle begins again in a new location eastward along the chain (*bottom*). (After McDougall, 1971, by permission of Macmillan Journals.)

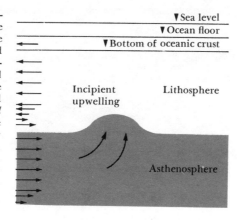
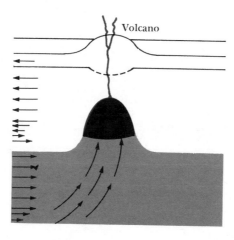
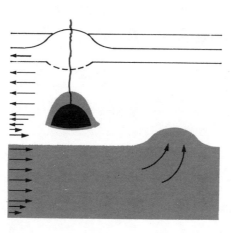
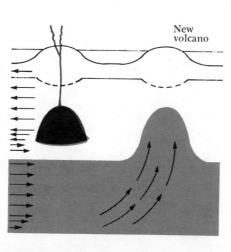

to decide among these various hypotheses, and it is probably premature to accept a single process as the final explanation for large-scale volcanic features of the Hawaiian type. The plume and propagating fracture hypotheses have some attractive features, but neither, as presently developed, accounts for the sigmoidal loci of volcanic centers, the close relation observed between the loci and the ages of the volcanoes, or the remarkably consistent chemical cycle followed by all of the accessible volcanoes. Although both hypotheses call for the generation of magma within or at the base of the lithosphere, geochemical and geophysical evidence indicates that nephelinitic lava is generated at greater depths than tholeiitic lava and that both come from the mantle.

Both hypotheses fulfill the requirement that the melting spot continually either tap fresh mantle material or be resupplied. However, McDougall's hypothesis presently is not consistent with the paleomagnetic evidence discussed above that Midway formed far south of its present latitude. Although Shaw's shear-melting model has many attractive features and seems potentially capable of explaining some of the distance–age–volume relations observed on the Hawaiian chain, it is not entirely clear why the melting should be confined to a single spot.

Finally, both Morgan's plume and Shaw's shear-melting models are in conflict with present knowledge about the rate of migration of volcanism. Morgan (1972a, 1972b) compared the motion vectors between the various plates and the melting spots (as predicted from the known relative motions of the plates) with the observed orientations and rates of formation of the volcanic chains. He found that the melting spots were satisfactory as a fixed reference frame for plate motions only if the average rate of volcanic progression on the Hawaiian chain was 9 cm/yr. Our best estimate, based on recent radiometric ages, yields an average rate of 12.5 cm/yr, in conflict with his prediction. Shaw (in press) predicted an age of about 50 m.y. for the formation of the southernmost part of the Emperor chain, on the basis of his shear-melting model and lava volume estimates, in good agreement with the measured age of about 46 m.y. for Koko Seamount. However, he also predicted an age of about 30 m.y. for Midway Island, which seems well dated now at 18 m.y.

Thus, all models have their strengths and their weaknesses. More detailed geochronological study of the Hawaiian chain and Emperor Seamounts should test whether (1) volcanism progresses linearly with random scatter at about an average rate of 9 cm/yr, as predicted by the plume–fixed spot models, or (2) volcanism is grossly episodic with accelerating rates along individual loci, as predicted from the shear-melting model. Many more paleolatitude determinations, especially on the Emperor chain, are needed before moving spot models can be completely precluded.

In attempting to define the cause of the volcanic activity and of plate motions, we should consider that causes and consequences are not necessarily separate entities. For example, convective motions in the mantle may just as well result from plate motions as cause them (McKenzie 1969). In the case of the Hawaiian volcanoes, we face the dilemma of determining whether plate motions cause or are caused by the processes inherent in the origin of volcanic chains, and we may have to conclude that the processes are interdependent. The prospect of discovering the real nature of this interdependence provides the impetus for continuing this exciting research.

References

Betz, F., Jr., and H. H. Hess. 1942. The floor of the North Pacific Ocean. *Geog. Rev.* 32 :99–116.

Chase, T. E., H. W. Menard, and J. Mammerickx. 1970. Bathymetry of the North Pacific. *Scripps Inst. Oceanogr. Charts 2, 7, 8.* LaJolla: Scripps Institution of Oceanography.

Clague, D. A., and G. B. Dalrymple. 1973. Age of Koko Seamount, Emperor Seamount chain. *Earth and Planetary Sci. Letters* 17:411–15.

Clague, D. A., and R. D. Jarrard. 1973. Tertiary Pacific plate motion deduced from the Hawaiian-Emperor chain. *Geol. Soc. Amer. Bull.* 84(4).

Dalrymple, G. B. 1971. Potassium-argon ages on the Pololu Volcanic Series, Kohala Volcano, Hawaii. *Geol. Soc. Amer. Bull.* 82:1997–2000.

Dana, J. D. 1849. *Geology, Volume 10 of United States Exploring Expedition, during the Years 1838–1839, 1840, 1841, 1842.* Philadelphia, Pa.: C. Sherman.

Dana, J. D. 1890. *Characteristics of Volcanoes.* New York: Dodd, Mead and Company.

Doell, R. R., and A. Cox. 1965. Paleomagnetism of Hawaiian lava flows. *J. Geophys. Res.* 70:3377–405.

Doell, R. R., and G. B. Dalrymple. 1973. Potassium-argon ages and paleomagnetism of the Waianae and Koolau Volcanic Series, Oahu, Hawaii. *Geol. Soc. Amer. Bull.* 84(4).

Eaton, J. P. 1967. Evidence on the source of magma in Hawaii from earthquakes, volcanic tremor, and ground deformation [abs.]. *Am. Geophys. Union Trans.* 48:254.

Fiske, R. S., and E. D. Jackson. 1972. Orientation and growth of Hawaiian volcanic rifts: The effect of regional structure and gravitational stresses. *Proc. R. Soc. London A* 329:299–326.

Green, D. H. 1971. Composition of basaltic magmas as indicators of conditions of origin: Application to oceanic volcanism. *Proc. R. Soc. London A* 268:707–25.

Grommé, C. S., and F. J. Vine. 1972. Paleomagnetism of Midway Atoll lavas and northward movement of the Pacific plate. *Earth and Planetary Sci. Letters* 17:159–68.

Hayes, D., and W. Pitman. 1970. Magnetic lineations in the North Pacific. In J. D. Hays, Ed., *Geological Investigation of the North Pacific. Geol. Soc. Amer. Mem. 126.*

Heirtzler, J. R. 1968. Sea-floor spreading. *Sci. American* 219:60–70.

Jackson, E. D., and T. L. Wright. 1970. Xenoliths in the Honolulu Volcanic Series, Hawaii. *J. Petrology* 11:405–30.

Jackson, E. D., E. A. Silver, and G. B. Dalrymple. 1972. Hawaiian-Emperor chain and its relation to Cenozoic circumpacific tectonics. *Geol. Soc. Amer. Bull.* 83:601–18.

Koyanagi, R. Y., and E. T. Endo. 1971. Hawaiian seismic events during 1969. *U. S. Geol. Survey Prof. Paper 750-C:*158–64.

Ladd, H. S., J. I. Tracey, Jr., and M. G. Gross. 1967. Drilling on Midway Atoll, Hawaii. *Science* 156:1088–94.

Larson, R. L., and C. G. Chase. 1972. Late Mesozoic evolution of the western Pacific Ocean. *Geol. Soc. Amer. Bull.* 83:3627–44.

Macdonald, G. A. 1949. Petrology of the Island of Hawaii. *U. S. Geol. Survey Prof. Paper 214-D:*51–96.

Macdonald, G. A., and A. T. Abbott. 1970. *Volcanoes in the Sea: The Geology of Hawaii.* Honolulu: University of Hawaii Press.

McDougall, I. 1964. Potassium-argon ages from lavas of the Hawaiian Islands. *Geol. Soc. Amer. Bull.* 75:107–28.

McDougall, I. 1971. Volcanic island chains and sea-floor spreading. *Nature* 231:141–44.

McDougall, I., and D. A. Swanson. 1972. Potassium-argon ages of lavas from the Hawi and Pololu Volcanic Series, Kohala Volcano, Hawaii. *Geol. Soc. Amer. Bull.* 83:3731–38.

McKenzie, D. P. 1969. Speculation on the consequent causes of plate motions. *Geophys. J. R. Astr. Soc.* 18:1–32.

McKenzie, D. P. 1972. Plate tectonics and sea-floor spreading. *American Scientist* 60:425–35.

Moore, J. G. 1970. Relationship between subsidence and volcanic load, Hawaii. *Bull. Volcanol.* 34:562–76.

Moore, J. G., and R. S. Fiske. 1969. Volcanic substructure inferred from dredge samples and ocean-bottom photographs, Hawaii. *Geol. Soc. Amer. Bull.* 80:1191–1202.

Morgan, W. J. 1971. Convection plumes in the lower mantle. *Nature* 230:42–3.

Morgan, W. J. 1972a. Deep mantle convection plumes and plate motion. *Amer. Assoc. Petrol. Geol. Bull.* 56:203–13.

Morgan, W. J. 1972b. Plate motions and deep mantle convection. *Geol. Soc. Amer. Mem.* 132.

Ozima, M., I. Kaneoka, and S. Aramaki. 1970. K-Ar ages of submarine basalts dredged from seamounts in the western Pacific area and discussion of oceanic crust. *Earth and Planetary Sci. Letters* 8:237–49.

Powers, S. 1917. Tectonic lines in the Hawaiian Islands. *Geol. Soc. Amer. Bull.* 28:501–14.

Scholl, D. W., J. S. Creager, R. E. Boyce, R. J. Echols, T. J. Fullam, J. A. Grow, I. Koizumi, H. Lee, H.-Y. Ling, P. R. Supko, R. J. Stewart, and T. R Worsley. 1971. Deep sea drilling project Leg 19. *Geotimes* 16:12–15.

Shaw, H. R. In press. Mantle convection and volcanic periodicity in the Pacific: Evidence from Hawaii. *Geol. Soc. Amer. Bull.*

Swanson, D. A. 1972. Magma supply rate at Kilauea Volcano 1952–1971. *Science* 175:169–70.

Wilson, J. T. 1963. A possible origin of the Hawaiian Islands. *Canadian J. Physics* 41:863–70.

Winterer, E. L. 1973. Sedimentary facies and plate tectonics of equatorial Pacific. *Amer. Assoc. Petrol. Geol. Bull.* 57:265–82.

Wright, T. L. 1971. Chemistry of Kilauea and Mauna Loa lavas in space and time. *U.S. Geol. Survey Prof. Paper 735.*

PART 4 *Igneous Activity*

Charles L. Rosenfeld, "Observations on the Mount St. Helens Eruption," **68**:494 (1980), page 128.

James G. Moore, "Mechanism of Formation of Pillow Lava," **63**:269 (1975), page 144.

J. R. Heirtzler, P. T. Taylor, R. D. Ballard and R. L. Houghton, "A Visit to the New England Seamounts," **65**:466 (1977), page 153.

Grant Heiken, "Pyroclastic Flow Deposits," **67**:564 (1979), page 160.

Donald Hunter, "The Bushveld Complex and Its Remarkable Rocks," **66**:551 (1978), page 168.

Charles L. Rosenfeld

Observations on the Mount St. Helens Eruption

An on-the-scene geologist presents the results of aerial reconnaissance of the events that led to the May 18 explosion and its immediate effects

On May 18, Mt. St. Helens shattered a 120-year dormant interval with an explosive eruption that devastated most of the immediate area in a 160° sector north of the mountain, killing an as yet unknown number of people, triggering numerous mudslides and destructive floods, and blanketing much of eastern Washington, northern Idaho, and western Montana with volcanic ash.

This article is a preliminary description of the events associated with the St. Helens eruption of 1980, the first eruptive activity in the Cascades since the eruption of Mt. Lassen in 1915, and the first to cause substantial impact in inhabited areas of the continental United States. This account of the events surrounding the present activity at Mt. St. Helens is based on numerous overflights that I made in OV-1 Mohawk reconnaissance aircraft and several ground verification visits by helicopter as part of the Cascade volcanoes surveillance project of the Oregon Army National

Charles L. Rosenfeld is an assistant professor of geography at Oregon State University, where he teaches geomorphology and remote sensing. He also serves as the imagery interpretation officer for the Oregon Army National Guard, conducting training and directing aerial surveillance projects such as monitoring the Cascade volcanoes. Dr. Rosenfeld has published widely on the applications of geomorphology to the assessment of environmental hazards and the use of remote sensing in geomorphic process studies. His work on the activity of Cascade volcanoes, which began in 1975 with the rapid increase in heating and gas venting exhibited by Mt. Baker, has enabled the National Guard reconnaissance aircraft to employ techniques yielding timely data to the scientists monitoring Mt. St. Helens. All photographs in this article were taken by the author. Address: Department of Geography, Oregon State University, Corvallis, OR 97331.

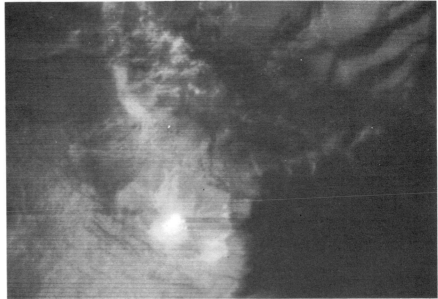

Figure 1. The explosion crater (*top*) on March 27, the day a 4.7 magnitude earthquake signaled the reawakening of Mt. St. Helens, was 80 m in length. This photograph looking down into the crater shows that it was flanked by fractures trending east-west. The thermal infrared image of the crater (*bottom*), taken from a different aircraft at about the same time, shows the heat flow from the crater. (Image courtesy Oregon Army National Guard.)

Figure 2. The greatly enlarged crater—400 m long—was formed by phreatic explosions on April 2 that expanded the crater to the full width between the fractures zones shown in Fig. 1. Annular fractures down the flanks of the cone outline the bulge area on the Forsyth glacier, as seen on April 15. Measurements made from April 23 to May 18 indicated the "Forsyth bulge," as it was labeled, was inflating at a rate of 1.5 m per day.

Guard. For the color photographs I used a UV-filtered 35-mm camera with a date-recording device and Ektachrome film. The thermal infrared and radar images were acquired by the sensor systems of the National Guard Mohawk aircraft.

An explosive history

Although Mt. St. Helens has been more active, and more explosive, during the last 4,500 years than any other volcano in the conterminous United States (1), its dormancy for the past 120 years belied its violent nature. The older volcanic center of the mountain predates 36,000 years B.P. (2); however, most of the visible part of the symmetrical cone has been formed by eruptions within the last thousand years. Beginning about 4,500 years B.P. Mt. St. Helens resumed activity after almost 4,000 years of dormancy, periodically forming domes and explosively ejecting large volumes of pumice and hot pyroclastic flows. During the last 2,000 years these eruptions have included lava flows and large-scale mudflows radiating tens of kilometers from the flanks of the volcano (1).

At 3:47 P.M. (2247 GMT) on 20 March 1980, an earthquake of 4.1 Richter magnitude was recorded in the St. Helens area by the Western Washington Seismic Network (3), the first of a series of unusual seismic events that preceded the eruption. Within 48 hours of the initial tremor, the University of Washington had installed four additional seismic stations near the volcano. These stations provided loci and magnitude data on swarms of microseisms, reaching magnitudes of 4.4 on March 25, which triggered minor avalanches along the flanks of the cone (3).

At 11:20 A.M. on March 27 an Army National Guard reconnaissance plane reported sighting a perforation through the ice near the summit area, with a light gray ash stain extending to the southeast. Later, at 1:32 P.M., an audible explosion accompanied by a seismic shock of magnitude 4.7 signaled the violent awakening of Mt. St. Helens. Figure 1 shows the initial explosion crater, approximately 80 m in length, together with the first thermal infrared image of the heat flow from the crater. The image reveals over a dozen fumaroles, or gas vents, in the bottom of the crater. The crater was also flanked by two fracture zones, separated by about 400 m, trending in an east-west direction across the summit area.

Phreatic explosions, caused by groundwater coming into contact with hot rocks and gases and flashing into steam, continued throughout the following week—first creating a second explosion crater 60 m east of the original site, then, as they enlarged, merging the two craters with a low saddle between them (Fig. 3). Harmonic tremors accompanied the eruptions during the first week of April. The parallel fractures that flanked the craters rifted during this period, allowing the central block to subside over 40 m. This isolated the crater on an active graben block.

Although the harmonic tremors continued until the middle of April, eruptive activity gradually tapered off. During this period, however, an area on the northeast flank of the mountain began to show signs of swelling; annular and radial fractures appeared on the upper part of the Forsyth glacier (Fig. 2), affecting an area about 1 km wide and nearly 2 km long. By April 23 an uplift of 100 m had been identified and labeled the "Forsyth bulge" (4); the Goat Rocks dome on the north flank was being moved laterally to the north-northeast at a rate of 1.5 m per day. There was virtually no eruptive activity from April 23 until May 6; steam emissions were confined to localized fumaroles, and a lake of meltwater formed on the crater's floor. On April 30 a National Guard helicopter made a radiometer survey inside the crater walls, allowing us to measure the radiant temperatures of the "hot spots" revealed on a thermal infrared image acquired by the National Guard on April 22. Several areas of thermal activity to the north of the crater were verified, and the convex Forsyth bulge was thoroughly examined.

The period from May 7 to 14 saw a dramatic return to phreatic eruptions, with steam and ash plumes frequently reaching an altitude of 4,000 m. The thermal areas outside the crater, imaged on April 22, were visibly issuing steam—as were several crevasses on the upper Shoestring glacier to the east. As the bulge expanded, the lower part became greatly oversteepened, resulting in a debris avalanche—triggered by an earthquake—which sent 50,000 m³ of rock and ice debris cascading down the north flank of the mountain on May 12 (Fig. 4). This incident may have been most fortunate, as the proximity of the avalanche to the Timberline parking area on the mountain's north flank quickly convinced the U.S. Geological Survey and media observers that future instability was imminent and that retreat to a more distant observation post was well advised.

The violent eruption of Sunday May 18

At 5:51 A.M. on Sunday May 18, a National Guard plane flying over the crater acquired what was to prove to be the last thermal infrared image prior to the major eruption (Fig. 9) (5). This imagery clearly shows an increase in heat flow throughout the crater area as well as several new hot spots spread throughout the bulge area. While the data would definitely indicate an increased risk of debris avalanche due to basal melting of the glacial mass, such an increase in heat flow would not of itself have signaled an urgent alarm of impending doom. This thermal infrared image was interpreted at 8:30 A.M.

At about the same time, Keith Stoffel, a geologist with the Washington Division of Geology and Earth Resources, was viewing the crater area from a private plane flying directly overhead at about 3,700 m. As the plane passed over the western end of the summit graben, he noticed landsliding of rock and ice debris inward toward the center of the crater, probably in response to earthquake activity. As he later wrote:

Within a matter of seconds, perhaps 15 seconds, the whole north side of the summit crater began to move instantaneously. As we were looking directly down on the summit crater, everything north of a line drawn east-west across the northern side of the summit crater began to move as one gigantic mass. The nature of the movement was eerie, like nothing we had ever seen before. The entire mass began to ripple and churn up, without moving laterally. Then the entire north side of the mountain began sliding to the north along a deep-seated slide plane. I was amazed and excited with the realization that we were watching this landslide of unbelievable proportions slide down the north side of the mountain toward Spirit Lake [6].

Figure 3. Phreatic explosions in the west crater on April 2 merged this crater with a smaller one to the east, forming the massive 400-m long crater pictured in Fig. 2. Three discrete ash columns are apparent. The vertical offsetting of the fracture zones, flanking a subsiding (graben) block, indicated a possible swelling of the stratovolcanic core.

Figure 4. An earthquake-triggered avalanche containing about 50,000 m³ of rock and ice debris from the oversteepened base of the Forsyth bulge divided into two lobes that cascaded down below the timberline on the mountain's north flank, alerting U.S. Geological Survey and other observers that danger was imminent. This photograph, looking to the east, was taken on May 12; the date on the photo is incorrect.

Figure 5. Clustered fumaroles at the base of the southeast wall of the crater vented steam and gases. When this photograph was taken on April 30, temperatures of 102°C were measured at the vents, and sulfur odors were evident.

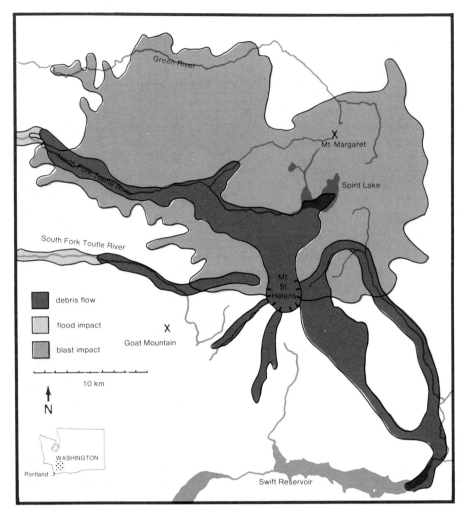

Figure 6. Impact of the May 18 eruption included debris flows and flooding which vastly increased the extent of damage beyond the area directly affected by the blast. Devastation was greatest along the valley bottoms; ridges provided limited shielding from the directional blast effects on the north side.

Figure 7. The May 18 eruption may have been triggered by (1) a landslide that detached a large portion of the bulge, (2) reducing pressure on the superheated groundwater and allowing it to flash into steam. (3) The steam explosion dislodged overlying rock, resulting in collapse of the summit graben and (possibly) explosive release of magmatic gases. (4) As the gases were released, an ash-laden plume issued from the breach, progressively enlarging the crater. The dashed brown line indicates the rim of the crater after the eruption.

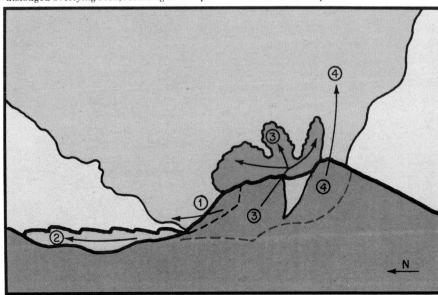

The Western Washington Seismic Network recorded a Richter magnitude 4.9 earthquake at 8:32 A.M. David Johnston, U.S. Geological Survey observer at Coldwater II—located 10 km north of the crater at 400 m above the floor of the Toutle River valley—radioed a brief, prophetic message: "Vancouver, Vancouver! This is it" (7).

I have attempted to reconstruct the following scenario from my personal observations, media coverage, and discussions with U.S. Geological Survey and other scientists. As the myriad of observations and ground evidence is sorted and analyzed, portions of this account will undoubtedly need revision, but it is hoped that it will serve to convey some initial impressions to the scientific community at large. The following sections are topically organized and do not necessarily represent a chronological sequence. Times and dates of events are noted where known.

The directional blast

Prior to the earthquake activity at 8:32 A.M. on May 18, the only visible steam activity on the mountain came from a small fumarole cluster on the southeast wall of the summit crater (Fig. 5). From eyewitness description and analysis of a series of photographs taken by a volunteer radio operator camped 15 km west of the peak (8), a triggering sequence, shown schematically in Figure 7, may be hypothesized:

1. A landslide, caused by the magnitude 4.9 earthquake, detached a large portion of the unstable bulge area, probably near its oversteepened base.

2. Detachment of the bulge sharply reduced the pressure on the superheated groundwater close to the magma, allowing the groundwater to flash into steam—sending a high-velocity flow of ash and steam north across the Toutle River valley.

3. As the steam explosion intensified (within seconds), overlying rock was broken up and traveled laterally as a high-velocity debris flow, clogging the Toutle valley and displacing much of the Spirit Lake basin. (I observed reddish rock at the base of Coldwater Ridge on my first overflight of the

Figure 8. The eruption plume on May 18, about 3 hours after the explosion, rises from the crater and the breach in the north flank. Pyroclastic flows, at lower left, descend into the north Toutle River valley. The elevation at the rim averaged 2,500 m, lowering to about 1,500 m at the base of the north flank.

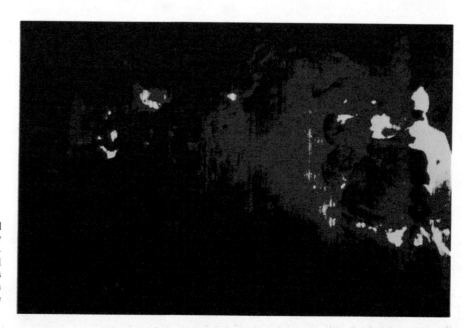

Figure 9. A color enhancement of a thermal infrared image taken at 5:51 A.M. on May 18—2½ hours before the directional blast—shows the heated areas within the crater and throughout the bulge area. Yellow highlights the hottest areas, including the crater area on the right and numerous hot spots in the bulge area on the left.

Figure 10. On the morning after the directional blast, the northwest flank of the volcano, which had been heavily forested, was devastated. The steam in the background was caused by phreatic activity in the north Toutle River valley. Black streaks are the remnants of the Toutle and Talus glaciers, whose surfaces were pitted and melted by numerous pyroclastic flows.

Figure 11. A ridge north of the Toutle valley shows the directional effects of the blast. The forest was completely removed from the up-wind slope, and the standing trees on the lee-ward side have been burned by the heat of the blast. Effects range from smouldering timber near the crest to scorched needles on the trees in the more protected sites.

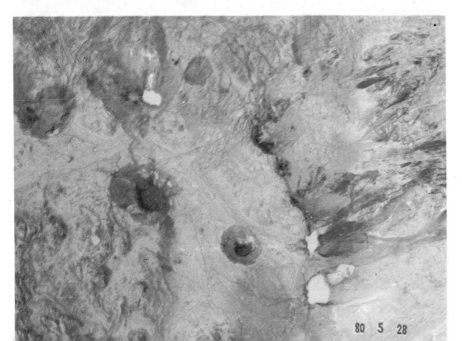

Figure 12. Numerous explosion craters with sharp rims mark the surface of the North Toutle River debris flow. A kettle hole, still containing some glacial ice, which is melting into a pool of turbid water, can be seen above the date. The remains of three trees in the bottom of the explosion crater at the center of the photograph provide an idea of the scale.

Figure 13. Large blocks of dacite, up to 20 m in diameter, were blasted from domes on the north flank of the mountain to the northwest shore of Spirit Lake, more than 10 km north of the peak. The water from Spirit Lake swept through this area as it was sloshed northward by the blast. The mud and log debris in the foreground was deposited by this wave of water.

mountain on May 18, and have subsequently collected similar rock from large blocks on the northwest shore of Spirit Lake. These were probably part of the Goat Rocks dome located at the 2,250-m level on the north side of St. Helens.) Collapse of the summit graben produced two large vertical blast columns, and may have triggered the explosive release of the magmatic gases.

4. As the gases were explosively released from the magma, a large vertical ash-laden plume issued from the breached crater, progressively removing the south wall of the crater, and lowering the summit elevation from 2,950 m to ~ 2,500 m (Fig. 8).

The lateral blast was probably a combination of steam and explosive gas releases mixed with pulverized rock material from the mountain's north flank, formed into a dense high-speed cloud and heated to about 500°C. Figure 6 illustrates the extent of the area affected by this blast. Several degrees of destruction occurred within the devastated area. Near the breached northern flank of the mountain nearly every exposed slope within a 10-km radius was completely denuded of all vegetation and covered by up to 2 m of ash and rock debris (Fig. 10). Beyond this (at about a 15-km radius), trees were snapped off at their bases and completely stripped of their branches (Fig. 11). Stumps were partially buried by heavy ash fall, and trees on the leeward sides of the slopes were singed but left standing. In a zone extending 100 to 300 m beyond this, needles and branches were either stripped from standing trees or the needles were singed to an orange color.

Several additional pieces of evidence support this scenario. Fishermen along the Green River, 26 km northwest of the peak, near the periphery of the devastated area, were badly burned. They survived only because they jumped into the river when they first became aware of the explosion, but were burned when they came up for air and after they left the water. They reported that the heat came upon them in a series of waves over a period of ten to fifteen minutes. The plastic parts of a blown-over truck at Ryan Lake, 21 km north-northeast of the mountain, were melted (4). Blocks of the more competent rock units on the mountain's north flank were hurled across the Toutle valley and Spirit Lake basin. Large blocks of

Figure 14. The debris avalanche immediately following the May 18 blast flowed up and over Coldwater Ridge—which rose 400 m from the floor of the valley—completely covering the north side of the ridge, shown here. The avalanche destroyed the forest, scoured the soil, and covered the slope with debris, which is currently sliding off the steeper portions of the slope, like the one in the foreground.

the Goat Rocks dome (up to 20 m in diameter) were found 8 km to the north, on the northwest arm of Spirit Lake, where they were apparently swept by the waters of the lake, which swashed nearly 300 m up the south side of Mt. Margaret. A trimline clearly marks this event on the side of Mt. Margaret, while piles of logs and a debris fan on the west arm of Spirit Lake attest to the power of this tsunami-like erosion (Fig. 13).

This phase of the eruption can best be described as a classic Peléean eruption—named after Mt. Pelée, a Caribbean volcano that killed 30,000 people in 1902. In general usage, Peléean refers to very violent eruptions of an explosive nature. Large quantities of pumice are erupted rapidly, owing to high concentrations of volatile gases in the magma chamber. *Nuées ardentes*, hot ash clouds that descend the volcano's flanks and adjacent valleys, are characteristic. In a strict sense, the term refers to the lateral eruption of such ash clouds: Mt. St. Helens began the May 18 eruption in precisely this manner. Although few close observations have been made of Peléean explosions, St. Helens's eruption closely resembles the description of Mt. Lamington in New Guinea in 1951 (9).

Landslides and debris flows

The initial landslide probably moved en masse to the bottom of the Toutle valley, where it broke up and was mixed with pulverized rock and ash debris being driven by the directional blast. The momentum of this flow carried debris over the crest of Coldwater Ridge, which rose over 400 m above the floor of the Toutle valley, removing the forest and scouring the

Figure 15. On the day after the blast, mudflows of melting glaciers mixed with ash and soil descended the east flank of St. Helens and flowed down the channels of Pine Creek and Muddy River, merging and entering the east end of Swift Reservoir. Mudflow debris filled the reservoir nearly a kilometer from its previous inlet, where the road toward the center of the photograph formerly bridged the river. The brown area beyond the mudflats is a large mass of floating log debris.

soils to a depth exceeding a meter in most areas (Fig. 14). A huge debris fan spread north from the breach in the crater wall into Spirit Lake, displacing the south shore nearly a kilometer northward and causing the lake level to rise almost 40 m. Farther west, the fan abutted against Coldwater Ridge, then deflected westward along the Toutle River valley.

The rock and ash debris, fluidized by groundwater and melting glacial ice, rapidly flowed nearly 28 km west along the valley floor, eventually coming to rest against a small hill that divides the channel. This stagnant flow filled the valley bottom to a depth of about 60 m, with trimlines along the upper end of the valley nearly 120 m above the former valley floor. Large blocks of glacial ice melted, forming kettle holes in the surface, or collapse pits when the ice melted just below the surface. Often, the meltwater or groundwater contacted pockets of hot ash within the flow, resulting in explosion pits (Fig. 12).

Beyond the terminus of the debris flow, mud-laden flood waters swept down the Toutle River, washing away the former channel, eroding the banks, and carving huge potholes in the bed. The flood crest floated away most of the log storage decks at Camp Baker, and the resulting log jams destroyed several bridges farther downstream. Rapid deglaciation of St. Helens's west flank produced similar flooding along the South Fork of the Toutle River. The combined floods entered the Cowlitz River near Castle Rock and then flowed south into the Columbia River at Longview, Washington. Within 24 hours the sediment from the Toutle River floods built a depositional delta at the

Figure 16. In the hours immediately after the May 18 blast, pyroclastic flows billowed from the crater. This photograph, looking southeast from 2,500 m altitude, shows three large pyroclastic flows descending the northwest flank. The steam plume on the left is rising from phreatic activity in the Toutle River valley. (From ref. *13*.)

mouth of the Cowlitz River, reducing the depth of the Columbia River navigational channel from 12 m to less than 4 m.

Pyroclastic and hot ash flows, pouring down the eastern side of the mountain, also caused mudflows. Heavy accumulations of hot ash melted the snowfields on the northeast flank, causing a flow that spread out over the Abraham Plains and then drained through Smith Creek into the Muddy River. Deglaciation of the upper Shoestring glacier cirque triggered an additional mudflow, which traveled down the Pine Creek drainage, merging with the Muddy River mudflow. These combined mudflows then entered Swift Reservoir. The displacement caused by the resulting delta raised the level of the water in the reservoir by over half a meter (Fig. 15).

The plume and pyroclastic activity

Immediately following the initial eruption, the vertical plume rose to heights of 20,000 to 25,000 m, trailing off to the east-northeast. Strong convective upwelling was observed in the vertical plume, and horizontal layers of steam appeared at several altitudes within the vertical column (Fig. 17). Intra-cloud lightning was frequently observed throughout the column, and a mushroom cap expanded outward at an altitude of about 15,000 m (Fig. 21).

Three distinct types of ash-plume activity were observed at different altitudes throughout the morning of May 18: a high, thin stratoform layer at about 25,000 m; a dense, nearly horizontal, serpentine plume dripping large lobes of ash at about 10,000 m;

and a layer of diffuse ash—possibly suspended particulates from the lateral blast—below 5,000 m. Data collected by a University of Washington atmosphere-sampling aircraft indicate a median particle size of approximately 1 μm in the dense plume about 12 km downwind of the crater (*10*). This is consistent with the particle diameters common in the areas of maximum ash fall in eastern Washington. It is probable that the low-level ash clouds had primarily a local downwind effect and that the main plume suspended most of the ash as far as eastern Washington before significant ash fall began. Satellite observations indicate that the leading edge reached the Idaho border by 11:00 A.M. and entered western Montana by 3:00 P.M. (*11*).

A side-looking airborne radar (SLAR) image of the mountain was obtained

Figure 17. Convectively rising eddies of ash swirl around a horizontal layer of steam in the vertical eruption plume at 5,000 m altitude, as viewed from 300 m distance. The mushroom cap (shown in Fig. 21) expanded outward about 5,000 m above the part of the vertical column shown here.

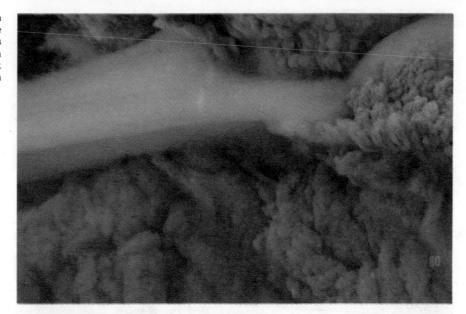

Figure 18. On May 30, the first day after the blast that the crater could be seen clearly, the interior was photographed through the breached north flank. The floor of the crater contained three steaming areas that corresponded to the eruptive centers identified on SLAR imagery during the May 25 eruption. Also visible was the wall-like pumice rampart (the somewhat lighter area immediately to the left of the date) built up during that eruption.

Figure 19. The lava dome that had formed within the crater following the eruption on June 12 is evident in this photograph looking to the south, taken on June 30. The pumice rampart is in the foreground. A line of steaming vents (fumaroles) can be seen radiating from the dome to the left side of the photo.

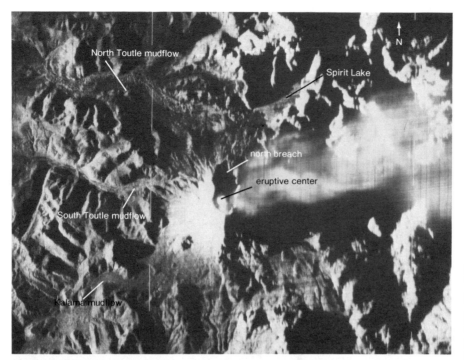

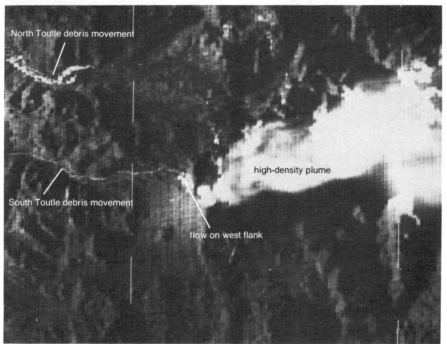

Figure 20. Side-looking airborne radar (SLAR) imagery was used to penetrate the heavy smoke and ash of the plume at 2:00 P.M. on May 18 during the main eruption. The image is divided into two channels: at the top, a fixed-target mode illustrates stationary objects; the moving-target mode below emphasizes motion, which is detected by Doppler shift. X-band (3-cm wavelength) radar was used to penetrate the eruption plume. The fixed-target mode shows the outline of the newly formed explosion crater and the mudflows in the valleys. On the moving-target mode the eruption vent area is evident in the southern part of the crater, as are the areas of dense particulate concentrations. The aircraft was 10 km west of the peak at an altitude of 8,500 m. The dashed lines which run vertically through the imagery are separated by 20 km. SLAR imagery provides the principal means of observation during eruptions when they are visually obscured by ash, clouds, or darkness. (Courtesy of Oregon Army National Guard.)

shortly after 2:00 P.M. (Fig. 20) (*12*). The fixed-target channel of the SLAR imagery shows the outline of the greatly enlarged crater, with the north breach clearly visible, and the debris flows descending the valleys. The bright area in the southern portion of the crater results from radar reflections of material being ejected at the eruptive center. The moving-target channel, which highlights detected motion, shows the direction of the densest portion of the plume as well as debris traveling down the north and south forks of the Toutle River.

Pyroclastic flows billowed out of the north breach of the crater throughout the first nineteen hours after the initial eruption on May 18, with several flows reported on the following day. These flows covered the mudflow materials immediately north of the mountain with more than 20 m of hot ash, causing the moisture contained in the mudflow debris to flash explosively into steam. Figure 16 shows several such flows descending the northwest flank, with a vertical plume of steam rising from a phreatic steam explosion in the Toutle valley.

Several nuées ardentes were observed on other flanks of the volcano. These incandescent ash clouds tended to be small and were primarily caused by partial collapse of the main vertical plume, which resulted in a spill over the remaining crater rim areas. One nuée ardente descended the south flank and flowed into the forest below in less than three minutes (*14*).

A second major explosive eruption occurred on May 25, one week after the first, at 1:32 P.M. An ash-rich plume rose, almost vertically, to an altitude of 15,000 m. Shortly thereafter a SLAR image indicated a constructional landform on the floor of the crater near three eruptive centers identified by SLAR during the eruption. On May 30 the crater was cloud-free for the first time since May 18, and a ridge of pumice, called the "rampart," and a flow of pumice blocks in the breach area were evident (Fig. 18). Numerous fumaroles were observed in the vicinity of the three eruptive centers. By June 8 thermal infrared images indicated a general heating of the eruptive center active on May 18 as well as the heat emitted by the pumice flow of May 25.

An explosive eruption on June 12 produced an even more extensive pumice flow. Pumice from this eruption reached the bottom of the Toutle valley, filling an extensive area of explosion craters formed on May 18. A SLAR image acquired on June 13, at 12:54 P.M., showed a dome forming south of the rampart within the crater. By June 30 the lava dome was clearly visible; it was nearly 200 m in diameter and about 40 m high (Fig. 19). Thermal images taken at about that time indicated heated fracture zones in the south and west wall of the crater. On July 19 heat traces along radial fractures seemed to indicate inflation of the dome.

The most recent chapter in this volcanic sequence (as of this writing) was a series of four powerful explosions that sent steaming ash plumes to altitudes of more than 15,000 m between 5:10 and 8:40 P.M. on July 22 (see Fig. 22 and cover). Once again SLAR imagery was used to pinpoint the eruptive centers within the crater. The activity was centered on the lava dome and elongated along the previously noted fracture zone. Pyroclastic flows were generated throughout the night, and two lobes of pumice flows

Figure 21. The ash-laden main eruption plume rose convectively, spreading into a mushroom cap at about 10,000 m. The aircraft was flying under the cap, shortly after the eruption, on a search and rescue mission. The photograph was taken from the north over Coldwater Ridge.

Figure 22. A powerful explosive pulse, at 8:39 P.M. on July 22, sent a plume of steam and ash to an altitude of more than 15,000 m. This photograph, looking down at the vent along the southwest rim, shows the ash rising rapidly immediately after the explosive burst. The gullied sides of the outward flank contain glacial remnants.

spilled out of the north breach toward Spirit Lake and the north Toutle valley.

Geothermal potential

Steam vents remain active in the explosion crater area at the base of the northwest flank, despite being buried by the pumice flow of June 12. The present venting appears to be fumarolic in nature, as well-developed sulfur crusts have formed around most of the vents. A helicopter survey on June 30 detected the odor of hydrogen sulfide and measured a general ground temperature of 39°C at a location that corresponds to the intersection of two geologic lineaments, observed on pre-eruption SLAR imagery. On this basis it is possible that the venting is not entirely phreatic in nature and that the north Toutle River valley may have some geothermal resource potential.

In addition to areas of pyroclastic deposition, the main cone may possess some accessible geothermal potential. The fracture systems that played active roles in the May 18 and July 22 eruptions are clearly visible on SLAR imagery, and could possibly be drilled into from the lower flanks of the cone at some future date.

Ancillary effects

The landslide-induced valley fills remain an area of continued concern. Tributary valleys, not directly affected by the debris flows, have been dammed, creating a series of lateral impoundments. Some of these lakes, such as the one at the mouth of Coldwater Creek, have begun to develop surface overflow channels along the margins of the valley fill. However, others appear to be infiltrating into the debris, causing saturation of

the mass. Thermal infrared images will be useful in monitoring the development of both surface drainage and material saturation as the winter rainy season approaches. High sediment yields, and possibly additional mudflows, are expected to develop during the reestablishment of the drainage network.

The erosional equilibrium of the hillslopes has also been seriously disturbed. Removal of the vegetation cover, followed in some areas by the scouring of the soil and deposition of ash, has resulted in extreme erosional instability on the midslope portions of most hillslopes in the devastation area. Although the area has been subjected to only minor rainfall thus far, the midslopes have developed long parallel rills—generally joining with one or two adjacent rills before ending in a small depositional fan at the base of the slope. In the dry period since the initial rains, the ash deposits began to develop desiccation cracks, which have further extended the surface rills. The fine particle size of the ash has variable infiltration characteristics, often producing runoff during precipitation of moderate intensity.

On flatter ground the ash has compacted over much of the downed organic debris, and where the accumulations are thicker than 50 cm, irregular surface depressions have formed. These depressions have already functioned as detention basins, and often a ring of light-colored pumaceous material may be found around their sides. It is unclear at this time whether these depressions will serve to detain runoff during higher intensity rainfalls or will overflow and add to peak runoff.

It is likely that the events described in this article are the first in a longer phase of eruptive activity of the Mt. St. Helens volcano. The impact of the eruption on transportation, agriculture, public health, and the economy of the Pacific Northwest are poorly understood at best. Although the magnitude and violence of the eruption of May 18 were certainly beyond expectations, the earth scientists who assessed the eruptive hazards of the area should be gratified with the effectiveness of their labors: the pre-

cautions and safeguards provided by science and society substantially reduced the loss of human life during an explosion that has been compared to a ten-megaton nuclear blast. The observations and experience gained should enable the scientific community to predict such occurrences with more accuracy in the future, and should teach society how to deal more effectively with major disasters.

References

1. D. R. Crandell and D. R. Mullineaux. 1978. Potential hazards from future eruptions of Mt. St. Helens volcano, Washington. U.S. Geological Survey Bulletin, 1383-C.

2. J. H. Hyde. 1975. Upper Pleistocene pyroclastic-flow deposits and lahars south of Mt. St. Helens volcano, Washington. U.S. Geological Survey Bulletin, 1383-B.

3. Personal communication, Dr. Stephen Malone, University of Washington, Geophysics Program, 1 July 1980.

4. M. A. Korosec, J. G. Rigby, and K. L. Stoffel. June 1980. The 1980 eruption of Mt. St. Helens, Washington. Wash. State Dept. Nat. Res., Div. Geol. and Earth Res., Info. Circ. 71.

5. The statement by H. Kieffer (*Science* 208, p. 1448, 27 June 80) that thermal infrared images were not available until Monday, May 19, is incorrect. In fact, the Oregon Army National Guard acquired, interpreted, and reported such data to the U.S. Geological Survey on May 16 and May 18, 1980.

6. Ref. *4*, pp. 9–11.

7. Personal communication, Dr. Dwight Crandell, U.S. Geological Survey, Vancouver, WA, 19 May 1980.

8. Interpreted from a sequence of photos taken by Ty Kearney, a volunteer radio operator for the Washington Dept. of Emergency Services, from a point 6 km north of Goat Mountain.

9. G. A. Taylor. 1958. The 1951 eruption of Mt. Lamington, Papua. *Bull. Bur. Mines Resour., Geol. Geophys.* Aust., v. 38.

10. P. V. Hobbs, L. F. Radke, M. W. Eltgroth, and D. A. Hegg. 1980. A preliminary report of airborne studies by the Univ. of Wash. of the effluents from the Mt. St. Helens volcanic eruptions. Seattle: Univ. of Washington.

11. Ref. *10*, p. 11.

12. For a description of the operations properties of this system see C. L. Rosenfeld and A. J. Kimerling. 1977. Moving target analysis utilizing side-looking airborne radar. *Photo. Eng. Rem. Sens.* 43(12): 1519–22.

13. C. L. Rosenfeld. 1980. Remote sensing of the Mount St. Helens eruption, May 18, 1980. *Oregon Geology* 42(6):103–14.

14. Personal communication, Dr. Guy Rooth, Geology Department, Oregon College of Education, who observed this nuée ardente from a plane 35 km west of the mountain.

James G. Moore

Mechanism of Formation of Pillow Lava

Pillow lava, produced as fluid lava cools underwater, is the most abundant volcanic rock on earth, but only recently have divers observed it forming

Much of the ocean floor is covered by lava of a distinctive character. The lava appears to be made up of closely packed ellipsoidal masses about the size and shape of pillows—hence the term *pillow lava*. Only within the last few years has the abundance of pillow lava on the ocean floor been fully recognized. Ocean-bottom photographs and dredge samples have shown that the great bulk of new ocean floor created at diverging plate boundaries (such as the Mid-Atlantic Ridge) is composed of pillowed basaltic lava flows. Closeup observations from submarines at depths of 2.7 km in the rift valley of the Mid-Atlantic Ridge have verified that virtually all the lavas erupted at this plate boundary are pillowed (Ballard et al., in press). The submarine portions of the great oceanic volcanoes, such as the ridge beneath the Hawaiian Islands, are also known to be built largely of pillow lava (Moore and Fiske 1969), and it is widespread in outcrops of uplifted ancient lava. Pillow lava is

Dr. Moore is a Research Geologist with the U.S. Geological Survey, which he joined in 1956, after receiving his Ph.D. from Johns Hopkins University. Currently associated with the Branch of Field Geochemistry and Petrology, Dr. Moore is the author of many publications on volcanic processes, including those taking place in the submarine environment. He was Scientist-in-Charge of the Hawaiian Volcano Observatory (1962–64). During eight oceanographic cruises in the Pacific and Atlantic oceans, the author participated in the mapping, sampling, and photographing of fresh submarine lava. In 1974 he was a diving scientist in the French-American Mid-Ocean Undersea Study (project FAMOUS), in which scientists from both countries studied the rift valley of the Mid-Atlantic Ridge by diving to depths of 2,700 meters in submersible vehicles. Address: U.S. Geological Survey, 345 Middle-field Rd., Menlo Park, CA 94025.

probably the most abundant form of volcanic rock on earth, though most of it is hidden beneath the world's oceans and mantled by younger sediments.

Most investigators agree that the pillows form when fluid lava chills in contact with water (Snyder and Fraser 1963), either when it erupts directly into water (or beneath ice) or when it flows across a shoreline and into a body of water. However, prior to our study, the process of pillow formation had never been directly observed (Macdonald 1972, p. 102).

The recent eruptions of Kilauea Volcano in Hawaii provided an unparalleled opportunity to study the movement and cooling of lava beneath the sea. In June 1969, lava from the new Mauna Ulu vent on the east rift zone of Kilauea spilled into the sea after flowing 12 km down the south flank of the volcano. This pattern was repeated, with lava flowing into the sea for a few weeks each year through 1973.

In April 1971 scuba divers for the first time investigated lava flowing underwater and learned that in favorable circumstances the lava could be approached closely. Despite heated water, explosive concussions, vigorous convective currents, and poor visibility due to suspended sediment, valuable observations were made (Moore et al. 1973).

Divers also studied the 1972 and 1973 flows at depths of 4–38 m and documented the process of pillow formation by still and motion picture photography. Much of the data in the present paper is based

on these observations and on analysis of motion pictures (Tepley and Moore 1974) showing the growth of Kilauean pillowed flows.

The actual shape of individual pillows in pillow lavas is a matter of dispute, leading to a controversy as to their mode of origin. Some writers (Snyder and Fraser 1963; Johnston 1969; Macdonald 1972) believe that most pillow lava is composed of discrete, independent, sacklike or ellipsoidal masses. Others (Jones 1968; Vaugnat and Pustaszeri 1965; and Moore et al. 1971) describe pillow lava as a tangled mass of cylindrical, interconnected flow lobes. In the recent extensive study of the rift valley of the Mid-Atlantic Ridge from submersible vehicles (Ballard et al., in press) most pillows were found to be interconnected, and independent sacks were rare. Pillows observed growing in the sea off Kilauea Volcano were invariably attached during growth, in rare instances a pillow would become detached from its feeding pipe on a steep slope and tumble downslope for a distance of a meter or so.

Much of the controversy over pillow shape is probably the result of misinterpretation of the three-dimensional configuration of pillow flows as seen in two-dimensional cross sections of eroded pillows in land outcroppings. Virtually every planar section through a tangled mass of interconnected flow lobes would give the appearance of a pile of discrete ellipsodial masses. Similar misinterpretation of three-dimensional shapes as seen in section has long been a problem in the biological sciences (Elias 1971). Adequate three-dimensional exposures

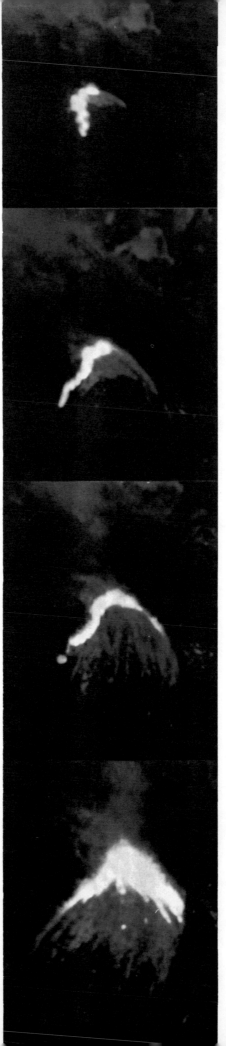

Figure 1. This sequence from a 16-mm film of an underwater lava flow from Kilauea Volcano in March 1973 shows a pillow lobe flowing downslope at a depth of about 15 meters. A layer of steam is adjacent to the incandescent crack. Timing of frames (*top to bottom*): start, 3 sec, 4 sec, 5 sec (film by L. Tepley). Scale: |‾10 cm‾|

are essential for defining the true geometry of pillowed flows (Jones 1968). The term *pillow* itself is in many cases a misnomer—it does not always accurately describe the shape—and that has contributed to the confusion.

All the pillowed flows studied by divers off Hawaii (Moore et al. 1973; Tepley and Moore 1974) are composed of interconnected, digitated, cylindrical lobes which advanced by the branching and enlarging of individual lobes, much as does a terrestrial fluid lava flow. This paper is concerned mainly with the mechanism of branching and enlargement of the individual flow lobes. The terms *pillow lobe* and *pillow bud* refer to lava flow units, a few decimeters in cross-sectional diameter, generally roughly circular or ellipsoidal in cross section, and generally connected upslope with other similar but larger flow units. The term *pillow* is used to refer to a part of or an entire flow lobe.

Pillow growth

Studies of the lava flows on the south coast of Hawaii from 1969 to 1973 (Moore et al. 1973; Tepley and Moore 1974) indicate that, at the shoreline, the flows tend to thicken and widen, entering the sea along a front characteristically much broader than that of the flow at a distance from the shore. The flow front then divides into many distributary lava tubes that feed a large number of flow lobes into the water, either over a sea cliff or, where there is no cliff, directly into the sea. The flow lobes develop most readily on a steep slope, producing a stack of foreset-bedded pillow tongues that slowly grows seaward. Eventually, the largest lava tubes divide and feed the ever-branching pillow lobes.

Where the surf is active and the flow front is steep, the new lava may break apart and slump down,

producing coarse rubble that becomes mixed in varying proportions with the coherent flow lobes. Pillowed flows fed from submarine vents in deep water—those along the mid-ocean ridges, for example—appear to have less intermixed rubble than the nearshore ones, but otherwise the structures produced are similar.

The surfaces of active pillow tongues are black and solid but hot to the touch; occasionally they crack open, exposing the core of incandescent red-orange lava within (Figs. 1 and 2). The solid black crust quickly forms on newly exposed lava as it cools below about 700–800°C, the minimum temperature of incandescence. The pillows are enlarged when fresh lava exudes from cracks as they open and begin to cool, depositing the hot lava on the diverging walls of the expanding cracks. During a period of steady divergence of the crack walls, the rate of spreading approximates the rate of cooling, and the width of the incandescent material remains about constant.

The shape and size of the spreading cracks can vary widely. Some are up to one meter long, and nearly straight, and parallel to the axis of a cylindrical pillow lobe; the expansion of this type simply increases the diameter of the lobe. Rapidly spreading cracks commonly have a more zig-zag shape than do slowly spreading ones. Other cracks are curved, or crescent-shaped, and enable a pillow lobe both to enlarge its diameter and to lengthen (Fig. 3). Still others, developing at a weak point on the side of a larger pillow, quickly develop into circular cracks that feed pillow buds from the side of the original pillow. The resulting pillow buds can grow in any direction, including straight up (Fig. 4), because the pressure of lava within the original pillow is maintained by the higher lava level in interconnected pillows farther upslope. As such upward-directed pillows grow larger, they eventually bend down under the effect of gravity until they rest on the pillow beneath them and continue growing downslope. Finally, some pillows enlarge at several branching cracks (see Fig. 2), all moving apart from each other as new crust forms adjacent to their diverging walls.

The width of incandescence in opening cracks ranges from 2 mm to 20 cm, and the cracks remain active spreading centers for a period ranging from a few seconds up to a few minutes. At several places we saw a crack revive one or more times, and the pillow enlarged, or the pillow bud grew, in surges.

A milky cloud (visible in Figs. 1, 2, and 4) rises from the incandescent cracks and is commonly carried horizontally by currents. The cloud is a mixture of water and tiny bubbles, generally much less than 1 mm in diameter. Presumably the incandescence is observed through a thin layer of superheated steam bubbles.

The steady growth of a pillow may be interrupted by implosions during which the pillow violently contracts, spalling off fragments of solidified glassy crust. The implosions are presumably produced when gases within the pillow cool and contract, and external water pressure crushes the thin outer crust which has been weakened by thermal stresses. Implosions may occur at intervals of about 5 seconds; the growth of the pillow is resumed immediately after the shock. Implosions produce concussions painful to divers within about 3 meters.

The molten lava inside a group of interconnected pillow lobes and buds drains downslope as it is fed from above. When the lava source is cut off by an obstruction or diversion, and fresh lava from upslope no longer reaches the lobes, water enters the crusted-over interconnected pillows through cracks and chills the inside, producing hollow pillow lobes and pipes (Moore et al. 1973). Broken hollow pipes of this type are common in oceanic dredge hauls. Many such curved fragments were collected on the Reykjanes Ridge: they ranged in thickness from less than 1 to 10 cm, averaging about 5 cm (Moore and Schilling 1973, p. 106).

Rates of spreading

From a study of motion pictures and timed still photographs, rates of spreading of accreting cracks have been estimated to range from about 0.05 to 20 cm/sec. Cracks on growing pillow buds are the easiest

Figure 2. During the same flow from Kilauea Volcano shown in Fig. 1, a pillow bud at a depth of about 7 meters expands on several accreting cracks, each moving away from the others. Timing of frames (top to bottom): start, 0.4 sec, 0.8 sec, 1.3 sec (film by L. Tepley). Scale: ⌐10 cm

to measure, because spreading in them is highly asymmetric, since the crack from which the bud grows is (to a first approximation) fixed to the main pillow lobe, and the great bulk of accretion is on the bud side. Hence the increase in distance from the tip of the bud to the crack axis over time is a measure of the spreading rate on one side of the crack—the half spreading rate —which is approximately equal to, though slightly less than, the full spreading rate.

The change in dimensions of two growing pillow lobes with time, plotted in Figure 5, shows the width of visible incandescence to increase as the spreading rate increases. In addition, the width of the thin layer of steam on the pillow surface adjacent to (and probably also on top of) the incandescent crack increases as the spreading rate increases.

No direct measurements were made of the spreading rates for symmetrically spreading cracks—although many such cracks were seen and photographed—because most symmetrical spreading is slow, and constant reference points on such pillows, especially on both sides of the crack, are difficult to see in a series of photographs. One slow-spreading crack on a pillow pipe which was photographed over a considerable length of time was estimated to have a half spreading rate of 0.1 ± 0.05 cm/sec; the incandescence was intermittent and discontinuous, but averaged about 0.2 cm in width.

The half spreading rate can be approximated by measuring the slope of the tangent to the curves, indicating the distance from the crack axis to the end of the pillow, as shown in Figure 5. The width of incandescence of spreading cracks relative to the half spreading rate shows a linear relationship, the slope of which indicates that the age of crust at the edge of incandescence is about 0.6 sec (Fig. 6).

The extremely rapid cooling effect of the surrounding water can be illustrated by comparing this age with the age of final incandescence along diverging plates of crust in a subaerial lava lake. At Mauna Ulu, Hawaii, Duffield (1972) found that the width of incandescence was 1 meter and the spreading rate of crustal plates was 20 cm/sec, indicating an age of the edge of incandescence of 5 seconds. Cooling is nearly 10 times faster underwater.

The age of crust at the edge of the visible steam layer on pillows is about 1.7 seconds (Fig. 6). In addition to the width of incandescence and steam, the general aspect of the spreading crack changes with the rate of spreading. When the rate is slow, the crack is quite straight and may be traced for distances up to 1 meter; when the rate is fast, the crack is markedly zigzag (Figs. 1 and 2). When the rate of spreading exceeds about 15 cm/sec, the plastic zone adjacent to the crack is so wide that a moderate-sized pillow may neck down at that point and pull off, particularly if it is forming on a steep slope. Moreover, as described in the next section, the character of surface ridges shows a definite relation to the rate of spreading.

Features of pillow crust

The outer crust of fresh basaltic pillows is glassy because of the drastic quenching of molten lava when it is exposed to water at the opening crack. The outermost glass is uniform, pale brown (in thin section) sideromelane, which contains only the large crystals (phenocrysts) and tiny crystals (microlites) present in the melt at the time of quenching. Generally, pillows fed from a terrestrial lava flow that enters a body of water contain more microlites in the outer glassy rim than do pillows erupted directly into the deep sea, because they have had more time to cool and crystallize during flow on the ground surface.

Just inside the glass zone is a zone in which microscopic incipient feathery crystals (crystallites), which develop due to slightly slower cooling, are mixed with the basalt glass. At a still deeper level, the crystallites replace glass, and

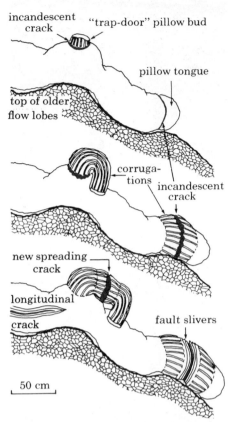

Figure 3. Pillow tongues enlarge by growth of new crust adjacent to incandescent cracks. *Top:* A pillow tongue growing downslope is connected to a larger feeder flow coming in from the left. Incandescent cracks in the outer thin crust form in response to tension induced in the crust by pressure of fluid lava within the tongue, and a small "trap-door" pillow bud grows out from a circular crack at a weak point. *Center:* New crust with corrugations perpendicular to the crack is produced adjacent to rapidly spreading cracks. The pillow tongue lengthens downslope, and the pillow bud continues to grow, eventually bending down under the influence of gravity. *Bottom:* Spreading on the crack at the end of the pillow tongue slows down, causing a shift from production of corrugations to production of fault slivers parallel to the spreading crack. A slow-spreading longitudinal crack opens on the thick part of the pillow tongue, increasing its diameter by producing fault slivers. A new spreading crack develops in the middle of the bud, causing it to lengthen still further.

the center of the pillow is mostly crystalline. At the base of the mixed crystallite-glass zone, the crystallites commonly organize into spherulitic masses up to a few millimeters in diameter, called varioles. Varioles are most fully developed in pillow rims that are poor in vesicles and phenocrysts or microlites. If these structures are present, crystallites tend to nucleate and crystallize around them rather than forming separate, more widely spaced spherulitic growths.

The glass and glass-crystallite zones are quite constant in thickness in basaltic pillows, averaging 4 and 5 mm thick, respectively, as measured in 50 samples of fresh, dredged oceanic basalt pillows. The glass zones are thicker in very small (less than 10-cm diameter) pillow buds because of the more rapid cooling of smaller masses. Glass thickness is also no doubt affected by the chemistry, and therefore the viscosity, of the basalt melt.

The glassy surface of pillows is commonly marked with small parallel ridges (Fig. 7), a ubiquitous feature in photographs of fresh pillows on the ocean floor (Cousteau and Dugan 1963; Moore and Fiske 1969). The ridges generally range from 0.5 to 10 cm in wave length and have a relief one-fifth to one-half their wave length.

Two very different types of ridges were observed during the growth of Kilauean pillows, here termed *corrugations* and *fault slivers*. Other types of small ridges are present on fresh pillows but have not been observed during formation. The ridges are rarely preserved in ancient pillows because the outer glass crust is easily destroyed. Corrugations are produced along fast-spreading zigzag cracks at right angles to the opening crack (Tepley and Moore 1974). Where spreading is symmetrical, corrugations are equally well developed perpendicular to the crack outward on both sides.

Fault slivers are produced along slow-spreading cracks, *parallel* to the opening crack. Generally less than 1 cm wide, they are thinner than corrugations and are bounded by small normal faults of a few millimeters displacement, downthrown on the crack side. The fault slivers are symmetrically arranged on each side of the crack and bound small fault blocks tilted away from the crack. The innermost faults encompass a small depressed area, or graben, at the site of the crack.

Fuller (1932), in a remarkably perceptive study, described these fault slivers in ellipsoids (pillows) on the Columbia River Plateau, near Vantage, Washington. He states that, "although separated by a distance of even a foot or more, these miniature opposing escarpments would coincide fairly closely if placed in conjunction." The fault scarps described by Fuller are 2–5 mm high, face inward toward the crack, and bound fault blocks about 0.5 mm wide that are evenly tilted 10–30° away from the crack. When well developed, the faulted pillow surface resembles a washboard in general aspect. Samples of pillows from Fuller's locality are quite similar to those observed adjacent to a slowly spreading crack on the Kilauean submarine lava flows (Fig. 8). Sections of these samples show that the inward-dipping faults curve downward at depth to become vertical or slightly overhanging, exactly as shown by Fuller.

Many opening cracks slow down in their spreading rate before they freeze shut, and as they do so, they begin to produce fault slivers instead of corrugations. At an intermediate spreading rate, both types of ridges are produced and intersect at about right angles.

In addition to corrugations and fault slivers, other small-scale ridges are present on pillows. Festoons of ropy wrinkles are rare (Moore et al. 1971), as they require very rapid development of the crust, so that a sizable area of thin plastic crust is exposed. So-called *toothpaste pillows* are relatively small cylindrical (~40-cm diameter) pillow buds with longitudinal striations or fine corrugations that average perhaps one cm in wave length (Ross and Schlee 1973). The pillow bud clearly exuded from a crack, but the striations may have a somewhat different origin than the corrugations parallel to them, which are thought to be produced by unequal cooling along a zig-zag crack. The striations may result from scraping of grooves in the plastic crust as it moves past the jagged upper lip of the adjacent crack.

Other small ridges transverse to the long axes of pillow buds are apparently not fault slivers, but form in different ways. One type probably forms by intermittent surging growth of a pillow bud from a crack, each surge apparently wrinkling the thin crust adjacent to the crack (Moore et al. 1971). Still another type of transverse wrinkle forms on the inner surface of an el-

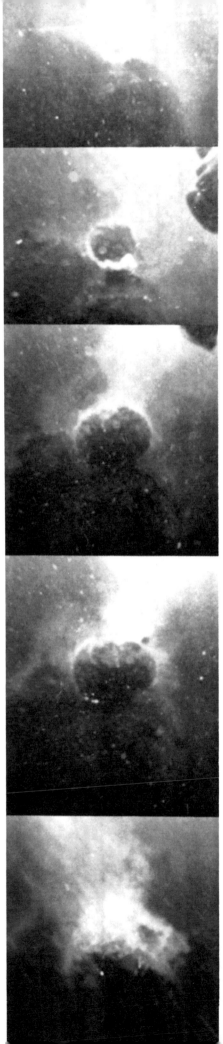

Figure 4. As these photographs demonstrate, a pillow bud can grow straight up from a larger pillow. Growth is interrupted in the last frame by implosion. Timing of frames (*top to bottom*): start, 1.7 sec, 4.2 sec, 5.8 sec, 7.5 sec. The sequence is from a 16-mm film made by L. Tepley during the Sept. 1972 flow from Kilauea Volcano, at a depth of 6 meters. Scale: ~10 cm

ongate cylindrical bud that is bent during growth (Ross and Schlee 1973).

Finally, new spreading cracks can open in previously ridged crust; they commonly form in the troughs between corrugations and may build new corrugations at right angles to the older ones. In addition, several spreading cracks can produce pillow crust simultaneously, as shown in Figure 2. Consequently, unraveling the growth history of a pillow that has enlarged by accretion on several cracks can be difficult, if not impossible.

Cooling of pillow crust

Pillows grow in volume, and their crust expands in area, as fresh lava is fed into them from a connecting pillow lobe which in turn is connected to an erupting vent by a larger flow unit or a shallow intrusive feeder. As lava is fed into a pillow bud, the crust cracks and opens. New pillow crust develops at the diverging crack walls, where fluid basalt is exuded, accretes onto the inner margin of the spreading crust, and solidifies as it cools. The inward cooling of new crust keeps pace with the outward movement of the crust, and thus the width of the incandescent part of the spreading center remains relatively constant.

The cooling and thickening of the pillow crust can be approximated by considering an infinite slab with an initial temperature, V, of 1,200°C, with the top surface held at 0°, and with cooling occurring entirely by conduction. The depth, x, that isotherms, v, will occupy after time, t, is expressed by the equation:

$$v = V\,erf\,\frac{x}{2\sqrt{Kt}}$$

where *erf* is the error function, K is the thermal diffusivity (assumed to be 0.007 cm²/sec); the latent heat

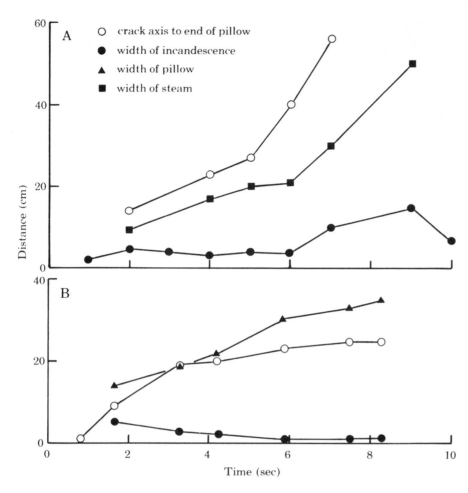

Figure 5. Graph A plots the growth of the pillow shown in Fig. 1 over time; graph B plots the growth of the pillow in Fig. 4. Dimensions are based on the estimated scales given in Figs. 1 and 4.

at 1,150° the viscosity is about 1,000 poises and at 1,065° it is greater than 10^6 poises.

If we assume that a pillow grows by steady spreading from a crack whose spreading rate is known, the model shown in Figure 9 can be considered a cross section of a growing pillow at a point in time. The figure shows the distribution of isotherms, and consequently the thickness of crust, for a slowly spreading pillow (0.1 cm/sec) and for a rapidly spreading pillow (5 cm/sec).

The thermal model does not take into account the fact that the pillow is not stagnant and new, hot melt is entering the system from below the crack area, retarding cooling near the crack. Nor does it consider radiant loss of heat from the incandescent crack, or insulation of the crack area by a layer of steam. The model suggests, however, that the crust of a pillow that grows at 5 cm/sec to a distance of 1 m has a thickness at its far end of about 8 mm. Presumably a crust of about this thickness is so rigid that it impedes free movement of the pillow and inhibits spreading at the crack and further expansion of the pillow, thereby back-throttling the spreading action at the feeding crack. Factors such as these are partly responsible for the uniform size of pillow lobes. The force required to split open a new crack eventually becomes less than that required to continue spreading on an old one.

Crystals grow very rapidly in basalt melt between the liquidus and solidus temperatures because of the relatively low viscosity of the melt as compared with more silicic melts. For this reason, the glassy crust on basaltic pillows is relatively thin, despite the drastic quenching it undergoes. The base of the pure glass layer (averaging 4 mm thick) is indicated in Figure 9. The graph shows that, in order for the melt to be quenched to glass without any crystallites forming, the

of crystallization is omitted because the outer part of the pillow is not crystalline but glass.

Two isotherms—1,000° and 1,150°—are plotted in Figure 9. The 1,000° isotherm is close to the solidus; Peck and his colleagues (1966)

determined by drilling into the basaltic Alae lava lake on Kilauea Volcano that the solidus is at 980°. The 1,150° isotherm is close to the liquidus; in the Alae lava lake, less than 5 percent of the crust was crystalline at 1,150°, but the true liquidus was at a somewhat higher temperature.

In both the Alae and Kilauea Iki lava lakes, the base of the crust (as defined by the depth at which, during drilling, the drill falls under its own weight) is at 1,065°. At this temperature about 40 percent of the rock is crystalline (Peck et al. 1966). Hence the base of the solid crust (and consequently the depth to which fracturing can occur) will lie about midway between the 1,150° and 1,000° isotherms. Shaw and his colleagues (1968) have shown that a strong viscosity gradient exists between these isotherms;

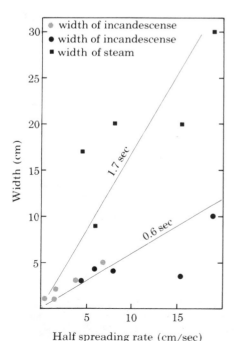

Figure 6. Change in the width of the incandescent zone and of the steam layer as a function of the half spreading rate is calculated from the data in Fig. 5: data from graph A in black, data from graph B in color.

Figure 7. This remote-camera photograph shows a pillowed lava flow on the Clarion fracture zone south of Baja California, 2,030 m deep (Moore 1970). The larger ridges on the pillow surfaces are corrugations that formed perpendicular to feeding cracks, which extend approximately from left to right in the photograph. Smaller fault slivers parallel to crack (poorly shown) apparently developed near waning stages of pillow growth when crack spreading had slowed. Diameter of compass is 10 cm.

basalt melt must cool from the liquidus to the solidus temperature in about 3 seconds. In order for any glass to be preserved, the cooling time from liquidus to solidus temperature must be less than about 13.5 seconds.

Implosions of pillows generally occur in rapidly spreading pillows that are only a few seconds old and have a crust 2–5 mm thick. At this stage, the crust is apparently thick and strong enough to maintain a pressure difference until the crust is crushed and broken by external water pressure, but thin enough so that it does not have the strength to resist fracture. The pressure difference is presumably created by cooling and contraction of gases within the crust. Hence implosions are no doubt less common at great depth, where the volume of a gas phase is smaller and less dependent on temperature changes.

Only in slowly spreading pillows is the crust thick enough close to the crack to be broken into tilted fault slivers, which are commonly 0.5 cm wide. If the faults first formed 1 cm from a crack with a half spreading rate of 0.1 cm/sec, the crust was about 10 seconds old and 0.6 cm thick. This is reasonable, since the fault scarp height is somewhat less than 0.6 cm.

These faults were apparently produced as molten lava within the pillow rose in the crack area and dragged up the solid crust some distance from the crack axis. The

thinnest crust adjacent to the crack was not uplifted because rising material was diverted laterally to accrete to the diverging crustal plates. Lachenbruch (1973) describes a similar model to account for the comparable (though vastly larger) tilted blocks adjacent to an oceanic spreading center.

The thermal model suggests that incandescence should be visible only in a hairline at the spreading crack, rather than in a zone up to 20 cm wide. Incandescence, as measured in Figure 6, passes outward from a golden yellow inner zone ($\sim 1,100°$) to an outer dull red zone ($\sim 700°$). The incandescence is probably viewed through a thin layer of steam bubbles that insulate the lava surface from the overlying cooling water. The layer of bubbles retards the depression of the iso-

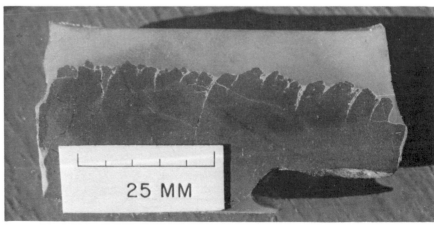

Figure 8. Samples from the Vantage area, in eastern Washington, show fault slivers on pillow surfaces. *Top:* Looking down on the fault slivers, the spreading crack is on the right. *Bottom:* The section is normal to length of fault slivers, with the spreading crack to the left. Note the steepening of normal faults with depth. (Samples collected by D. A. Swanson.)

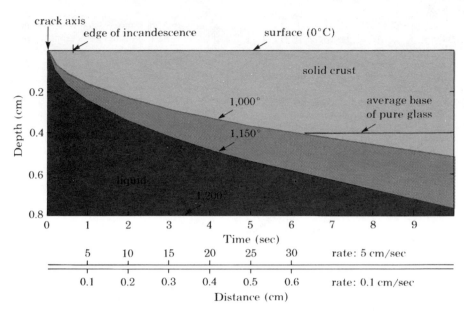

Figure 9. In this theoretical model, iso-thermal surfaces within a cooling, infinite slab of basalt are calculated for the following conditions: original temperature of 1,200°C, top surface chilled to 0°C, cooling within the slab by simple conduction. 1,150°C is the approximate temperature at which crystals first appear in the liquid (liquidus); 1,000°C is the temperature at which the last liquid crystallizes (solidus). The temperature at which the crystal-liquid mixture acts as a solid body (defining the base of the solid crust) thus lies between the liquidus and solidus temperatures. Lower scale shows dis-tance from the crack axis at left for a. fast-spreading crack (half spreading rate of 5 cm/sec) and for a slow-spreading crack (0.1 cm/sec).

therms and also provides a trans-parent window through which in-candescence can be viewed.

Photographs indicate that the bub-bles above incandescent cracks are very small, clearly less than 1 mm in diameter. Bubbles of 0.2 mm di-ameter would rise at a rate of 0.2 mm/sec (Stokes law), and hence even those originating at the crack axis would have risen only 0.34 mm above the crust surface at the edge of the steam zone at a crust age of 1.7 sec.

In addition to the insulation pro-vided by the steam layer, the width of incandescence is wider than pre-dicted by the thermal model be-cause of the movement of new, hot lava from within the pillow up to the spreading crack axis. The time required to solidify the outer 5 cm of a stagnant pillow is about 20 minutes (10 cm, 75 minutes), but the time could be greatly extended if fresh, hotter lava were entering and flowing through the tube sys-tem. Hence the hollow pillows with a crust 5–10 cm thick, common on the ocean floor, were probably part of a feeding system that was active for one to several hours.

Sea-floor spreading analogy

The growth of pillows by accretion on the trailing edge of plates of crust that move away from spread-ing cracks is a process that can be compared to sea-floor spreading, in which oceanic lithosphere is formed

at, and diverges from, oceanic spreading ridges (Le Pichon 1968). The process differs from the total picture of global plate tectonics be-cause the pillows are expanding as the crust grows and no zones of crustal destruction (subduction) are necessary, or were observed. Also, no clear-cut transform faults were noticed in the limited obser-vations made, but the zig-zag spreading cracks seen in fast spreading can be regarded as sim-ple spreading cracks modified by many small transforms.

The structures developed on pillow crust can be related to the rate of spreading and show interesting comparisons with the structures as-sociated with oceanic spreading ridges. Fast-spreading pillow crust develops corrugations parallel to the direction of plate motion, and the spreading crack itself is broad and not depressed. Slow-spreading pillow crust develops fault slivers perpendicular to the direction of plate motion, and the spreading crack is narrow and depressed, oc-curring within a graben.

Fast-spreading oceanic ridges are relatively smooth and display a central positive area, or horst, whereas slow-spreading ridges ex-hibit rough topography, frequent earthquakes, and a central graben (Anderson and Noltimier 1973). The fault slivers paralleling slow-spreading pillow cracks are down-tilted away from the crack in the same fashion as the fault blocks flanking the slow-spreading Gorda

Rise (Atwater and Mudie 1968) and Mid-Atlantic Ridge (Aumento 1972). The fault slivers in pillows apparently are restricted to slow-spreading cracks because only ad-jacent to them is the crust thick enough to fracture close to where the stresses of crustal uplift are highest. The fact that the faults bounding the slivers of pillow crust steepen with depth (see Fig. 8) may be useful in interpreting the shape of the oceanic faults.

Formation of pillow lava

Most of the actively growing pil-lowed lava flows examined by di-vers, as well as pillowed flows in the deep sea studied from photo-graphs or by observations from sub-marines, are composed of elongate, interconnected flow lobes that are elliptical or circular in cross sec-tion. The flow lobes are fed from upslope by larger, connected lava tubes that maintain lava pressure within the growing lobes. Isolated pillow sacks are rare but may form when a flow tube pinches off on a steep slope and tumbles a meter or so downhill.

The individual pillow lobes do not grow by the stretching of an outer plastic skin. Rather they expand, branch, and lengthen as fresh lava feeds into the pillow, and distends and cracks the outer crust. New crust is continuously formed adja-cent to the incandescent, opening cracks. The pillow lobe grows spo-radically downslope as a given crack stops spreading and a new one forms closer toward the distal end of the growing pillow tongue.

Fast-spreading cracks (opening at

about 5 cm/sec) are commonly zig-zag in shape and produce corrugations perpendicular to the crack. Slow-spreading cracks (opening at about 0.1 cm/sec) produce smaller fault slivers parallel to, and tilted away from, the crack. Ridges of both types account for the parallel-ribbed appearance of fresh pillows in ocean-bottom photographs.

The development of pillow crust adjacent to opening cracks is analogous to sea-floor spreading in which oceanic lithosphere is formed at, and diverges from, oceanic spreading ridges. The fault slivers in pillow crust can be compared to the outward tilted fault blocks that bound slow-spreading ocean ridge systems.

References

Anderson, R. N., and H. C. Noltimier. 1973. A model for the horst and graben structure of midocean ridge crests based upon spreading velocity and basalt delivery to the oceanic crust. *Geophys. Jour. Royal Astrophys. Soc.* 34:137–47.

Atwater, T. M., and J. D. Mudie. 1968. Block faulting on the Gorda Rise. *Science* 159:729–31.

Aumento, F. 1972. The oceanic crust of the Mid-Atlantic Ridge at 45° N. In *The Ancient Oceanic Lithosphere*, vol. 4. Ottawa, Canada: Dept. Energy, Mines, and Resources, pp. 49–53.

Ballard, R. D., W. B. Bryan, J. R. Heirtzler, G. Keller, J. G. Moore, and Tj. van Andel. In press. Manned submersible observations in the FAMOUS area, Mid-Atlantic Ridge. *Science*.

Cousteau, J. Y., and J. Dugan. 1963. *The Living Sea*. N.Y.: Harper and Row, 325 pp.

Duffield, W. A. 1972. A naturally occurring model of global plate tectonics. *Jour. Geophy. Research* 77(14):2543–55.

Elias, Hans. 1971. Three-dimensional structure identified from single sections. *Science* 174:993–1000.

Fuller, R. E. 1932. Tensional surface features of certain basalt ellipsoids. *Jour. Geology* 50(2):164–70.

Johnston, W. E. Q. 1969. Pillow lava and pahoehoe: A discussion. *Jour. Geology* 77:730–32.

Jones, J. G. 1968. Pillow lava and pahoehoe. *Jour. Geology* 76:485–88.

Lachenbruch, A. H. 1973. A simple mechanical model for oceanic spreading centers. *Jour. Geophys. Research* 78(17):3395–417.

Le Pichon, X. 1968. Sea-floor spreading and continental drift. *Jour. Geophys. Res.* 73:3661.

Macdonald, G. A. 1972. *Volcanoes*. Englewood Cliffs, N.J.: Prentice-Hall, 510 pp.

Moore, J. G. 1970. Submarine basalt from the Revillagigedo Islands region, Mexico. *Marine Geology* 9:331–45.

Moore, J. G., R. Cristofolini, and A. Lo Giudice. 1971. *Development of Pillows on the Submarine Extension of Recent Lava Flows, Mount Etna, Sicily*. U.S. Geol. Survey Prof. Paper 750-C, pp. C89–C97.

Moore, J. G., and R. S. Fiske. 1969. Volcanic substructure inferred from dredge samples and ocean-bottom photographs, Hawaii. *Geol. Soc. America Bull.* 80:1191–202.

Moore, J. G., R. L. Phillips, R. W. Grigg, D. W. Peterson, and D. A. Swanson. 1973. Flow of lava into the sea 1969–1971, Kilauea Volcano, Hawaii. *Geol. Soc. America Bull.* 84:537–46.

Moore, J. G., and J. G. Schilling. 1973. Vesicles, water, and sulfur in Reykjanes ridge basalts. *Contr. Mineral. and Petrol.* 41:105–18.

Peck, D. L., T. L. Wright, and J. G. Moore. 1966. Crystallization of tholeiitic basalt in Alae lava lake, Hawaii. *Bull. Volcanol.* 29:629–56.

Ross, D. A., and J. Schlee. 1973. Shallow structure and geologic development of southern Red Sea. *Geol. Soc. Am. Bull.* 84:3827–48.

Shaw, H. R., D. L. Peck, T. L. Wright, and R. Okamura. 1968. The viscosity of basalt magma: An analysis of field measurements in Makaopuhi lava lake, Hawaii. *American Jour. Sci.* 266:255–64.

Snyder, G. L., and G. D. Fraser. 1963. *Pillowed Lavas, II: A Review of Selected Literature*. U.S. Geological Survey Prof. Paper 454-C, 7 pp.

Tepley, L. (prod.), and J. G. Moore (sci. consult.). 1974. Fire under the Sea: The Origin of Pillow Lava. (16-mm sound motion picture.) Mountain View, Calif.: Moonlight Productions.

Vaugnat, M., and L. Pustaszeri. 1965. Réflections sur la structure et mode de formation des coulées en coussins du Montgènevre. *Archives des Sciences* Genève 18(3):686–89.

J. R. Heirtzler
P. T. Taylor
R. D. Ballard
R. L. Houghton

A Visit to the New England Seamounts

Seamounts, one of the largest topographic features of the ocean floor, are largely volcanic, yet their origin is obscure

Seamounts are one of the few prominent features of the deep-sea floor, but unlike the mid-ocean ridges, we know little about them. From what we now know, if the seawater was drained away, their peaks would look like isolated volcanic peaks on land. But although they are volcanic, their origin is different from most of the volcanic peaks on land, such as those ringing the Pacific Ocean.

The sea floor spreads and cools away from the axis of the mid-ocean ridges. The volcanic activity associated with this process seems to be confined to a very thin axial line and is due to crustal stretching and tension. The deep-Earth forces that cause the sea floor to spread also break it into a rugged profile near the mid-ocean ridge axis, but they have produced no observable folding. It is not characteristic of these forces to produce seamounts near the ridge axis.

Across the vast oceanic abyssal plains and away from the mid-ocean ridges thick layers of sediment cover the hard volcanic rock basement. Occasionally the volcanic rock protrudes above the sea floor to form seamounts and sometimes rises above the sea

J. R. Heirtzler and R. D. Ballard are on the scientific staff of the Woods Hole Oceanographic Institution and were both associated with Project FAMOUS—Heirtzler as the U.S. Chief Scientist and Ballard as one of its principal diving scientists. Ballard is just completing a submersible study of the Cayman Trough in the Caribbean. P. T. Taylor is a member of the scientific staff of the U.S. Naval Oceanographic Office and has made several cruises studying Gilliss Seamount. R. L. Houghton is a student in the joint M.I.T.– Woods Hole graduate degree program; his thesis work is on the petrology of the New England Seamounts. Address: Woods Hole Oceanographic Institution, Woods Hole, MA 02543.

surface to form islands. Seamounts may occur as isolated peaks or as chains that extend for thousands of miles. The chains show no continuity with land forms and are obviously a product of the forces that construct the sea floor.

Some Earth scientists have suggested that seamounts mark lines where the crust is thin or weak, enabling molten volcanic material from deep in the Earth to seep through. Others have noted that certain seamount chains increase in age along their length and thus may mark the trail of material that rose over a single "hot spot" in the mantle as the crust moved over it (Wilson 1963; Morgan 1971; Vogt 1971). It has also been proposed that the rising material provides the driving force for sea-floor spreading.

The only seamounts that have had detailed study are those forming island chains, especially the Hawaiian in the Pacific Ocean, the Comores in the Indian Ocean, and the Azores in the Atlantic. There have been no detailed studies of submerged seamount chains, and although there may be no essential geologic difference between submerged and island seamounts, they may have a different history of vertical elevation or subsidence and were subjected to different weathering, erosive, and depositional actions.

The New England Seamounts

In the Atlantic Ocean there is only one extensive linear chain of submerged seamounts—the New England Seamounts. They extend for about 1,600 km southeast from the

coast of New England, and a somewhat separate group of peaks called Corner Rise continues this chain for another 500 km (Fig. 1).

The first exploration of the New England Seamounts was undertaken by Ziegler (1955), who located many of the peaks (and named them after ships of the Woods Hole Oceanographic Institution), made simple bathymetric charts, and collected samples. Ziegler's report initially carried a security classification and was not widely circulated. Some of the more recent surveys have been made by Walezak (1963), Uchupi (1968), Uchupi et al. (1970), Taylor and Hekinian (1971), McGregor and Krause (1972), McGregor et al. (1973), Taylor et al. (1975), and Johnson and Lonsdale (1976). During these investigations bathymetric surveys, seismic studies, dredges, bottom photographs, and bottom current measurements were made. In addition, there were two unreported submersible dives on Bear Seamount by Milliman and Emery in 1968.

The more than 30 major peaks in the New England seamount chain are interspersed with numerous smaller unnamed peaks and hills which have not been looked at closely because the usual oceanographic research ships cannot resolve small features at such great depths. The sea floor varies in depth from about 2 km on the continental rise to about 5.5 km in the Sohm Abyssal Plain on the eastern end of the chain. Some of the peaks rise 4 km above the sea floor—twice the height of Mt. Washington in the White Mountains of New England and comparable to the major peaks in the Alps. None of the seamounts, however, comes within 1 km of the sea surface.

The chain from Bear Seamount on the west to Nashville Seamount on the east follows a smooth line that is not quite in the direction of sea-floor spreading for the area. Manning, Greg, and Rehoboth seamounts lie just to the north of this line and Mytilus just to the south. The depth contours of the individual peaks seem to be elongated along the direction of the chain. Corner Rise, an apparently separate feature to the east of Nashville, is composed of two short chains running northeast and southeast from a common point, in the shape of a ∨, and covers a large area nearly equal to all the New England states combined.

The geologic ages of these seamounts are not known with any degree of certainty. Ziegler recovered shallow-water limestones from his dredges on Bear and Mytilus seamounts and tentatively dated them as Eocene (45 million years ago). Presumably these peaks sank to their present depth since their formation; they could be older than the limestone but not younger. The sea floor immediately to

the south has been dated as Jurassic (about 150 m.y. ago) by Pitman and Talwani (1972) using magnetic anomalies. Near Nashville the magnetic anomalies give an age of Cretaceous (about 60 m.y. ago) while deep-sea drilling at the base of Nashville recovered rocks that radiometric dating showed to be 80 to 90 million years old. It is thus uncertain whether the seamounts are contemporaneous with the adjacent sea floor or whether the age increases along the chain.

The submersible program

In the summer of 1974, upon returning to Woods Hole from the Azores, the submersible *Alvin* had the opportunity to make brief dives on Corner Rise and the New England seamount chain. Because the operating-depth limit of *Alvin* was 3,300 m, only upper parts of peaks could be reached (Fig. 2). Nonetheless this was the first time man had directly viewed the expanse of the Earth between the Mid-Atlantic Ridge and the North American continent.

Single dives were made on seven seamounts: Corner Rise and Nashville, Gilliss, Rehoboth, Manning, Balanus, and Mytilus. The operations were hindered by charts that mislocated some of the seamounts as well as by the lack of bathymetric charts to give the configuration of the features. Since these were reconnaissance dives, the sites were selected for logistic reasons rather than for some particular scientific mission. Dives were made early in the day and when the weather was good (it was hurricane season). Before daybreak the seamount was located by the escort vessel *Knorr* and surveyed briefly to determine its depth and general configuration. A transponder—an acoustic navigation aid—was dropped at a place that had water depths within *Alvin*'s operating limits. The tender *Lulu* determined local surface currents and assumed a launch position that would allow *Alvin* to drift toward the transponder during descent. Three persons—a pilot and two of the authors of this article, on a rotating basis—made each dive. Those who were not diving

Figure 1. The New England Seamounts and Corner Rise are located off the northeast coast of North America and north of Bermuda. *Alvin* was able to dive to a depth of 3 km and to reach the tops of several of the seamounts.

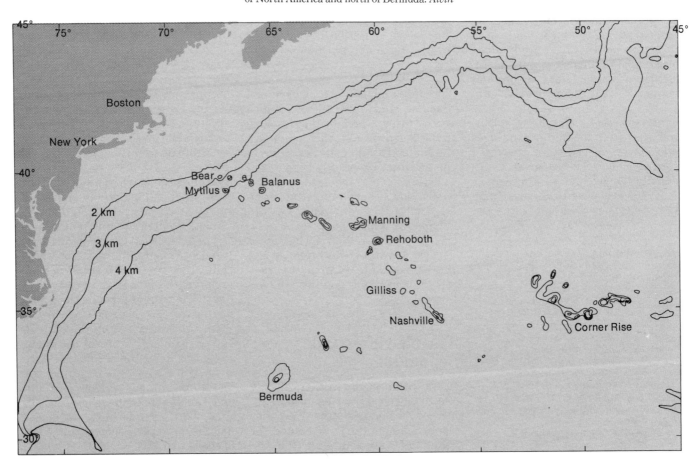

assisted in a brief regional survey and conducted nearby dredging operations from the *Knorr*.

The round trip to the bottom took about three hours, and an additional three to four hours were spent on the bottom covering a distance of one to a few hundred meters. We obtained numerous photographs and television video tapes and collected rock samples. During the ice ages, glaciers picked up rocks from North America and carried them to sea on icebergs, which eventually melted and dropped the rocks. Since the last ice age was only about 14,000 years ago, those continental rocks have not been covered by the slow rain of sediments to the sea floor, and their composition is quite different from the seamount rocks. Only rocks that were clearly broken from a native rock were collected.

Work on the bottom was terminated when the supply of electrical energy was exhausted or when darkness was approaching at the surface. After each day's work was finished *Knorr* would acoustically release the transponder and recover it while *Alvin* was surfacing and being secured on *Lulu*. The vessels would then proceed to the next dive location.

The dives provided many surprises.

Of course, until we could go down to examine them, we had never had any real knowledge about the smaller features on the bottom, and we were unprepared for what we found. The seamounts proved to be much more rugged and structurally complex than expected. We had had a close look at the bottom volcanic terrain near the Azores during Project FAMOUS, but many of the volcanic forms of the seamounts are different. Some interesting aspects of the bottom sediments and several erosional features had not been seen before. The diversity of biological specimens and the presence of an increasing number of glacially transported rocks on the western end of the seamount chain made each dive a unique experience. Only a few brief comments on each of the dives can be given here. A complete report of the diving investigations is available elsewhere (Heirtzler et al. 1977).

What we found

The first dive took place on Corner Rise on the western side of one of the easternmost peaks. The submersible landed at a depth of 2,035 m on native rocks heavily encrusted with manganese. Manganese accretes from the seawater onto bottom objects by a process that is not well understood. The rate of accretion is very slow and

not more than one to a few millimeters per million years. The heavy manganese encrustation indicates that the rocks are many millions of years old.

These rocks were quite blocky and fractured, and a sample, later examined in the laboratory, showed that they were composed of basaltic lava like the rocks of the mid-ocean ridges. Eighty-five meters above our landing point, a break in the upward slope exposed a surface of characteristic pillow-shaped lavas.

For the next 100 m upslope the bottom was mostly sediment covered with some relatively recent (probably Pleistocene) mollusk shells. There were a few scattered rocks, some extremely vesicular and scoriaceous. On previous reported occasions oceanographic ships had dredged rocks of this type, but they were thought to be cinders from the furnaces of old steamships. We now think they may be indigenous and deserve more serious study.

Coming up to a depth of 1,800 m the first major cliff was encountered. It was 260 m high and had several small benches covered with sediment that showed ripple marks caused by persistent ocean currents. This rugged cliff, which was characteristic of al-

Figure 2. This generalized elevation of the seamounts shows the depths at which the dives were made. The Gulf Stream usually sweeps the westernmost of these peaks with the current at least occasionally reaching completely down to the regional sea floor. Rocks transported by icebergs from the last ice age were especially evident on Balanus and Mytilus seamounts.

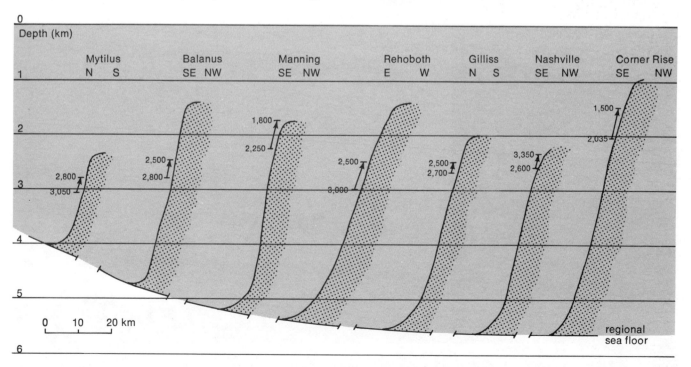

most all the seamounts visited on the cruise, was made up of steps that appeared to be small faults.

Near the base of this cliff *Alvin* passed quickly over a series of very striking topographic features that can best be described as small buttes. They are no doubt erosional in nature and, in fact, resemble eroded basaltic structures in central Iceland. Because they have dimensions of a few meters it was awkward to photograph them with the narrow field of view of our cameras; Figure 3 is a sketch of them.

Some well-worn lava flows and lava pillows were encrusted with manganese. Young lava forms are coated with a glassy surface that is later replaced with material called palagonite, and still later replaced by a man-ganese coating. The steep cliffs may be due to tectonic activity after formation, but no major earthquakes have been recorded on the seamounts recently. A detailed visual inspection of the rock face itself is made difficult by the pervasive manganese coating.

Nashville Seamount, the easternmost peak along the main part of the seamount chain, is divided into four prominent and equally spaced summits along the northwest-southeast axis of the chain. The dive took place on the southeasternmost and largest summit, beginning at a depth of 2,582 m and moving upslope a vertical distance of 260 m.

Massive rock outcrops with vertical faces, the main feature here, have a sharp, angular exterior, suggesting the faces are the result of faulting. Ravines a few meters in width and depth were too large to photograph but were worthy of inspection. On the interior of one of them, there were five or six separate lava flows, suggesting that the seamount was built in several discrete stages. Since there were several lava pillows elongated in the downslope direction, we know the rocks have not been tilted since their formation.

In this section of Nashville, sediments are nonexistent or exist only as a thin veneer, with some thickening behind ledges, suggesting downslope movement. There is much manganese in the form of a pavement, and there are broken coral fragments, sometimes mixed with the manganese. In places the pavement is tilted 20–30° and cut by gullylike features. The rocks of this

Figure 3. An artist's sketch of some unusual volcanic features found on Corner Rise. At no other location did we encounter such a series of small buttes. The 7-meter *Alvin* is illustrated for scale; its manipulator arm at the front places rocks in the "basket" beneath it.

area have a usual manganese coating of 5–6 cm and sometimes several times that.

The third dive, on Gilliss Seamount, some 190 km west of the Nashville dive site, began at a depth of 2,689 m and proceeded upward, covering a vertical distance of 200 m and a horizontal distance of 750 m. At the end *Alvin* was 500 m below the southern peak of this twin-peaked seamount.

The most commonly observed feature was the, by now, nearly ubiquitous lava outcrops covered with manganese and broken and whole pillows. In addition, there was a most unusual linear wall, or ridgelike feature, first observed at a depth of 2,600 m. Its height is 18 m, width 40 m, and length 475 m, and it follows the general northwest-southeast direction of the seamount chain. The orientation of ridges is as previously mapped by Taylor et al. (1975) by a special precise bathymetric mapping system. Taylor and his colleagues assumed that this was an erosional feature; however, lava pillows observed in the wall indicate it is an emplaced igneous feature, perhaps like the radial dikes sometimes found in certain types of volcanoes on land.

Several "tongues" of sediment, observed at 2,610 m, appeared to be sediment-debris flows partially covering older brown pelagic sediments. The height of these features was 10–20 cm, and they appeared to be oriented northeast-southwest. They generally resemble features that Heezen and Hollister (1971, their Fig. 9.18) called "megaripples" and seem to have moved downslope from the northwest. Ripple marks, leaning corals, and other organisms showed that the local water-current direction was from the southeast, against the apparent sediment-tongue movement.

Located nearly halfway along the New England seamount chain, Rehoboth Seamount comprises two peaks oriented northeast-southwest. The northeastern one rises from an abyssal depth of 5,000 m to about 1,600 m. Its neighbor, some 55 km to the southwest, has a peak at about 2,400 m depth. *Alvin* dived on the southern face of the northern peak, starting at 3,000 m (Fig. 4).

The sea floor in this vicinity has a 30°

Figure 4. (*top*) A map of the depth contours, in meters, of Rehoboth Seamount shows its two peaks. The small line shows the approximate dive track of *Alvin*. (*bottom*) The time-depth history of this dive shows the two igneous outcrops observed. One was at stations 1 and 2 and the other at stations 3 and 4. In this figure the flat portions of the curve show where the submersible briefly stopped climbing. The manipulator arm collected rock samples at the stations.

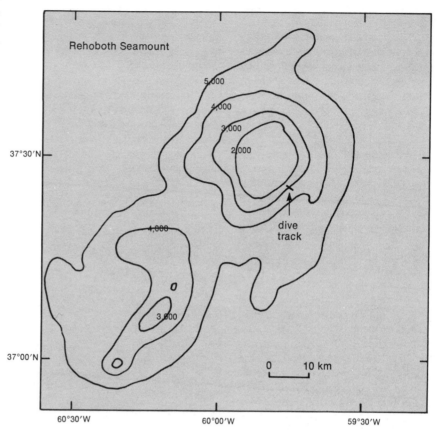

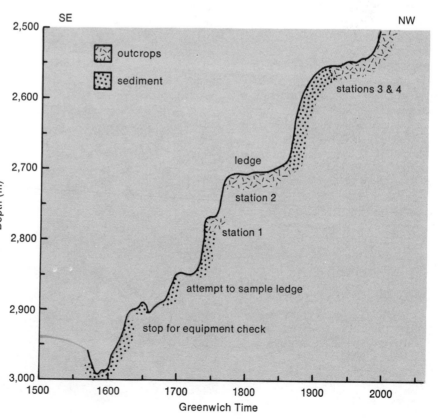

slope and is sediment covered. However, two very distinct vertical rock exposures were encountered, each with a face about 100 m high. Broken rock fragments from these two outcrops were scattered downslope, making it possible for the observers to anticipate an outcrop before they could see it. The sediment, apparently a calcareous ooze not unlike that seen at similar depths in the ocean, had numerous depressions one-half to one meter in length and 10–20 cm in depth and width (Fig. 5), which were most likely made by some fish or animal.

Many sponges, sea urchins, and fish were observed. No live corals were seen here, but in some areas manganese covers dead coral. It would seem unlikely that environmental conditions in the deep ocean have changed sufficiently over the recent geological ages to make large coral populations extinct, but that cannot be ruled out.

Manning Seamount has two peaks, and the dive, on the western one, began at a depth of 2,252 m and proceeded entirely up the face of a cliff with a 60–90° slope, which was covered with manganese. A mixture of foraminiferal ooze and manganese-coated coral was seen here and there on local ledges. There was considerable evidence of downslope movement of sediments, probably because of the steep angle of the slope. In one place a sediment slump 60 cm high and a few meters wide was observed. Ravines and gullies were carved into the sediments, the largest of which was 10 m long, 3 m wide, and a meter deep. The rim of one gully was raised 5 cm. In contrast to the other seamounts, there was no indication of bottom-current activity on Manning. Oceanographers know little about the flow of water around these seamounts. Our observations indicate that there may be many local current regimes and that these may vary with time.

No pillow lava forms were observed here—only massive outcrops cut by jointlike features. The small vertical gullies in the rock face apparently act as chutes for falling rock fragments and sediments which in some places have given a smooth polish to the inside surface of the chutes (Fig. 6).

The dive on Balanus Seamount

commenced at 2,800 m on the southwest peak and proceeded in a generally northerly direction. The slope was gentler than we had found on other seamounts and was entirely covered with sediment except for a single rock outcrop at the upper end of the traverse at 2,500 m depth. For the first time we encountered rocks of continental origin; they were granites and were conspicuously rounded from glacial action.

The sediment had many small depressions measuring about 100 cm long and 50 cm wide and deep. Unlike those on Rehoboth, these holes were

Figure 5. Between stations 2 and 3 on Rehoboth we observed these sediments with gouge marks one-half to one meter long and 10 to 20

bridged across the middle, creating a burrow with two exits, which crustaceans of some type occupied—their tentacles were frequently visible. Ripple marks in the sediment near the top of the traverse and the movement of standing coral indicated bottom water flowing to the west or southwest—the same as on the surface.

Small native rocks—that is, those that have resulted from volcanic action, as opposed to those moved by glacial action—became more numerous upslope until we reached the outcrop, which was vertical for about

cm deep and wide. They were probably made by some fish or animal.

Figure 6. This photograph of the nearly vertical face of Manning Seamount shows a gully (*center, top to bottom*) with sides polished smooth by rocks and sediments that have fallen through it.

10 m and contained no ledge on which *Alvin* could rest. The flat top was heavily covered with glacially transported rocks. *Alvin* spent three hours working on the face of this outcrop, moving to the southwest. Each time sampling was attempted the little sub would push itself away from the cliff; it was necessary, therefore, for the pilot to drive the submersible gently into the face while a second person operated the mechanical arm for sampling. Many of the native rocks, although igneous, were unusually crumbly and could not have been recovered by dredging. Loose native rocks were collected near the base and, upon subsequent inspection, were found to be parts of lava pillows. Several rocks were composed of more than one lava flow. A "cave" about 2 m into the outcrop's face housed an eel. This was the only place where anything resembling an overhang was found.

The last dive was on Mytilus Seamount, located on the continental rise 70 km south of the main axis of the New England seamount chain. It has probably been the most thoroughly studied and, in many ways, is unique in the chain. Continuous seismic profiles across it show that it has a 300 m thick cap of relatively acoustically transparent material (Uchupi 1968; Emery et al. 1970), which, because of its acoustic similarity to shallow-water material that had been previously dredged, Uchupi interpreted to be a coral reef. The contact between this cap and the underlying material is at a depth of approximately 3,000 m—close to *Alvin*'s operational depth limit.

The *Alvin* traverse began on the north side of the seamount at a depth of 3,057 m and moved upward in a southerly direction, terminating at a depth of 2,772 m. Rocks sampled at 3,009 m support the reef hypothesis; they contained coral fragments, a mollusk fragment, foraminifera, and certain elongated microfossils that could not be precisely identified.

At a second sampling station, the rocks contained prominent algae strands in a calcite matrix. The algae has been identified as Melobesia, a family that now grows on the outermost ridges of reef breccia platforms in less than 100 m of water (R. Johnson, pers. comm.). Its occurrence at the greater depth offers firm evidence

that this seamount has subsided by 3,000 m.

Twenty years ago Ziegler (1955) dredged carbonate rocks from both Bear and Mytilus seamounts that he believed came from shallow-water environments; he estimated them to be Upper Cretaceous to Eocene in age. Thus we might assume a subsidence of 3,000 m in 50 to 70 million years, or an average rate of subsidence of 40 to 60 m per million years. Menard (1964) has estimated about 30 m per million years as the rate of subsidence for seamounts of the western Pacific—a rate comparable to that of Mytilus. These rates are also not excessive for certain places on land, but no land volcano is known to have subsided 3,000 m.

It is difficult to offer a reasonable hypothesis for the subsidence of Bear and Mytilus. Perhaps it might be related to the fact that these two are closest to the continental margin, which has undergone a major subsidence due to its load of sediment. Perhaps because of its proximity to the continental margin, there may be some special tendency for the crust to be weaker and to sink.

The bottom topography we observed here ranged from near flat to vertical, and many rock faces were covered with manganese. There were many glacially transported rocks including granites. Some of the rocks observed on the face of a small ledge about 30 cm high were tinted yellow and reddish-yellow. Since we were unable to sample the material, we don't know the origin of the coloration. There were abundant corals and sponges—some types we had not seen before.

The dives of the New England seamount chain have suggested much about this relatively unexplored ocean territory. Although *Alvin*'s studies were limited, they support the usual hypothesis that the seamounts are volcanic in origin. They are covered by less sediment than expected, but have more manganese encrustations than previously thought. Those seamounts closest to the continental margin, however, have been affected by geological phenomena—such as subsidence and glacial transportation of rocks—that are not evident on the farther-out seamounts. Deep-sea exploration has just begun, and the peaks of the New England Seamounts

offer interesting locations for further study of the biological, oceanographic, and geologic phenomena of the deep sea.

References

Emery, K. O., E. Uchupi, J. D. Phillips, C. O. Bowin, E. T. Bunce, and S. T. Knott. 1970. Continental rise off eastern North America. *Bull. AAPG* 54:44–108.

Heezen, Bruce C., and Charles D. Hollister. 1971. *The Face of the Deep.* Oxford Univ. Press.

Heirtzler, J. R. 1975. Where the Earth turns inside out. *Natl. Geogr.* 147:586–603.

Heirtzler, J. R., P. T. Taylor, R. D. Ballard, and R. L. Houghton. 1977. The 1974 *Alvin* dives on Corner Rise and the New England seamounts. Tech. Rept. WHOI-77-8, Woods Hole Oceanographic Inst.

Johnson, David A., and Peter F. Lonsdale, 1976. Erosion and sedimentation around Mytilus Seamount, New England continental rise. *Deep Sea Res.* 23:429–40.

McGregor, B. A., and Dale C. Krause. 1972. Evolution of the sea floor in the Corner Seamount area. *J. Geophys. Res.* 77:2526–34.

McGregor, B. A., P. R. Betzer, and D. C. Krause. 1973. Sediments in the Atlantic Corner seamounts: Control by topography, paleowinds, and geochemically detected bottom currents. *Mar. Geol.* 14:179–90.

Menard, H. W. 1964. *Marine Geology of the Pacific*, p. 94. McGraw-Hill.

Morgan, W. Jason. 1971. Convective plumes in the lower mantle. *Nature* 230:42–43.

Pitman, Walter, III, and Manik Talwani. 1972. Sea-floor spreading in the North Atlantic. *Bull. Geol. Soc. Amer.* 83:619–46.

Taylor, P. T., and R. Hekinian. 1971. Geology of a newly discovered seamount in the New England seamount chain. *Earth Plan. Sci. Ltrs.* 11:73–82.

Taylor, P. T., Daniel Jean Stanley, Tom Simkin, and Walter Jahn. 1975. Gilliss Seamounts: Detailed bathymetry and modification by bottom currents. *Mar. Geol.* 19:139–57.

Tucholke, Brian, and scientific party. 1975. *Glomar Challenger* drills in the North Atlantic. *Geotimes* 20:18–21.

Uchupi, E. 1968. Long lost Mytilus. *Oceanus* 14:1–7.

Uchupi, E., J. D. Phillips, and K. E. Prada. 1970. Origin and structure of the New England chain. *Deep Sea Res.* 17:483–94.

Vogt, P. R. 1971. Asthenosphere motion recorded by the ocean floor south of Iceland. *Earth Plan. Sci. Ltrs.* 13:153–60.

Walczak, J. E. 1963. A marine magnetic survey of the New England seamount chain. Project M-9, Tech. Rept. TR-159, U.S. Naval Oceanographic Office, Washington, DC.

Wilson, J. T. 1963. A possible origin of the Hawaiian Islands. *Can. J. Phys.* 41:863–70.

Ziegler, John M. 1955. Seamounts near the eastern coast of North America. Tech. Rept. 55-17, Woods Hole Oceanographic Inst.

Woods Hole Oceanographic Institution, Contribution no. 3856, carried out under contract with the Office of Naval Research, Washington, DC.

Pyroclastic Flow Deposits

Grant Heiken

Hot particles mixed with gases flow from volcanic centers at terrific speeds, leaving thick deposits that become useful to man in many ways

Pyroclasts are fragments of rock that have been explosively ejected from volcanoes. Extensive plateaus consisting of thick pyroclastic deposits are scattered over the continents, evidence of short-lived catastrophic eruptions on a scale that has never been observed by man. Such eruptions would be awesome, burying hundreds to thousands of square kilometers under pyroclastic flows that move away from the vent at hurricane velocities. Eruption clouds may deposit fallout over millions of square kilometers. A safe distance from which to watch such an event might be the earth orbit of a space station.

Voluminous pyroclastic flow deposits are found throughout the geologic record, from Precambrian times until only a few thousand years ago. Some of the younger, more familiar deposits are located in and around Yellowstone National Park in the United States; in the high plateaus, or altiplano, of the Andes; and on the Anatolian Plateau of Turkey. These deposits have proved enigmatic to geologists; until recently, many were mistaken for lava flows or sedimentary rock units. Over the last twenty years, however, numerous field and laboratory studies have begun to explain the eruption phenomena that created them.

But the deposits have been of interest to man for longer than there have been scientists. Portions of these deposits have provided singularly useful building materials, and ancient dwellings were actually carved into pyroclastic deposits on several continents. Today, young deposits provide evidence for underlying geothermal reservoirs, which may prove to be valuable resources. Some rare elements with commercial and strategic value, such as lithium, are made recoverable through large-scale pyroclastic eruptions and subsequent weathering processes. Uranium and thorium, as well as lithium, are concentrated at the surface in this way. On the other end of the nuclear fuel cycle, older pyroclastic deposits have many properties that suggest their use for the storage of radioactive wastes.

Formation of the deposits

The process by which extensive pyroclastic deposits are produced starts when a large silicic magma body rises through continental crust toward the earth's surface. (Silicic magmas are those that contain 60–75 weight % SiO_2 and are rich in alkalis.) The bodies may rise either along subduction zones or within continental plates—through different sets of processes, though in both cases bodies of magma are believed to rise buoyantly in the same manner that salt diapirs rise through a sedimentary basin, owing to their relatively low density.

What we know of the nature of the magma bodies that produce these spectacular eruptions comes from three sources of data: the petrology of associated lavas and pyroclastic rocks; the geophysical characteristics of the volcanic field; and the visible features of old, deeply eroded intrusive-volcanic complexes such as the coastal batholith of Peru and the older igneous complexes of the Cascade Range of the United States.

A classic study of one such system, the San Juan volcanic field of Colorado, was made by T. A. Steven and his co-workers (1974, 1975). Volcanic activity in the San Juan field culminated between 35 and 26 million years ago, as large silicic magma bodies approached the earth's surface, erupting thousands of cubic kilometers of silicic pyroclastic rocks. The volcanic field is nearly coincident with a large negative gravity anomaly, which has been interpreted as indicating the presence of a shallow body of intrusive rock, or batholith (Plouff and Pakiser 1972). This batholith may be about 50×100 km (Lipman et al. 1978), comparable in size to the well-exposed Boulder batholith of Montana.

The San Juan field is marked by many large circular depressions known as calderas. The formation of these features is a complex subject worthy of a book (see Fig. 1). In general, smaller domical or diapiric intrusions, cupolas, rise toward the earth's surface from the main batholith. Why cupolas form is not understood; they may be related to regional structure, instabilities within the main body, or both. In any case, a cupola generates a concentric and radial fracture system over its apex. Then, as magma is erupted from the chamber, there is collapse along the ring fractures to form a depression,

Grant Heiken is a staff member in the Geosciences Division of the Los Alamos Scientific Laboratory. Previously, he worked at the NASA Johnson Space Center from 1969 to 1975, involved mainly with research on volcanism and evolution of regoliths on other planets. His main research activities are in the field of volcanology, as applied to geothermal resource evaluation, pyroclastic rocks, remote sensing of volcanic fields, and hazard analysis. This paper was written under the auspices of the U.S. Department of Energy. Address: MS 978, Los Alamos Scientific Laboratory, Los Alamos, NM 87545.

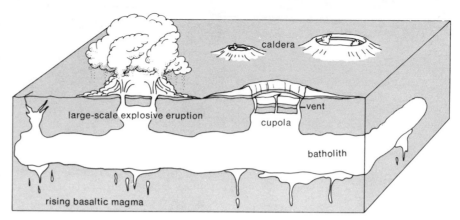

Figure 1. This drawing illustrates the relationship between a batholith, individual cupolas, and large-volume silicic volcanic eruptions. Eruptions generally occur along circular vents formed above the cupolas as they approach the earth's surface. During these eruptions, rapid emptying of the upper part of the magma chamber results in the collapse of the overlying rock units and the formation of a cylindrical caldera, which is usually partly filled with material from the eruption. It is believed that the batholith may be kept hot with fresh magma from underneath. This drawing is based mainly on data and interpretations of Lipman et al. (1978).

and much of the erupted pyroclastic debris falls back into the caldera itself. Circular vents above silicic magma bodies have been observed in older, well-exposed igneous complexes in Britain, Australia, and Peru. If the San Juan calderas are approximately the same width as the underlying cupolas, the cupolas are nearly circular bodies 10–40 km in diameter.

Prior to eruption, the shallow silicic magma bodies forming the cupolas develop gradients in volatile components such as H_2O and CO_2, with enrichment in the uppermost parts. Such gradients have been observed and calculated petrologically for many deposits—for example, the Bishop tuff in California. Eruption begins when the pressure of volatile phases within the magma has exceeded the pressure of the overlying rocks of the crust. The depth to which the body rises before this occurs depends on many factors but is believed to be between 2 and not more than 7 km (Lipman et al. 1966; Eaton et al. 1975). The volatile content of these bodies at this stage is believed to be between 3 and 5 weight % (Sommer 1977; Hildreth, in press).

Historical observations of eruptions much smaller in scale than those that produced the San Juan field show that prior to many eruptions seismic activity increases. This may be due to expansion of the magma as volatile phases come out of solution, leading to the hydraulic cracking of overlying rock units along fractures developed over the apex of the magma body. Magma moving along these features then reaches the surface and the eruption begins. Small eruptions of tephra—the collective term for all the pyroclasts ejected—herald the be-

ginning of activity leading to larger catastrophic eruptions. Most large-volume pyroclastic deposits have thin ash-fall deposits at the base of each set of beds that represents a single eruption sequence.

Once vents are widened by early activity, volatile-rich portions of the magma body are tapped, resulting in a highly explosive eruption of pumice pyroclasts full of vesicles, or bubbles. During this violent "Plinian" phase, enough volatiles are released to support a large column of tephra, which is then nearly all deposited as fallout from the air. Major eruption clouds are produced, depositing measurable tephra layers over large areas and injecting finer-grained tephra into the stratosphere, where it circulates over much of the earth's surface. These deposits, often dispersed over vast regions, are extremely useful as stratigraphic markers within sedimentary deposits on the continents and in the ocean basins. They have been used to correlate sedimentary sections and as markers within archaeological sites (see, e.g., Wilcox 1965).

Vesiculation—the formation of bubbles in magma (i.e. foaming)—is an important step in the eruption process. What little we know of the vesiculation of the magma is derived from study of the individual pyroclasts and from theoretical models. It should be noted here that eruption temperatures, calculated from mineralogical geothermometers, range from about 720°C to 900°C, and magma viscosities are in the range of 10^6 to 10^8 poises. The depth at which vesiculation occurs and the rate of vesicle growth have been modeled, but the models are qualitative at best. We do know that vesiculation gener-

ally occurs before eruption commences, for the vesicles in pyroclasts are drawn out into long, thin bubbles. This deformation must have taken place during movement of the highly viscous melt toward the surface.

Fragmentation of the vesiculated magma has also been the subject of some debate. Some investigators believe that a spray is erupted and individual globules continue to vesiculate to form glassy pyroclasts. Others favor the theory that rapid decompression of the melt leads to shock waves that pass down the vent, disintegrating the highly viscous, vesiculated mass into angular pumice fragments and shards—glassy fragments of thin bubble walls (Rittman 1962; Bennett 1974; McGetchin et al. 1976). The latter explanation best fits the angular, blocky nature of glassy pumice and shard pyroclasts characteristic of pyroclastic deposits. The deposits are composed mostly of silica-rich (~70–75 wt. %) glassy pyroclasts, minerals formed in the magma chamber, and pieces of the vent walls.

As the eruption proceeds, the vent is widened and eroded by the erupting magma, and lower, less volatile-rich portions of the magma chamber are tapped. A pyroclastic flow occurs when the volume of gas being released from the magma decreases to the point where it can no longer support the eruption column. The column collapses and flows radially downhill from the vent at extremely high velocities (Sparks et al. 1978). This turbulent mass of hot pyroclasts and gas follows existing canyons or valleys on the flanks of the volcanic field until enough gas has been lost from the suspension for the flow to stop. (For a thorough analysis of flow phe-

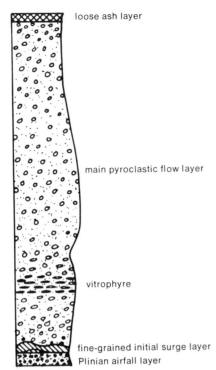

loose ash layer

main pyroclastic flow layer

vitrophyre

fine-grained initial surge layer
Plinian airfall layer

Figure 2. Texture and mineralogy vary within a thick silicic pyroclastic deposit. A typical sequence begins with a nonwelded, pumice-rich airfall deposit. A pyroclastic flow then makes up the bulk of the deposit, usually as a single massive layer, but sometimes with oblique layers at the base. The degree of welding within the flow increases to a maximum at the vitrophyre, a dense, glassy zone of low porosity. Imprinted on the welding are zones of alteration caused by devitrification and alteration of glassy pyroclasts by hot gases. A thin pyroclastic airfall deposit is present at the top. This diagram is abstracted mainly from the work of Smith (1960).

nomena, see Smith 1960, or the soon-to-be-published Geological Society of America Memoir on pyroclastic flows.) The flows fill valleys and result in a leveling of the region to form the large plateaus which are a geomorphic characteristic of pyroclastic flow deposits.

Deposits range in thickness from a few tens of meters to hundreds of meters, thinning from source to terminus. Individual pyroclastic flows may have traveled <10 km up to 120 km and cover thousands of square km. Total volumes of tephra from these short-lived catastrophic eruptions range from tens of cubic km to over a thousand. Eruption sequences must occur over a short period of time, because those consisting of several flow units—each representing an individual explosive event lasting

minutes? hours? days?—rarely show evidence of any erosion or weathering between them. Many flows, erupted at closely spaced intervals, cool together and form a single "cooling unit" (Fig. 2).

Owing to loss of heat by adiabatic cooling as the magma vesiculates and expands, by radiation, and by mixing of the flow with air, emplacement temperatures of pyroclastic flow deposits, it is believed, range from 500°C to 700°C—although temperatures have not been measured in large deposits immediately after the eruptive activity. Within a cooling unit, be it composed of one or several pyroclastic flows, a number of processes modify the glassy pyroclasts *after* emplacement (for an extensive review, see Smith 1960 or Ross and Smith 1961). In the hot, well-insulated interior of the flow, individual glassy pyroclasts are welded together and flattened by weight of the overlying deposit to form a dense rock. Larger, blocky pumice pyroclasts are flattened into lens-shaped disks, giving the characteristic texture of these deposits. Welding of pyroclasts and compaction of the deposit are generally restricted to the interior; the top and bottom are quite often composed of loosely bonded, uncompacted pyroclasts that cooled quickly.

After deposition, volatiles continue to escape from the cooling unit and may interact with the upper portions of the deposit, altering their original character. Within these "vapor-phase" zones, most of the original pyroclast textures are present only as relicts within a friable, crumbly deposit. Postdepositional processes produce many of the physical and chemical characteristics of large silicic pyroclastic deposits which make them so useful to man.

Geothermal energy

The "highest-grade" geothermal resources now being developed are located over large silicic magma bodies. Geothermal areas of New Zealand are located in the Taupo volcanic zone, where over 16,000 km³ of mostly silicic pyroclastic rocks were erupted during the late Pliocene and Quaternary time (Ellis and Mahon 1977). A large silicic magma body (or bodies) still underlies this zone (Healy 1962), which is characterized by large vol-

canic depressions or calderas, boiling springs, geysers, and steaming ground. Steam wells drilled into these geothermal areas are used in the generation of electricity for the North Island of New Zealand.

One of the best-known geothermal areas of the United States, the Yellowstone Plateau, is characterized by voluminous young silicic pyroclastic flows, calderas, extremely high heat flow, and active surface manifestations in the form of hot springs and geysers. Eaton and his colleagues (1975) have concluded that a large silicic magma chamber exists a few kilometers below Yellowstone Park. During the last 2 million years—and as recently as 600,000 years ago—the roof over this body has broken several times, resulting in the eruption of large volumes of silicic volcanic ash and in the formation of calderas. Smith and Shaw (1975) estimate that this large magma body (or bodies) has a volume of 45,000 km³, and although it may have cooled significantly during the last million years, the thermal energy still remaining is about $16,000 \times 10^{18}$ calories (6.7×10^{22} J).

The large volume of a silicic magma body and its close approach to the earth's surface makes it an ideal thermal source. Prospecting for these high-grade geothermal resources is fairly simple. A young volcanic field, say less than 2 million years old, with a large caldera surrounded by plateaus consisting of silicic pyroclastic flow deposits, is an excellent target for detailed exploration. Smith and Shaw (1975), in an evaluation of the geothermal potential of the United States, conclude that large (10^2–10^5 km³) magma chambers located in the upper 10 km of continental crust are the most attractive targets for geothermal development.

For heat loss by conduction alone, Smith and Shaw demonstrate that a magma chamber with a horizontal slablike geometry, 5 km thick, takes 2 million years to cool from an initial temperature of about 850°C to ambient temperatures. A larger system, such as the one hypothesized beneath Yellowstone, may be preserved for 10 million years. Moreover, it is believed (see, e.g., Lachenbruch et al. 1976; Eichelberger 1978) that many large silicic magma bodies are kept hot by continued intrusion of hotter basaltic magmas into the base of the bodies.

Ground water interacts with shallow magma bodies, especially when the ground-water system is confined within a volcanic depression. Numerical models for cooling by conduction and water convection (e.g. that of Norton and Knight 1977) indicate that the cooling rate for a large magma body is not greatly shortened if the body has a permeability of less than 1 microdarcy. Ground water does, however, effectively transfer heat from the magma body to the surface, where it is visible as hot springs or geysers.

At present, the Union Oil Company is developing a natural hydrothermal steam field for the generation of electricity within the 22 km-diameter Valles caldera of the Jemez volcanic field, in New Mexico. Taking another approach, the Los Alamos Scientific Laboratory is experimenting with the use of heat conducted from the cooling silicic magma body located under the Valles caldera into impermeable Precambrian gneisses and granitic rocks that underlie the volcanic field. They are using a technique in which water is circulated through a man-made fracture in hot basement rock units to generate hot water and steam.

Shelter

Man has made use of many extensive young (late Tertiary through Holocene) pyroclastic flow deposits for shelter. The appeal of easily quarried, light-weight, strong materials of low thermal conductivity has been universal.

Pyroclastic flow deposits vary in degrees of welding—annealing or bonding—of glassy pyroclasts, of deformation of hot, glassy pyroclasts under the weight of the overlying deposit, and of alteration by devitrification and interaction with gases escaping from the cooling deposit during the vapor phase activity. These variations make some portions more useful than others.

The less-compacted, nonwelded or partly welded zones near the base of a pyroclastic flow deposit are ideal in their strength and ease of carving or quarrying. This is due to devitrification and growth of minerals by interaction with a vapor phase in and between glass pyroclasts (Fig. 3). The glass is replaced by parallel or radial

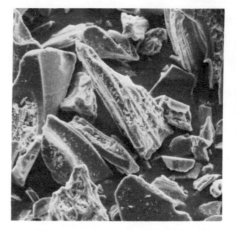
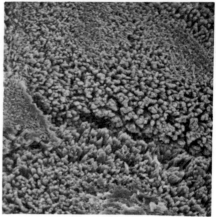
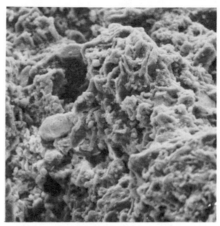
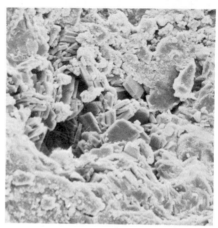
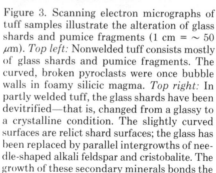

Figure 3. Scanning electron micrographs of tuff samples illustrate the alteration of glass shards and pumice fragments (1 cm = ~ 50 μm). *Top left:* Nonwelded tuff consists mostly of glass shards and pumice fragments. The curved, broken pyroclasts were once bubble walls in foamy silicic magma. *Top right:* In partly welded tuff, the glass shards have been devitrified—that is, changed from a glassy to a crystalline condition. The slightly curved surfaces are relict shard surfaces; the glass has been replaced by parallel intergrowths of needle-shaped alkali feldspar and cristobalite. The growth of these secondary minerals bonds the pyroclasts, resulting in a cohesive but light-weight, easily carved rock. The mineral tridymite is present in spaces between pyroclasts. *Bottom left:* Partly welded tuff has been altered during vapor-phase activity. Visible are intergrowths of alkali feldspar, cristobalite, and tridymite. There is little relict pyroclast texture left in this sample; grain boundaries have been destroyed by vapor-phase alteration. *Bottom right:* In this sample of welded tuff, the glassy pyroclasts have been replaced and the interstices partly filled by the zeolite mineral clinoptilolite. The tuff was altered mainly by ground water flowing through the deposit.

growths of fibrous crystalline silica (cristobalite) and alkali feldspar, usually only a few μm wide. Void space between pyroclasts is partly or completely filled with growths of somewhat coarser-grained (10–50 μm) silica (tridymite) and alkali feldspar. It is the intergrowth of minerals which bonds the altered pyroclasts together yet allows the rock to be easily cut, carved, or scraped with even primitive tools.

Owing to the lack of welding, the rock—called tuff—is still porous (50%–60%) and is an excellent insulator. The thermal conductivity of

tuffs from these partly welded zones ranges from 0.2 to 0.4 W/m°K at 25°C (W. L. Sibbitt, pers. comm.). In contrast, granite has a conductivity of about 3.5 W/m°K. Within this zone the tuff is light (1–1.7 g/cm^3) and of reasonable strength to be used for construction of dwellings (its crushing strength is 70–500 kg/cm^2) (Zalessky 1961). Devitrified tuffs are resistant to weathering, particularly after slight weathering leaves a thin (a few tens of μm) patina composed of microcrystalline carbonate and sulfates. This patina forms rapidly and is visible on joint surfaces, canyon walls, and on carved tuff block surfaces. In a study of longevity of tuffs as construction materials, Atzagortzian and Martirosian (1962) determined that there are at least 400 architectural monuments in Armenia that are more than 1,000–1,500 years old.

Other portions of thick pyroclastic deposits are less useful. The unbonded, loose pumice of Plinian deposits can serve as lightweight filler for concrete, but not as carved shelters or blocks. The more densely

Figure 4. Cone-shaped "tent rocks" are a landform characteristic of eroded nonwelded or partly welded pyroclastic flows. In Cappadocia, a region in central Turkey, villages have been carved into this soft rock. (Photo courtesy of the Turkish Tourism and Information Office.)

welded portions are difficult to quarry; they are brittle, have a high fracture density, are less frost resistant, and are poor insulators, being dense and less porous (10%–20%).

The central Anatolian Massif in Turkey has been the site of widespread, voluminous eruptions from early Pliocene through Quaternary times. These plateaus consist of extensive tuff deposits, with pyroclastic flow layers up to 400 m thick in places (Milanovsky and Koronovsky 1963; Pasquare 1968). These plateaus are characterized by distinctive coneshaped "tent rocks," formed by differential erosion. Within these layers homes, churches, and whole villages have been excavated (Figs. 4 and 5). The strength of this rock is demonstrated by the fact that churches carved during the fourth century A.D. have survived many earthquakes and remain today.

Cone-shaped tent rocks, favored for ease of carving and structural strength by the troglodytic communities of the Anatolian Plateau, are located in nonwelded or poorly welded portions of pyroclastic flows and in some pumice-fall deposits below them. Within a cooling pyroclastic flow, hot gases trapped between pyroclasts and released from vesicular pyroclasts may migrate toward the surface of the deposit along cracks or thin, tubelike conduits. The more intense alteration and growth of vapor-phase minerals within a thin zone adjacent to these conduits produces an erosion-resistant column within the more easily eroded, poorly welded portion of the pyroclastic flow. Erosion of deposits containing these conduits or "fossil fumaroles" produces these coneshaped landforms.

The earliest occupants of the plateau are not known, but a few Roman tombs remain. At Goréme, seventh-century monks excavated cells and churches into the pyroclastic flows. During the eighth and ninth centuries, Byzantine churches, monasteries, chapels, and hermitages were carved into cliffs and tent rocks. At the onset of Arab invasions the troglodytic villages were expanded into defensible underground cities, consisting of as many as 10 levels, connected by tunnels.

After the population of Anatolia by Seljuk Turks in the twelfth century, the troglodytic settlements were occupied by Turkish farmers. Homes excavated in tuff were extended by masonry houses attached to the cliff face—made with bricks carved from tuff! Utilization of tuff continues: in 1966 the Kaya Hotel at Uchisar was excavated in a pyroclastic deposit. Most modern carved spaces are, however, used for storage rooms and dovecotes (Ozkan and Onur 1975).

Carved dwellings similar to those of the Anatolian Plateau exist in the lesser Caucasus of Georgia and Armenia, where silicic pyroclastic deposits cover thousands of square kilometers and individual pyroclastic flow units, generally 10 to 20 m thick, may reach thicknesses of 50 to 100 m (Milanovsky and Koranovsky 1963). During the twelfth century, the city of Vardzia in Soviet Georgia was quarried out of the lower, poorly welded portion of a thick pyroclastic flow unit. Vardzia has 500 rooms and apartments, a chapel, banqueting halls, and stables, connected by a labyrinth of stairs and tunnels (Lang 1966), all cut into a vertical cliff face.

The use of silicic tuff as a building stone has also continued in the USSR. Fitted blocks cut from porous, poorly welded tuffs are used in masonry construction, and unwelded Plinian-phase pumice deposits are used as concrete filler (Zalessky 1961). These building materials are characterized by resistance to weathering and low thermal conductivity.

In Mexico and South America, silicic pyroclastic deposits of late Tertiary to Holocene age form plateaus within the Cordillera that are parallel to the Pacific Coast and cross Mexico from west to east. In Latin America little use has been made of tuff deposits for carved shelters, but there was, and continues to be, extensive use of quarried tuff blocks for construction. Arequipa, commercial center of southern Peru—known as the "white city"—is located at the base of Volcan El Misti on a plateau consisting of thick pyroclastic flow units of Pleistocene age (Fenner 1948). Arequipa's buildings are constructed mostly of blocks cut from pyroclastic flows, quarried from nonwelded tuff units. Building blocks of this tuff are easily cut and trimmed and are characterized by their low density, high porosity, and resistance to weathering. Young silicic pyroclastic flows that may be sources for building materials are found along the western Andes intermittently for about 3,200 km, from northern Peru to central Chile (Jenks and Goldich 1956).

Along with limestone and basaltic cinders, one of the most common building materials in Mexico is nonwelded to slightly welded tuff from pyroclastic flow deposits located throughout central Mexico and along the Sierra Madre Occidental. Moderately welded tuff is used not only as building stone but also, since it is

easily carved, in columns, elaborate carved facings, and statues. As an architectural and artistic material, silicic tuffs have been used since pre-Columbian times, most extensively in the construction and decoration of Spanish colonial cities. Today it is still used in architecture, for paving stones, and for sculpture.

The Pajarito Plateau of New Mexico consists of massive pyroclastic-flow units erupted during the last 2 million

years. This plateau has been dissected by drainage systems radial to the central part of the volcanic field. The region was populated by native Americans during the thirteenth century and abandoned during the mid to late sixteenth century (Steen 1977). Tuff units characteristic of the plateau were used in two different ways for shelter. On mesa tops and in the broader canyons, blocks of tuff from talus slopes were shaped with basalt scrapers and picks and used for masonry construction, bonded with adobe mortar. Natural caves and depressions weathered into cliff faces in the lower part of pyroclastic flow

Figure 5. Though the rock from which the villages in Cappadocia are carved is soft, it is extremely durable. Dwellings and churches dating from the fourth century have survived weathering and earthquakes and remain today. (Photo courtesy of the Turkish Tourism and Information Office.)

units were expanded into rooms used for storage and ceremonial purposes (see cover). Masonry structures were usually built against cliff faces in front of the carved shelters, the stone mainly consisting of tuff blocks, as in Turkey.

Modern uses of the tuff in the region are varied. Buildings at park headquarters for the Bandelier National Monument are constructed of cut tuff blocks, built during the labor-intensive years of the W.P.A. The loose pumice deposits of the Plinian phases which preceded ash-flow eruptions are now used as aggregate for lightweight cement cinder blocks. If the labor costs within this region ever decrease, quarrying and use of lightweight, strong, weather-resistant tuff blocks as a major building material should become attractive.

Lithium resources

An increasing demand for the element lithium will evolve from its use in thermonuclear plants and in lithium batteries, in addition to its present use in glass and ceramic industries. Currently, the two main sources of lithium are lithium-rich pegmatites and brines and clays from evaporative basins called playas. Basins containing lithium-rich brines usually are surrounded or partly surrounded by volcanic rocks (Vine 1975).

Through careful stratigraphic and petrologic analysis of the Bishop tuff, a large-volume silicic pyroclastic deposit that erupted from the Long Valley caldera, in California, Hildreth (in press) has determined that trace elements, including lithium, were concentrated in the silica-rich, volatile-rich upper portion of the large, shallow magma body prior to eruption. The first, volatile-rich phase erupted as a Plinian airfall and produced glassy pumice pyroclasts which contain anomalously high amounts of uranium, fluorine, and lithium. For the Bishop tuff, the earliest erupted samples contain 37 ppm lithium.

These pyroclasts, which cooled by radiation during fallout from eruption clouds, form a highly permeable deposit at the base of the pyroclastic sequence. Ground water, passing through similar Plinian deposits, could easily leach lithium from the glass and carry it, in solution, into

aquifers and surface streams. Water flowing through an unwelded Plinian deposit comes into contact with enormous surface areas of glassy pyroclasts—between 500 and 700 cm^2 of surface area per cm^3 of unconsolidated tephra—and can easily hydrate and leach the deposit. There may also be leaching and concentration by hot springs. If these are part of an open drainage system the lithium will be lost. If, however, the waters are trapped in a closed arid basin where evaporation exceeds rainfall, the lithium may be concentrated as a brine (Vine 1975; Erickson et al. 1976).

The major known sources for lithium are some of the closed saline basins of the western United States and the central Andes of Bolivia, Chile, and Argentina. Brines at Clayton Playa, in Nevada, a playa bounded on several sides by voluminous silicic pyroclastic rocks, contain 300 ppm lithium (Kunasz, Ph.D. thesis). Within the Andes are 75 playas, which range in size from 1 km^2 to 9,000 km^2. Most of the mountain regions draining into these playas contain thick silicic pyroclastic deposits, of Pliocene and younger age (Erickson et al. 1976). Preliminary sampling of brines from one of these playas (Salar de Atacama) indicates that they contain between 2,000 and 4,000 mg/l, making this one of the world's largest lithium deposits.

Erickson et al. (1976) and Vine (1975) indicate that favorable conditions for the concentration of lithium brines are (1) large, old, closed sedimentary basins; (2) minimal loss of surface or ground water from the basin; and (3) an abundance of silicic pyroclastic deposits subject to leaching and fumarolic activity surrounding the basins. Similar lines of reasoning can be followed for the leaching of uranium from silicic volcanic glass with subsequent deposition in a sedimentary basin—a subject worthy of another review paper and an entire research program.

Storage of radioactive waste

Thick silicic pyroclastic deposits may offer one solution to the contemporary problem of radioactive-waste storage. Of interest are not the very young deposits described so far, but older (early to mid-Tertiary age) tuff

units. Use of these formations is being examined by J. R. Smyth and his co-workers for the U.S. Department of Energy (Smyth et al., in press). Present targets for exploration are the voluminous tuff deposits of the Timber Mountain–Oasis Valley volcanic field in southern Nevada. The 9.5- to 16-million-year-old tuff deposits cover 11,000 km^2 and have a total volume of over 5,000 km^3 (Byers et al. 1976). Some of the basins contain tuff deposits that may be over 1,200 m thick.

These thick pyroclastic deposits originally consisted primarily of glassy silica-rich pyroclasts or such pyroclasts that had been transformed, in many cases to mineral phases, by vapor-phase activity soon after deposition. After deposition and burial, the interaction of glass and aluminosilicate minerals with ground water produced a variety of minerals, including zeolites. Within both closed and open basins similar to those in southern Nevada where the thick tuff sections are buried, the tuff may be wholly altered in 10^3 to 10^4 years (Hay 1977). In zeolite-rich zones, the pyroclasts may be replaced by poorly developed zeolite minerals and void spaces partially or completely filled with well-developed coarser phases of the same minerals (see Fig. 3).

How does all of this relate to the storage of radioactive waste? Zeolite minerals have structures consisting of open frameworks with sites for large-radius cations; natural zeolites are now used commercially as molecular sieves and have been used in the United States, the USSR, and Italy to filter radionuclides from contaminated effluents (Mumpton 1978). A zeolite-rich, thick tuff deposit would thus act as a natural barrier against liquid loss from a nuclear-waste repository. There are other good reasons for using thick deposits of zeolitized tuff located in the western United States for radioactive-waste storage: many of these basins are located far from centers of population; the climate is arid, with low rainfall and low rate of recharge of aquifers; and many of the basins with buried tuff deposits are closed, with no outlets for surface or ground water. Research on the physical and mineralogical properties of tuff units as hosts for radioactive waste is currently being supported by the U.S. Department of Energy.

Our understanding of large-volume silicic pyroclastic flow deposits has grown rapidly during the last 25 years, but there are many problems yet to be solved concerning their genesis and eruption phenomena. No such large-scale eruption has been observed, and observation would tell us a great deal. The eruption frequency is low enough that we have no way of predicting the next one; it will most likely occur sometime during the next several tens of thousands of years.

When such an eruption occurs, almost everyone on the planet will be aware of it. But since the skies are not yet darkened by volcanic ash, we must be satisfied with our interpretations of eruption phenomena based on old pyroclastic deposits. As if to counteract the gloom and catastrophe one of these eruptions would bring, the benefits of this type of volcanic activity are many, ranging from thermal sources for geothermal energy to a natural supply of molecular sieves.

References

Atzagortzian, Z. A., and O. S. Martirosian. 1962. *Tuffs and Marbles of Armenia* (in Russian). Erevan: Armenian Inst. of Materials.

Bennett, F. D. 1974. On volcanic ash formation. *Am. J. Sci.* 274:648–61.

Byers, F. M., W. J. Carr, P. O. Orkild, W. D. Quinlivan, and K. A. Sargent. 1976. *Volcanic Suites and Related Cauldrons of the Timber Mountain–Oasis Valley Caldera Complex, Southern Nevada.* USGS Professional Paper 919.

Eaton, G. P., R. L. Christiansen, H. M. Iyer, A. M. Pitt, D. R. Mabey, H. R. Blank, Jr., I. Zietz, and M. E. Gettings. 1975. Magma beneath Yellowstone Park. *Science* 188:787–96.

Eichelberger, J. C. 1978. Andesitic volcanism and crustal evolution. *Nature* 275:21–27.

Ellis, A. J., and W. A. J. Mahon. 1977. *Chemistry and Geothermal Systems.* Academic Press.

Erickson, G. E., D. G. Chang, and V. G. Tomás. 1976. Lithium resources of salars in the central Andes. In *Lithium Resources and Requirements by the Year 2000,* ed. J. O. Vine, pp. 66–74. USGS Professional Paper 1005.

Fenner, C. N. 1948. Incandescent tuff flows in southern Peru. *Geol. Soc. Am. Bull.* 59:879–93.

Hay, R. L. 1977. Geology of zeolites in sedimentary rocks. In *Mineralogy and Geology of Natural Zeolites,* ed. F. A. Mumpton, pp. 53–64. Mineral. Soc. Am. Short Courses and Notes, 4.

Healy, J. 1962. Structure and volcanism in the Taupo Zone, New Zealand. In *Crust of the Pacific Basin,* pp. 151–57. Am. Geophys. Union Monogr. no. 6.

Hildreth, W. In press. The Bishop Tuff: Evidence for the origin of compositional zonation in silicic magma chambers. *Geol. Soc. Mem.*

James, D. E. 1971. Plate tectonic model for the evolution of the central Andes. *Geol. Soc. Am. Bull.* 82:3325–46.

Jenks, W. F., and S. S. Goldich. 1956. Rhyolitic tuff flows in southern Peru. *J. Geol.* 64:156–72.

Kunasz, I. A. Ph.D. thesis. *Geology and Geochemistry of the Lithium Deposit in Clayton Valley, Esmeralda Co, Nevada.* Penn. State Univ., 1970.

Lachenbruch, A. H., J. H. Sass, R. J. Munroe, and T. H. Moses, Jr. 1976. Geothermal setting and simple heat conduction models for Long Valley Caldera. *J. Geophys. Res.* 81:769–84.

Lang, D. M. 1966. *The Georgians.* Praeger.

Lipman, P. W., and R. L. Christiansen. 1964. *Zonal Features of an Ash-Flow Sheet in the Piapi Canyon Formation, Southern Nevada.* USGS Professional Paper 501-B.

Lipman, P. W., R. L. Christiansen, and J. T. O'Connor. 1966. *A Compositionally Zoned Ash-Flow Sheet in Southern Nevada.* USGS Professional Paper 524-F.

Lipman, P. W., B. R. Doe, C. E. Hedge, and T. A. Steven. 1978. Petrologic evolution of the San Juan volcanic field, southwestern Colorado: Pb and Sr isotope evidence. *Geol. Soc. Am. Bull.* 89:59–82.

McGetchin, T., M. T. Sanford III, and E. M. Jones. 1976. Eruption mechanism of large silicic calderas (abstracts). *Geol. Soc. Am. Abstracts with Programs* 8:610–11.

Milanovsky, E. E., and N. V. Koronovsky. 1963. Ignimbrite-tuff lava formations in the alpine belt of southwestern Eurasia. *Trudy Paleovolc. Lab. Alma-Alta Univ.* (Kazakhstan), Issue 2, pp. 38–53.

Mumpton, F. A. 1978. Natural zeolites: A new industrial mineral commodity. In *Natural Zeolites,* ed. F. A. Mumpton, pp. 3–30. Oxford: Pergamon.

Norton, D. L., and J. E. Knight. 1977. Transport phenomena in hydrothermal systems: Cooling plutons. *Am. J. Sci.* 277:937–81.

Ozkan, S., and S. Onur. 1975. Another thick wall pattern: Cappadocia. In *Shelter, Sign, and Symbol,* ed. P. Oliver, pp. 95–106. London: Barrie and Jenkins.

Pasquare, G. 1968. Geology of the Cenozoic volcanic area of central Anatolia. Acad. Naz. dei Lincei. Roma. Class di Sci. fis., mat. e. Nat., Atti Memoire S.9, no. 3, pp. 53–204.

Plouff, D., and L. C. Pakiser. 1972. *Gravity Study of the San Juan Mountains, Colorado.* USGS Professional Paper 800-B, B183-B190.

Rittman, A. 1962. *Volcanoes and Their Activity.* Wiley.

Ross, C. S., and R. L. Smith. 1961. *Ash-Flow Tuffs: Their Origin, Geologic Relations and Identification.* USGS Professional Paper 366.

Smith, R. L. 1960. Ash Flows. *Geol. Soc. Am. Bull.* 71:796–842.

Smith, R. L., and H. R. Shaw. 1975. Igneous-related geothermal systems. In *Assessment of Geothermal Resources of the United States,* ed. D. F. White and D. L. Williams, pp. 58–83. USGS Circ. 726.

Smyth, J. R., B. M. Crowe, and P. M. Halleck. In press. An evaluation of the storage of radioactive waste within silicic pyroclastic rocks. *Environmental Geol.*

Sommer, M. A. 1977. Volatiles H_2O, CO_2, and CO in silicate melt inclusions in quartz phenocrysts from the rhyolitic Bandelier air-fall and ash-flow tuff, New Mexico. *J. Geol.* 85:423–32.

Sparks, R. S. J., L. Wilson, and G. Hulme. 1978. Theoretical modeling of the generation, movement, and emplacement of pyroclastic flows by column collapse. *J. Geophys. Res.* 83:1727–39.

Steen, C. L. 1977. *Pajarito Plateau: Archeological Survey and Excavations.* Los Alamos Sci. Lab. Report LASL-77-4.

Steven, T. A. 1975. Middle Tertiary volcanic field in the southern Rocky Mountains. In *Cenozoic History of the Southern Rocky Mountains,* ed. B. F. Curtis, pp. 75–94. Geol. Soc. Am. Mem. 144.

Steven, T. A., P. W. Lipman, W. J. Hail, Jr., F. Barker, and R. G. Luedke. 1974. *Geologic Map of the Durango Quadrangle, Southwestern Colorado.* USGS Map I-764.

Vine, J. D. 1975. Lithium in sediments and brines: How, why and where to search. *J. Res. USGS* 3:479–85.

Wilcox, R. E. 1965. Volcanic-ash chronology. In *The Quaternary of the United States,* ed. H. E. Wright, Jr., and D. G. Frey, pp. 807–16. Princeton Univ. Press.

Zalessky, B. V. 1961. Use of volcanic tuffs and tuff lavas as building material. *Trudy Volc. Lab. Acad. Sci.,* USSR, Issue 20, pp. 220–22.

Donald Hunter

The Bushveld Complex and Its Remarkable Rocks

The Bushveld Complex in South Africa, which contains some of the world's most precious resources, is itself a precious resource for geologists studying the crystallization of magmas

Selection of the seven geological wonders of the world is not a pursuit most geologists favor. If they did, foremost among their choices would almost surely be the Bushveld Complex, an extraordinarily large pile of layered igneous rocks in the Transvaal Province of the Republic of South Africa. Layered igneous rocks, resembling giant layer cakes, are known in many parts of the world. In each case it is apparent that the layering is somehow produced as a result of the way in which a parent magma cooled and crystallized to form the igneous rocks we see. What makes the Bushveld Complex unique is its extraordinary size—66,000 km^2 (see Table 1)—the remarkable perfection of its layering, and the chemical complexity of its rock layers.

Why the complex exists at all is an unanswered puzzle. How the layering came about and made the complex one of the world's great treasure

Donald Hunter is Professor and Head of the Department of Geology and Mineralogy and Dean of the Faculty of Science at the University of Natal. He is also president-elect of the Geological Society of South Africa and program director of the geology program of the South African Scientific Committee for Antarctic Research. After graduating from the University of London, he served with the Colonial (subsequently Overseas) Geological Surveys in Africa until 1970, when he relinquished his post as director of the Geological Survey and commissioner of mines in Swaziland. A period of four years with the Economic Geology Research Unit at the University of the Witwatersrand in Johannesburg preceded his appointment to his present position. His research interests have been primarily concerned with the evolution and geochemistry of Archean granitoids in southern Africa and with the structural setting of the Bushveld Complex. Address: Dept. of Geology, University of Natal, P.O. Box 375, Pietermaritzburg 3200, South Africa.

houses of unusual mineral resources are equally difficult and still unanswered questions. Yet the answers are clearly worth seeking, because the Bushveld Complex may well be one of the keys to the youthful years of the earth and the beginning of continents.

Igneous rocks of widely differing composition in the Transvaal were first reported by Adolph Hübner during his travels in 1869–71. Hübner did not realize that his samples came from a single geological entity; that realization came in 1901 when the distinguished geologist G. A. F. Molengraaff recognized the common parentage of all igneous rocks in the complex. Rock compositions range from exceedingly mafic (meaning rich in iron and magnesium silicates; one example is olivine) to highly salic (meaning rich in quartz and in alkali-alumino silicates; an example is potassium feldspar). The rock layers have clearly been formed by processes that segregated certain minerals during the cooling and crystallization of the parent magma. The layers (Fig. 1) resemble, at least superficially, those seen in common sedimentary rocks, but the lengths of the layers and the uniformity of their thickness would be surprising in a sedimentary rock.

How, then, could such remarkable features have formed in a cooling mass of hot, viscous magma intruded into the earth's crust? Furthermore, why should layers of certain rock types be observed again and again in puzzling repetitive patterns? And why in this layered intrusion in particular should some of the layers prove to hold uniquely valuable mineral resources, among them platinoid metals, copper, nickel, chrome,

vanadium, and iron ore? By the end of 1973, $3,679,104,276 worth of minerals had been mined (Pretorius 1976). And the best is clearly still to come, as can be seen in Table 2, where production totals for 1975 are compared with the estimated total resources for the complex.

Age and structure

The time frame within which the Bushveld Complex was formed is summarized in Figure 2. The magma from which the rocks of the Bushveld Complex crystallized rose into the crust of the earth soon after the deposition of sedimentary rocks called the Transvaal Supergroup came to an end. Intrusion of this magma was largely confined within the Transvaal sedimentary basin, which contained a maximum thickness of about 12 km of older sediments and volcanic rocks. The Transvaal Supergroup consists of a lower unit of volcanic and sedimentary rocks (the Wolkberg Group), overlain by a unit of chemically precipitated rocks (the Chuniespoort Group), and an upper unit of coarse

---

Table 1. Areal extent of some mafic layered complexes

	Km^2
Skaergaard, Greenland	104
Stillwater, U.S.	194
Sudbury, Canada	1,342
Great Dyke, Rhodesia	3,265
Muskox, Canada	3,500*
Duluth, U.S.	4,715
Dufek, Antarctica	>8,000†
Bushveld, South Africa	66,000

* Includes 2,000 km^2 inferred from geophysical evidence to exist beneath cover rocks.
† This figure is inferred from geological evidence concerning rocks underlying the Antarctic ice.

Figure 1. The layers of chromite (*dark*) and anorthosite (*white*) in this spectacular outcrop of the Bushveld Complex, exposed through erosion by the Dwars River in the eastern lobe, superficially resemble those seen in common sedimentary rocks, but they actually comprise an intricate system of igneous rock. (Photograph by Brian J. Skinner.)

and fine-grained sedimentary rocks (the Pretoria Group). The Pretoria Group constitutes the roof and floor of the layered intrusion throughout most of its areal extent.

The Bushveld Complex underlies so large an area, and there are so few good outcrops, that the detailed geometry of the complex and its individual intrusions is still uncertain. In recent years gravity measurements have been made over the complex to enable geophysicists to estimate the distribution at depth of rocks of different densities. The measurements indicate that there are four separate concentrations of more dense, mafic rocks (Fig. 3). The presence of these four areas refutes the previously held belief that the magma from which the Bushveld rocks were formed rose to the upper levels of the crust through one centrally located feeder or channel. Rather, it appears that four feeder channels enabled the magma to enter the crust of older rocks.

The gravity measurements provide no evidence of connections between the four outcrops of mafic rocks. One authority on the complex (Coertze 1970) believes that a thin but continuous mafic rock layer links the large western lobe to a smaller, roughly circular body north of Zeerust (see Fig. 3), but another authority (Vermaak 1970) disagrees. Acceptance of both Vermaak's opinion and the gravity data would give the Bushveld Complex an approximately cruciform shape made up of four lobes.

Bushveld granite, which overlies the mafic portion of the complex, leaves only some of the mafic rocks exposed, and the gravity measurements are insufficiently detailed for an accurate reconstruction to be made of the precise configuration of the individual lobes. Biesheuval (1970) has provisionally interpreted his data for the Zeerust intrusion to indicate a lopolith, which is a sill-like intrusion that is saucer shaped and depressed in the center, with a stalklike feeder channel. A few exposed rock layers in the eastern lobe, together with gravity data, suggest that the eastern lobe has an asymmetric funnel-shaped profile, with an eastern limb dipping at a low angle of about 10° toward a dikelike feeder, and a more steeply inclined (45°) western limb (see Fig. 3). A similar geometry for the northern lobe has been deduced from geophysical and borehole data (van der Merwe 1976). No valid conclusions can be reached regarding the shape of the western lobe; although it is possible that it may be similar in general to the eastern or northern lobes, it would be presumptuous to extrapolate from the less than conclusive evidence for these lobes to structural arrangements elsewhere in the complex.

Although all four lobes are apparently discrete intrusions, remarkable similarities in the sequence of layers, particularly between the eastern and western lobes, imply that each intrusion was derived from a common magma source and was subjected to similar physicochemical conditions during cooling and crystallization.

Layering of the mafic rocks

The overall composition of the mafic portion of the complex approximates that of basalt, the most common magma type on earth. The thickness of the mafic rocks alone in the two main lobes is about 8 km. Subdivision of the sequence into the five zones shown in Figure 4 is based on the appearance or disappearance of easily recognized, diagnostic minerals. The Lower Zone consists mainly of rocks in which the ferromagnesian minerals, olivine—$(Fe,Mg)_2SiO_4$—and pyroxene—$(Fe,Mg)SiO_3$—are predominant. Toward the top of this zone the mineral plagioclase—ideally, $CaAl_2Si_2O_8$—makes its appearance. In the succeeding Transition Zone plagioclase is no longer an important constituent, so the base of this zone is taken as the level where plagioclase disappears. The boundaries of the higher zones are marked by the appearance of chromite—$(Fe,Mg)Cr_2O_4$—clinopyroxene—$Ca(Mg,Fe),Si_2O_6$—and magnetite—Fe_3O_4.

Individual zones are made up of a number of repetitive layers of rocks of widely different compositions. Such layering reaches its most spectacular development in the Critical Zone, so called by Arthur L. Hall, one of the greatest of the early students of the complex. Hall selected the name Critical Zone to emphasize the delicate adjustments in the physicochemical conditions that he believed prevailed during the crystallization of this batch of magma to account for the variety and rapid repetition of the different rock layers.

Zones may be subdivided according to the concept of a cyclic unit. *Cyclic unit* refers to a specific fixed sequence of cumulus layers that is repeated within a zone; it represents a separate batch of magma that has solidified to

Table 2. Mineral production and resources of the Bushveld Complex

Mineral	1975 production	Estimated total resource	Depth
Platinoid metals and gold	~85,000 kg	62.89 kg × 10⁶	to 1,200 m
Nickel	~18,250 t	22.80 t × 10⁶	to 1,200 m
Copper	~10,800 t	9.95 t × 10⁶	to 1,200 m
Chromite	2,075,378 t	156.0 t × 10⁶	to 300 m
Vanadium pentoxide	19,002 t	16.8 t × 10⁶	to 30 m
Magnetite iron ore	1,561,670 t	1,030.5 t × 10⁶	to 30 m
Magnesite	27,000 t	~10.0 t × 10⁶	
Fluorspar	202,583 t	~40.0 t × 10⁶	
Tin	5,232 t	~0.05 t × 10⁶	

SOURCE: von Gruenewaldt 1977

give the characteristic sequence of minerals. An idealized cyclic unit in the Critical Zone would start with crystallization of chromite, and be followed in sequence by the silicate minerals olivine, pyroxene, and plagioclase, thus giving rise to successive layers in which each of these minerals in turn is predominant. A chromite-rich layer would be followed by a layer distinguished by the presence of olivine, and so on, until finally the top layer would be composed entirely of the mineral plagioclase, the rock it forms being called anorthosite.

A zone consists of a number of these cyclic units, many of which are frequently incomplete due to the nondevelopment of layers at the base or top of a particular cyclic unit. On this cyclicity is superimposed a gradual change in the overall composition of

the mafic rocks from base to top of the 8-km-thick pile. For example, the rock layers in the Lower Zone are composed of pyroxene that contains more magnesium—ideally, $MgSiO_3$—than the pyroxenes in rock layers in the Upper Zone, where the composition of the pyroxenes approaches $FeSiO_3$. The net result is that the bulk composition of the rocks containing pyroxene also changes from magnesium-rich in the bottom zones to iron-rich in the higher zones. Similar changes in composition are found in other minerals in the mafic rocks.

The change in bulk composition of the rocks both within cyclic units and through the entire thickness of the mafic rock pile implies that the magma changes in composition between the forming of the lower layers

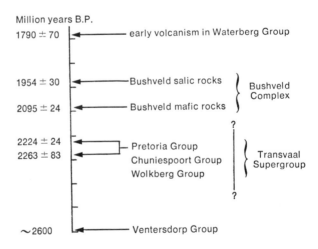

Figure 2. The timing of events immediately surrounding the intrusion of the Bushveld Complex is shown schematically in this diagram. Following deposition of the Transvaal Supergroup, the Bushveld Complex was intruded about 2,095 to 1,954 m.y. ago, along planes of weakness within the uppermost unit of the Transvaal Supergroup—the Pretoria Group, consisting of sediments deposited between 2,200 and 2,300 m.y. ago.

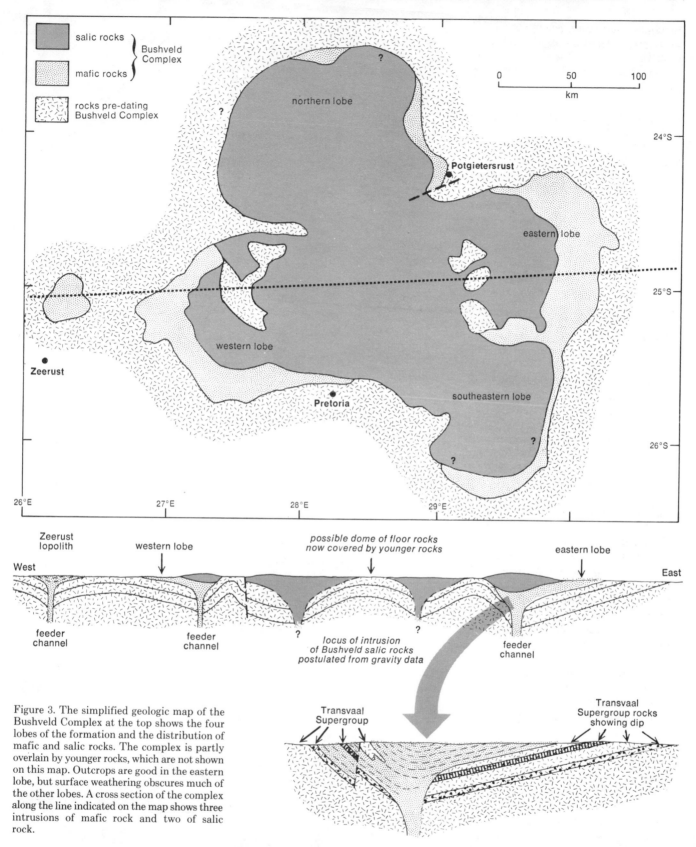

Figure 3. The simplified geologic map of the Bushveld Complex at the top shows the four lobes of the formation and the distribution of mafic and salic rocks. The complex is partly overlain by younger rocks, which are not shown on this map. Outcrops are good in the eastern lobe, but surface weathering obscures much of the other lobes. A cross section of the complex along the line indicated on the map shows three intrusions of mafic rock and two of salic rock.

and the upper layers. The process by which an originally homogeneous magma changes its composition, either within the cyclic unit or in the larger sequence of zones, is called differentiation. In terms of a cyclic unit, as the originally wholly molten magma starts to cool, pyroxene and olivine will be the first crystals to form, resulting in a mush of crystals and magma. The molten portion of the mush becomes depleted in the iron and magnesium required to make the minerals pyroxene and olivine, until ultimately insufficient iron and magnesium remain for pyroxene and olivine to form. New mineral species such as plagioclase will then crystallize, removing calcium and aluminum from the residual magma. The magma thus continues to change its composition until solidification is complete. This concept of differentiation is important in understanding the origin of the layering in the Bushveld Complex.

What caused the layering?

How did this layering originate? There is no question that the entire mafic sequence of rocks crystallized from a magma, but how many intrusions of magma were there? Does each layer represent a discrete intrusion of magma having the appropriate chemical composition, or was a single magma intruded that cooled in such a way that the layers formed by differentiation as the intrusion solidified? If the latter, one must ask what caused the magma to crystallize so that compositionally similar rock layers are repeated several times within a zone. Most workers accept at least some intrusion of new magma periodically during the formation of the mafic rocks. It is the frequency and magnitude of these additions that are most controversial. The ideas that have been proposed to account for the formation of a layered igneous rock sequence like the Bushveld Complex can be classified into two main groups.

The first group of hypotheses has differentiation of an originally homogenous magma occurring at great depths in the earth. Fractions of the magma with different compositions are then intruded into the rocks of the Pretoria Group. Adherents to this hypothesis believe that each layer seen in the Bushveld Complex represents a separate intrusion of magma. A modification of this hypothesis of differentiation in depth

Figure 4. This generalized stratigraphic column of Bushveld Complex is based on the mean thicknesses of layers throughout the complex. The layers in the northern lobe are significantly thinner than those in the rest of the complex, except for the Basal Subzone, which is about 10 times thicker. Levels at which new intrusions of magma may have occurred are marked by the boldface numbers on the left. A portion of the Lower Zone is drawn at an expanded scale to show the repetitive or cyclic layers of olivine and orthopyroxene of which it is composed. These layers become less mafic in composition toward the top of the zone, where plagioclase of the Basal Subzone becomes cumulus. *Cumulus* refers to rock material that has accumulated as a result of gravitational settling.

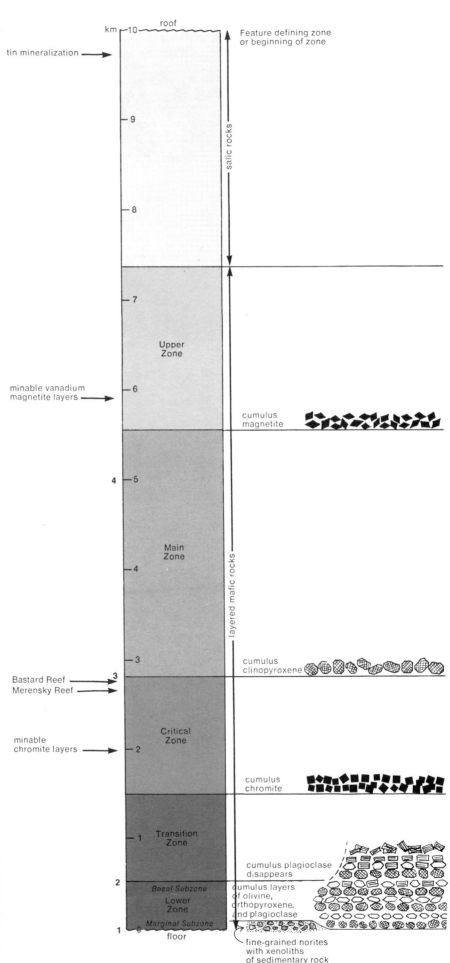

has both primary differentiation at depth and secondary differentiation at the level of intrusion. Once a batch of magma has been intruded, cooling of the magma initiates crystallization of the high-temperature minerals, which, being more dense than the enclosing molten magma, sink and accumulate to form a discrete layer. As the temperature continues to fall, minerals of lower crystallization temperature form and also sink. This process continues until the magma is entirely solidified. Thereafter another batch of magma is intruded and the sequence of crystallization is repeated. In simple terms it may be said that an individual cyclic unit would represent a discrete batch of intrusive magma according to this theory.

The second group of hypotheses proposes that an entirely homogeneous magma was intruded into the crust and that differentiation occurred only at the site of intrusion. Repetition of rock layers is considered a result of convection within the crystallizing magma. Consider the situation where a thick batch of magma is intruded at shallow levels of the crust; in fact, the top of an intrusion may be as little as 4 km below the surface of the earth. The coolest parts of this magma are at the top, so crystallization starts there, again taking place in descending order of crystallization temperature. And again the solidified portions have a greater density than the molten magma and consequently sink under gravity. Sinking of cooler solid matter sets up convection currents in the magma chamber, with the result that hot, molten, undifferentiated magma rises upward from the bottom of the magma chamber to the level at which crystallization is taking place. This magma has the undepleted, original chemical composition, and crystallization of a new set of high-temperature minerals is repeated.

This process, called convective overturn, could have been repeated several times, until the magma was no longer molten and crystallization was complete. Some authorities think that this process of differentiation at the site of intrusion may have been modified by the intrusion of relatively small volumes of new magma from depth. Such additions of new magma would have had the effect of recharging the original magma with elements that had been removed by

crystallization of early-formed minerals.

Coertze (1970) has argued in favor of the hypothesis of differentiation at depth and of separate intrusion of each layer on the basis of the presence of inclusions derived from stratigraphically lower layers in overlying layers. His proposal represents one extreme in the range of opinions regarding the number and frequency of intrusions.

Vermaak (1976) has suggested tentatively that the Basal Subzone of the Lower Zone may represent a discrete intrusion of magma. He points to the presence of plagioclase-bearing layers at the top of the subzone as evidence for complete differentiation of this magma batch. In addition, the presence of a huge inclusion of rocks belonging to the Lower Zone in the Main Zone at Potgietersrus in the northern lobe (see Fig. 3) is taken as evidence for the intrusion and crystallization of the Lower Zone magma prior to the intrusion of the magma from which the remaining zones crystallized. Vermaak's conclusion is supported by van der Merwe (1976), who found evidence for the intrusion of the Lower Zone magma prior to the emplacement of the Transition and succeeding zones in the northern lobe. There appears to be an abrupt change in the bulk chemistry of the rocks above the base of the Main Zone, however, and a new intrusion of magma to form the Main and Upper zones is postulated, but more information is required to test this suggestion.

There is as yet no unequivocal evidence on the side of any of these hypotheses, but it seems reasonable to assume that the 8-km-thick pile of igneous rocks was not intruded in one stupendous magmatic heave. It seems equally improbable that each layer could represent a separate intrusion of magma, for it is difficult to visualize how the delicate layering illustrated in Figure 1 could be so consistent over long strike lengths. Undoubtedly, the truth lies somewhere between these extremes.

The Merensky Reef cyclic unit

The rock layers immediately above and below the Merensky unit have been studied in greater detail than

other cyclic units in the Critical Zone, because the Merensky Reef contains the world's greatest reserves of platinum metals and because the cyclic unit of which the Merensky Reef forms a part is complete, in contrast to most other units in the Critical Zone. Consideration of the detailed information obtained from the extensive underground workings provides clues to the possible origin of the mafic layered rocks.

The Merensky Reef cyclic unit and the overlying Bastard Reef cyclic unit are the topmost pair of units in the Critical Zone. The Merensky Reef unit is about 10–25 m thick and contains economically significant concentrations not only of platinum metals but also, at its base, of copper and nickel sulfides. Although the succeeding Bastard Reef cyclic unit contains an assemblage of rock layers identical to that of the Merensky Reef unit, mineralization is insignificant. The name applied to this unit accurately expresses the despair of early exploration geologists seeking continuations along strike and dip of the Merensky Reef.

Below the Merensky Reef the Critical Zone consists of cyclic units in which thick layers composed almost exclusively of chromite are found. The proved and potential reserves of chromite in these cyclic units constitute the world's largest deposits of the mineral. Several of these strata are mined, particularly in the eastern lobe. Underlying the thick chromite layers are layers of anorthosite mingled with thin chromite layers. An example of the mixed anorthosite/chromite layers is shown in Figure 1, in which it can be seen that the two components have a complex relationship. In underground workings at the Rustenberg platinum mine the full complexity is magnificently displayed.

Frequently the anorthosite and chromite show features that in a sedimentary rock would be called cross-bedding, slumping, troughs, or cut-and-fill. In fact, these pseudo-sedimentary structures can be interpreted as a result of "sedimentation" in an igneous environment. Instead of grains of sand or silt settling in water subjected to current action, mineral grains settled in molten magma that was affected by currents. The existence of currents is implied by the

Figure 5. The Merensky Reef, as it is seen in the Bleskop Mine, Rustenberg, South Africa, forms a dark layer overlying a massive layer of white anorthosite. The dimpled base of the reef is clearly visible in this photograph. The reef itself is layered on a fine scale: at the base, a centimeter-thick layer of chromite; above this, to the point of the pick, a thicker layer of pyroxene and olivine; then another thin chromite-rich layer; and finally a pyroxene-olivine-plagioclase layer of variable thickness. The rock unit overlying the Merensky Reef is again anorthositic and appears white. (Photograph by Brian J. Skinner.)

way in which certain minerals are consistently aligned.

The layers beneath the Merensky Reef are further distinguished by the presence of a so-called boulder layer, located about 30 m below the base of the Merensky Reef. This layer consists of scattered rounded but flattened aggregates of mafic minerals set in anorthosite. Individual boulders have a mean diameter of about 20 cm in the plane of the layering, but perpendicular to the layering their axes measure about 10 cm. The boulders consist of pyroxene, amphibole—ideally, $Ca_2(Mg,Fe)_5Si_8O_{22}(OH)_2$—and biotite—$K_2Mg_6(Mg,F,Al)_8$ $(Si,Al)_8O_{20}$. Amphibole and biotite are not common in the mafic rocks generally but are found sparingly in the Merensky Reef.

The Merensky Reef cyclic unit consists of a paper-thin chromite layer at the base, followed by a coarse-grained layer composed of pyroxene and olivine, overlain by an olivine-pyroxene-plagioclase layer, and finally capped by plagioclase-rich layers. Several features, apart from the rich concentrations of platinoid metals and copper/nickel sulfides, distinguish the reef. First, individual mineral grains in the reef attain dimensions of 2 to 3 cm in length and 1 cm in width, unusually large compared to those found in other layers. Second, the surface on which the reef rests is peculiarly dimpled (Fig. 5). Third, in shallow mine workings where weathering has been active the reef appears to be composed of a series of closely

packed spheres. And fourth, there is a reversal of the theoretical differentiation trend. For example, olivine shows an increase in magnesium content upward through the cyclic unit rather than a decrease.

When the interface between the top of the Merensky Reef cyclic unit and the base of the Bastard Reef unit is taken as a horizontal datum (Fig. 6), there is an obvious variation in thickness of the mafic layers of the Merensky unit. This thickening occurs on the flanks of gentle arches of the rock layers below the reef, which Vermaak (1976) attributes to gentle folding of these underlying layers (accompanied by erosion, caused by currents or turbulence in the magma, over the crest of the arch) prior to the crystallization of the Merensky unit. On a larger scale, an overall dipping of the mafic layers toward the center of the complex is observed. Paleomagnetic evidence strongly suggests the observed relationship of the basal layers of the Merensky unit to the underlying rock layers must be a consequence of sagging and arching of the floor while the Merensky unit was accumulating.

The presence of erosional breaks, disconformities, and scattered inclusions of rock types from lower layers in overlying layers in other zones of the complex suggests some movement (i.e. currents) in the unsolidified magma. Moreover, anomalous relationships between the more gently dipping layers of the complex and the steeply inclined floor rocks of

the Pretoria Group in the eastern lobe indicate deformation of the sedimentary rocks prior to the solidification of the Bushveld magma. The evidence seems to point firmly toward the occurence of periodic tectonic adjustments throughout the crystallization of the Bushveld mafic magma. We will discuss the important implications of this hypothesis after some of the other clues supplied by our knowledge of the Merensky unit have been examined.

The confinement of the main platinum and copper/nickel sulfide mineralization to the penultimate cyclic unit in the Critical Zone led to early suggestions that the Merensky unit represented a new addition of magma enriched in these elements. Vermaak (1976), however, is of the opinion that this enrichment is the normal result of differentiation whereby sulfur, platinoids, nickel, and copper are progressively concentrated in the residual magma, and that crystallization of copper and nickel sulfide minerals occurs when sulfur enrichment reaches an appropriate level. The lack of sulfide mineralization in the Bastard Reef would be a direct consequence of the depletion of sulfur in the magma following the formation of sulfides in the Merensky Reef.

It has also been noted that both biotite and apatite—$Ca_5(F,OH)$- P_3O_{12}—are found in the Merensky Reef. The lack of apatite and the lower abundance of hydrous minerals in the immediately overlying Main Zone is taken as further evidence that

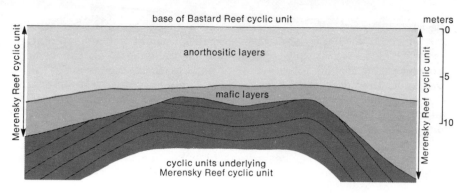

Figure 6. This schematic representation of the Merensky cyclic unit shows variations in its thickness and the relationship of the unit to layers in preceding cyclic units along a 100-km-long section around part of the periphery of the western lobe. Only the general distribution of the mafic and anorthositic rocks is shown: the Merensky unit contains more layers than are seen here. The basal 0.5 to 3 m of the mafic layer is the platiniferous section of the reef. (Data from Vermaak 1976.)

the Merensky Reef unit formed late in the crystallization history of the magma that formed the Critical Zone. In addition, the extensive crystallization of the oxide mineral chromite below the Merensky Reef would have had the effect of increasing the relative sulfur content of the residual magma. The state of oxidation of the residual magma would have been decreased so that only limited amounts of chromite could have crystallized at the level of the Merensky Reef, and hence only a paper-thin layer formed.

The boulder bed underlying the Merensky Reef may provide a clue to the origin of the reef and the dimpled floor on which it rests. The boulders are thought to have resulted from the aggregation of mafic minerals, by a mechanism that is at present uncertain, into clusters that settled under gravity toward the floor. The enclosing magma solidified around the aggregates before the boulders could collect as a continuous layer. The mineral composition of the boulders closely resembles that of the Merensky Reef—even including a paper-thin chromite layer at the base of each boulder. The bottom layer of the Merensky Reef thus seems to consist of boulders that did accumulate as a continuous layer on a floor consisting of the preceding cyclic unit, which was not completely solid. As the boulders settled they would have dimpled the floor of partly molten anorthosite. Subsequent settling of further mafic minerals would have filled the interstices between the boulders.

The reversal of the theoretically expected differentiation trend was identified by Vermaak (1976), who suggests that, early in the crystallization history of the batch of magma from which the Merensky unit formed, plagioclase grains, which were less dense than the magma, floated upward to form an impermeable layer below which residual magma was trapped. It is not certain whether crystallization and settling of chromite into the thin basal layer of the Merensky Reef preceded or overlapped the floating of the plagioclase. Crystallization of plagioclase and chromite would have removed CaO and FeO, thus causing a relative and progressive increase in concentration of MgO of the residual magma. An impermeable plagioclase covering could also have trapped the volatile components of the magma, the presence of which would have provided conditions favorable for the crystallization of large mineral grains.

Although this hypothesis provides a neat explanation of the observed facts, the problem here is to understand the mechanism that arrested the upward migration of the first-formed plagioclase crystals. Vermaak suggests some kind of compositional, density, and/or temperature inversion, but precise details are not clear. If Vermaak's hypothesis is valid the real difficulty lies in the apparent necessity of isolating the batch of magma that forms a particular cyclic unit from the main volume of the magma.

Vermaak sees an interplay between the top of one cyclic unit with the base of the succeeding unit in the mixed anorthosite/chromite layers shown in Figure 1. The relationship of the plagioclase-rich anorthosite to the chromite layers is inconsistent with any mechanism of intrusion. Vermaak interprets these layers as being a consequence of mixing of low-temperature plagioclase crystallized from a preceding magma batch with the high-temperature chromite from the first crystallization of a succeeding magma batch. Turbulence-induced current action within the partially solidified magma batches, probably related to the inferred tectonic adjustments, would have caused the winnowing of the less-dense plagioclase grains and the higher-density chromite grains into interspersed layers which exhibit structures, such as cross-bedding, that are more typical of sedimentary rocks. The plagioclase-rich layer would still act as an impermeable mat, trapping residual magma below it, though remaining sufficiently molten to interact with the chromite from a new phase of crystallization.

At the level of the Merensky Reef, the low state of oxidation of the magma inhibits the formation of chromite, and thus the complex relationships of the mixed anorthosite/chromite layers seen at lower levels in the Critical Zone are absent at the base of the Merensky Reef. It is also probable that in some earlier units the plagioclase-rich mat was less efficient as a trap for the volatile components of the magma. The escape of these components, together with nickel, copper, and platinum, would result in their increased concentration in the final residual magma. Thus in the magma from which the Merensky unit crystallized, these components would have been highly concentrated. The coincidental formation of an impervious mat created a combination of circumstances that provided the ideal conditions for the crystallization of platinum, nickel, and copper minerals.

The meaning of the complex

In the present state of our knowledge of the Bushveld Complex it seems reasonable to postulate that a first discrete intrusion of the magma crystallized to form the Lower Zone, about 500 m thick. Following tectonic movements, a second intrusion of magma occurred that solidified as the Transition and Critical zones, altogether about 2,100 m thick. The Main and Upper zones were formed from either one or two intrusive magmas. Von Gruenewaldt (1973) has proposed that an addition of new magma occurred at a level approximately 2,200 m above the top of the Critical Zone, which would make the remainder of the Main Zone and the Upper Zone about 2,600 m thick. The bulk chemical compositions of these zones reflect broad differentiation of the Bushveld magmas from magnesium-rich to magnesium-poor.

Although evidence for intricate controls of crystallization has been found in the Bushveld Complex, no effective mechanism has yet been identified that permits relatively small batches of magma to crystallize in isolation from the main volume of the magma. Detailed study of the Merensky cyclic unit has shown a complex history of crystallization. It is not known whether other cyclic units in the Critical Zone or in other zones exhibit the same complexity. Until more detailed studies have been completed, no satisfactory model can be proposed to account for the origin of the layering and indeed the origin of the complex as a whole. Many questions must remain unanswered until new data are available.

The problem of the origin of the cyclic repetitions of rock layers is not unique to the Bushveld Complex. Similar sequences are found in other layered complexes, such as Stillwater in Montana, the Great Dyke of Rhodesia, Muskox in Canada, and Skaergaard in Greenland. Similarities among these complexes have been stressed by Jackson (1970), and he concluded that similar processes must have been operative during the crystallization history of all these intrusions, the observed slight differences being attributed to variations in the initial bulk chemistry of their magmas.

Study of the Bushveld Complex has wide economic implications for our metal-hungry society, because of the enormous reserves of metals such as platinum and chrome present in the formation. Understanding the crystallization history of this remarkable intrusion might well aid exploration for metal deposits in the layered complexes in other parts of the world. But detailed investigation of the complex has other applications as well.

The Bushveld magmas were derived from the mantle that underlies the crust of the earth. At the present time scientists are studying the chemistry of magmas, also derived from the mantle, that well up today along mid-oceanic ridges. Variations in the chemistry of these magmas are well documented, but the reasons for these variations are not fully understood. Many possibilities present themselves—a chemically inhomogenous mantle, variations in depth of origin of the extruded magma, or differences in the nature of extrusion. Study of the Bushveld magmas will help resolve these questions and will contribute to studies concerned with the evolution of the mantle and the continents on which we live.

Nearly 100 years of geological endeavor have provided a sound basis for the more detailed investigations that are now required to elucidate the problems presented by the Bushveld Complex. The complex provides geologists with almost unrivaled opportunities to study a wide spectrum of igneous processes.

References

Biesheuval, K. 1970. An interpretation of a gravimetric survey in the area west of the Pilanesberg in the western Transvaal. *Geol. Soc. S. Afr., Spec. Pub.* 1: 266–82.

Coertz, F. J. 1970. The geology of the western part of the Bushveld Igneous Complex. *Geol. Soc. S. Afr., Spec. Pub.* 1: 5–22.

Gough, D. I., and C. B. van Niekerk. 1959. A study of the palaeomagnetism in the Bushveld Complex. *Phil. Mag.* 4, part 38: 125–36.

Jackson, E. D. 1970. The cyclic unit in layered intrusions: A comparison of repetitive stratigraphy in the ultramafic parts of the Stillwater, Muskox, Great Dyke, and Bushveld complexes. *Geol. Soc. S. Afr., Spec. Pub.* 1: 391–423.

Pretorius, D. A. 1976. The stratigraphic, geochronologic, ore-type, and geologic environment sources of mineral wealth in the Republic of South Africa. *Econ. Geol.* 71: 5–15.

van der Merwe, M. J. 1976. The layered sequence of the Potgietersrus limb of the Bushveld Complex. *Econ. Geol.* 71: 1337–51.

Vermaak, C. F. 1970. The geology of the lower portion of the Bushveld Complex and its relationship to the floor rocks in the area west of the Pilanesberg, western Transvaal. *Geol. Soc. S. Afr., Spec. Pub.* 1: 242–62.

———. 1976. The Merensky Reef: Thoughts on its environment and genesis. *Econ. Geol.* 71: 1270–95.

von Gruenewaldt, G. 1973. The Main and Upper zones of the Bushveld Complex between Tauteshoogte and Pardekop in the eastern Transvaal. *Trans. Geol. Soc. S. Afr.* 76: 207–27.

———. 1977. Mineral resources of the Bushveld Complex. *Minerals Sci. Engng.* 9: 83–95.

Index

Adagdak volcano, 99
aerosols, 43–44
Africa, 81–82, 168–76
age, *see* dating
Age of the Earth, The
 (Holmes), 21
Aja Fault, 70
Alae lava lake, 149
Alaska, faults in, 70–71
Aleutian trench, 100
algae, 32, 159
alkali feldspar, 163
allyl diglycol carbonate, 41
Alps, 11
Amak Island, 101
amphibole, 174
andesites, 102
andesitic basalts, 103
anorthosite, 170, 173, 175
Antarctica, 65–67
Appalachian Mountains, 11
Arequipa, Peru, 165
argon, 23
ash clouds, 136, 140
 See also volcanic ash
asthenosphere, 104–6, 120
Atlantic Ocean, 97
Atlas Mountains, 11
atmosphere
 C^{14} in, 59–60
 of Earth, 11, 31, 32
 and nuclear tracking, 43
Australia, 65–67, 98
avalanche, on Mount St.
 Helens, 130

Baja California, 82, 97
Balanus Seamount, 158
Ballard, R.D., 153
Baltica, 88–89, 93–95
Bambach, Richard K., 86
Barus, Carl, 20
basalt
 hydration measurement of, 53
 island-arc lavas composed
 of, 103
 magnetization of, 68
 and mantle source volume,
 120
 on Moon, 7
 on ocean floor, 11, 12, 28, 68
 in pillow lava, 144–52
 on planets, 12–13
 seamounts composed of, 155
 tholeiitic, 101–2, 107, 115,
 121
 See also lava

basaltic andesites, 103
batholith, 160
Baxter, Gregory P., 22
Bear Seamount, 154
Becquerel, Henri, 21
Benioff zone, 102–4, 108–10
birds, and nuclear tracking, 45
Black, Joseph, 18
Bogoslof Island, 101
Boltwood, Bertram, 21
boron, 41, 43
bristlecone pine, 57–59, 61
Bull Lake glaciation, 52
Bushveld Complex, 168–76

C^{14} dating
 and obsidian dating, 49
 reliability of, 51, 55–61
 and tree ring dating, 57–59
caldera, 160–62
California, 80, 82, 98
Caloris Basin (Mercury), 8
Cambrian era, 89, 93–94, 97
carbon dioxide ice, 9
Carboniferous period, 94–96
Caspian Sea, 33–34, 35
Chamberlin, T.C., 20
channels, Martian, 9, 10
chemical differentiation
 of Earth, 28, 31
 of planets, 15
China (paleocontinent), 88–89,
 93–96
chromite, 170, 173, 175
Chuniespoort Group, 168
climate
 continental drifting affecting,
 81–82
 deposits indicating, 29, 91,
 95–96
 of modern world, 92
 during Paleozoic era, 87, 91,
 94–97
 impact of Panthalassic Ocean
 on, 97
clinopyroxene, 170
coal, 91, 95–96
Coldwater Ridge, 132, 136
Columbia River, 137–38
Columbia River Plateau, 147
cone, volcanic, 130
continents
 ages of rocks on, 69
 drifting of, 23, 25, 32, 64–66,
 76, 81–82, 86–98
 erosion of, 76
 margins of, 83

during Paleozoic era, 86–98
 and plate tectonics, 64–74
convection, and plate tectonics,
 25, 32, 72–74, 124
core
 deep sea drilling and, 80
 of Earth, 27, 68
 of Mercury, 13
Coriolis effect, 92
Corner Rise, 153–55
Cowlitz River, 137–38
craters
 of Earth, 11
 on Mars, 9–10
 on Mercury, 8
 on Moon, 7
 morphology of, 8
 on Mount St. Helens, 130,
 140–41
 planetary comparisons of, 12
 on Venus, 10
cristobalite, 163
Cromwell current, 81
crust
 continental collisions
 affecting, 86–98
 geologic time told by, 27–28
 of Mars, 9
 of Mercury, 8
 of Moon, 7–8
 oceanic, 25, 77–78, 80,
 115, 153
 around pillow lava, 145–47
 and plate tectonics, 11, 13,
 64–74
 silica rich rocks from, 12
 See also earthquakes;
 volcanism
cupola, 160–61

dacite, 54
Daicel cellulose nitrate, 41
Dalrymple, G. Brent, 112
Darwin, Charles, 18–19
dating
 of continental rocks, 69
 of Hawaiian volcanoes, 112,
 115, 119–20
 using impact cratering, 10, 14
 of lunar mare, 7–8
 of New England Seamounts,
 154
 using obsidian, 47–54
 of ocean floor, 12, 69
 of planetary surfaces, 9–10
 using radiometrics, 7, 21–22,
 25–26, 29–31, 55–61
day, Devonian, 30

Deep Sea Drilling Project, 75–84
dendrochronology, 49, 57–59,
 61
deuterium, 41–42
Devonian period, 30, 93–94
dikes, 68, 107, 114, 119
dosimetry, 38–39
drilling project, 75–84
dust storms, Martian, 9

Earth
 age of, 14, 18–23, 25–36
 Bushveld Complex on,
 168–76
 continental drifting on, 23,
 25, 32, 64–66, 76, 81–82,
 86–98
 cooling of, 19
 density of, 27
 differentiation of, 28, 31
 elevations on, 13
 magnetic polarity, reversal of,
 23, 25, 60–61, 68–71,
 76–77, 86
 Paleozoic era of, 86–98
 Planetary Evolution Index
 for, 12
 polar regions of, 12
 See also plate tectonics;
 volcanism
earthquakes
 between Australia and
 Antarctica, 65, 66
 from lava transport, 115
 and plate tectonics, 65–74,
 102
 and radon, 44
 recording of, 64
 in Mount St. Helens area,
 130, 132
 and transform faults, 82
 trenches and, 71
Easter Island, 122
elevations, planetary, 13–14
Emperor Seamounts, 112,
 114–15, 118–25
Engel, A.E.J., 25
eolian erosion, 9
equatorial belt, 92
erosion
 of continents, 11, 76
 dating of, 51–52
 on Mars, 9
 on sea floor, 29
 on seamounts, 156
 and sun's energy, 19
 of volcanoes, 103, 135–38,
 143

Faul, Henry, 18
faults
    from continental collision, 97
    of Mars, 10
    See also transform faults
fault slivers, 147–48, 150–51
fission track method, 42
Fleischer, Robert L., 37
flooding, from Mount St.
    Helens, 128, 137
Forsyth glacier, 130
fossil records
    climate revealed by, 29,
        81–82, 95–96
    isotopic age determinations
        using, 21–23, 28–31, 120
    of Paleozoic era, 89–90,
        94–96
    in strata, 18
Franciscan formation, 69
Friedman, Irving, 47
fumaroles, 130, 132, 141, 143
fusion track method, 42

Geikie, Archibald, 20
geologic time, 18–24, 25–36,
    86
geothermal energy, 162–63
geysers, 162, 163
Gilliss Seamount, 157
glaciers
    in Antarctic, 81
    on Mount St. Helens, 130,
        137–38
    in North America, 32, 51–52,
        130, 137–38
    in Paleozoic era, 94, 96
    rocks transported by, 155
    tillites left by, 91
glass, volcanic *see* obsidian
Glomar Challenger, 76, 79–84
Goat Rocks dome, 136
Gondwana, 88, 93–97
Gorde Rise, 151
grabens
    on Mars, 10, 14
    on Moon, 13
    on Mount St. Helens, 130,
        135
granite
    in Bushveld Complex, 169
    in continental crust, 28, 30–32
    and obsidian, 48
    on seamount, 158
gravity, on Mercury, 8
Great Magnetic Bight, 72
Great Sitkin volcano, 99
Green River, 135

Halley, Edmond, 21
Hawaiian Islands, 112–25
Hawaiian Ridge, 112, 144
Head, James W., 7
Heiken, Grant, 160
Heirtzler, J.R., 153
Holmes, Arthur, 21–23
hornblende, 103
"hot" spot hypothesis, 123
hot springs, 162, 163, 166
Houghton, R.L., 153
Houtermans, F.G., 23
Hunter, Donald, 168
Hutton, James, 18
Huxley, Thomas, 19
hydrosphere, 11

Imbrium Basin (Moon), 7
impact cratering, 7–8, 10,
    13–14
International Phase of Ocean
    Drilling (IPOD), 83
iron
    in Earth's core, 31
    heavy nucleus of, 40
    in obsidian, 50
    in pyroxene, 170
iron oxides, 94
island arcs, 71–73, 99–110

Jackson, Everett D., 112
Jameson, Robert, 18
Joint Oceanographic Institutions
    (JOI), 84
Joint Oceanographic Institution
    Deep Earth Sampling (JOIDES),
    75
Joly, John, 21
Juan de Fuca Ridge, 70, 82

Kauai volcano, 114
Kazakhstania, 88–89, 93–94
Kelvin, Baron, 19–21
Kermadec trench, 71, 73
Kilauea volcano, 112, 116–18,
    114–52
King, Clarence, 20
Knopf, Adolf, 22
Koko Seamount, 120
Kurile Islands, 101

lakes, from glacial ice, 32, 34
landslide, at Mount St. Helens,
    130, 132, 136
Lane, A.C., 22
Laurentia, 88, 93–94
Laurussia, 94, 96
lava
    from Hawaiian volcanoes, 115
        120–25
    from island-arc volcanoes,
        99–110
    lunar, 7, 8
    magnetization of, 68
    from Mount St. Helens, 130
    obsidian dating of, 50–51
    pillow, 68, 144–52, 155–57
    and plate tectonics, 68
    potash in, 103–4, 106, 110
    seamounts composed of,
        155–59
    from shield volcanoes, 13,
        115, 120–25
    See also volcanism
lava channels, 12
lava dome, 141
lava tubes, 145
lead
    and nuclear tracking, 41
    radiometric ages determined
        using, 26, 28, 30, 31
    uranium decaying into, 23
life, early evolution of, 31, 32
limestone, 90
lineaments, lunar, 8, 13
lithium
    nuclear tracking of, 41, 43
    from pyroclastic lows, 160
        166
lithosphere
    continental collision affecting,
        97

earthquakes restricted to, 71
    and Hawaiian volcano
        formation, 112, 123
    and plate tectonics, 11, 13
    study of, 83
    viscosity of, 108
Los Alamos Scientific Laboratory,
    163
Lyell, Charles, 18, 23

Macdonald Seamount, 122
McKenzie, Dan P., 64
mafic rocks, 168–73
magma
    ascent of, 107–8
    asthenosphere corner
        formation of, 104–5
    near Benioff zones, 103–4
    forming Bushveld Complex,
        168–76
    from island-arc volcanism,
        99–110
    and obsidian, 48
    and pyroclastic flows, 160
    radioactive elements in, 30
    silica content of, 12, 103,
        115, 160, 162
    source of, 103–7
    See also andesite; basalt;
        lava
magnesium
    in mantle, 31
    in obsidian, 50
    in pyroxene, 170
magnetic field
    and C[14], 60
    measurement of, 64
    and pole reversal, 23, 25,
        60–61, 68–71, 76–77, 86
    of sun, 61, 68
magnetic reversal stripes, 68–71,
    78, 82, 86
magnetite, 103, 170
manganese, 50, 155–59
Manning Seamount, 154, 158
mantle
    Bushveld magmas derived
        from, 176
    continents evolved from, 25
    convective cells within, 25,
        32, 33
    magnesium in, 31
    melting spot in, 114–15,
        120–25
    and plate tectonics, 67, 72,
        78, 101
    silica-poor rocks from, 12
mantle source volume, 120
mare, lunar, 7, 12, 13
Mars, 9–10, 12–14
Marsh, Bruce D., 99
Mauna Loa volcano, 112
Mauna Ulu vent, 144
Mediterranean Sea, 82
Meiji Seamount, 120
Melobesia, 159
melting spot, 121–25
Mendocino escarpment, 78
Mercury, 8, 12–14
Merensky Reef, 173–75
meteorites
    used in dating, 27, 30, 60
    in nuclear tracking, 39–40,
        45
Michael, Henry N., 55
Mid-Atlantic Ridge, 80, 144
Midway Island, 83, 125
minerals, at Bushveld Complex,
    168–76

Mohorovicic discontinuity, 75
Moon
    dating of, 14
    during Devonian era, 31
    and nuclear tracking, 39–40
    Planetary Evolution Index
        for, 12
    surface of, 7–8
    tectonism on, 13
    volcanism on, 7, 12
Moore, James G., 144
mountains
    on Earth, 11, 128–43
    on Moon, 8
    during Paleozoic era, 92–95,
        97
Mt. Lassen, 128
Mt. Margaret, 136
Mount St. Helens, 128–43
Muddy River, 138
mudflows, 128, 130, 138, 140
Mutch, Thomas A., 7
Mytilus Seamount, 154, 159

Nashville Seamount, 154, 156
natural selection, 19
nephelinitic basalt, 115
neutrons, in radiocarbon
    dating, 55
New England Seamounts, 153–59
New Zealand, geothermal areas
    in, 162
Nier, Alfred O., 22
Nierenberg, William A., 75
Nihoa volcano, 121
Niihau volcano, 121
North America
    Atlantic coast of, 97
    continental drifting of, 32, 81
    glaciers covering, 32
    Pacific coast of, 80, 82, 98
nuclear tracking, 37–46
nuees ardentes, 136, 140

obsidian
    chemical composition of, 50
    as a dating tool, 47–54
    and nuclear tracking, 41
ocean floor
    age of, 25, 29, 115
    basaltic plains on, 12
    magnetic anomalies on,
        68–71, 78, 82, 86
    and nuclear tracking, 43
    pillow lava on, 144–52
    plate boundaries on, 68
    ridges on, 11, 64, 68–71,
        77, 80, 82, 101, 144
    spreading of, 11, 32, 64–74,
        77–78, 80, 153
    trenches consuming, 67, 69,
        71
oceans
    C[14] turnover in, 60
    covering continents, 31
    evolution of, 31
    during Paleozoic era, 93–95,
        97
    See also ocean floor
oil, 80, 83
olivine, 103, 108, 168, 170–71,
    174
Olympus Mons (Mars), 9
Ordovician period, 90, 93–94
Orientale Basin (Moon), 7
Origin of Species (Darwin), 20

orogenies, 31
oxygen, 30, 31, 50

Pacific plate, 80–81, 99, 112, 114, 120, 122
Pajarito Plateau, 165
palagonite, 156
paleocontinents, 87–98
paleogeography, 86–98
paleomagnetism, 89
Paleozoic era, 86–98
Pangea, 86, 89, 94, 97
Panthalassic Ocean, 97
periodotite, 101–2, 104, 106–9
permafrost, on Mars, 9
Permian period, 96, 97
photosynthesis, 55
pillow lavas
    formation of, 144–52
    magnetization of, 68
    on seamounts, 155–57
Pinus aristata, 57–59
plagioclase, 103, 170–71, 174–75
plains, volcanic, 7–9, 12
Planetary Evolution Index (PEI), 12
planets, evolution of, 7–15
plate tectonics
    continental drifting caused by, 86–98
    description of, 13, 14, 64–74
    and Hawaiian Islands, 120–25
    island-arc volcanism and, 99–110
    mid-ocean ridges caused by, 11, 68–71, 77, 82, 101, 144
    plate subduction and, 82, 92, 95, 100–106, 109
    sea floor spreading caused by 32, 64–74, 77–78, 80, 153
    See also earthquakes; volcanism
plates
    boundaries of, 12, 64–71, 144
    movement of, 64–74, 82, 88, 92–98, 100–106, 109
    names of, 69
    See also plate tectonics
platinum, 173, 175, 176
plume hypothesis, 123–25
plutonium, 39, 42–43
plutonium dioxide, 43
plutons, 107
polar regions
    of Earth, 12, 92
    of Mars, 9
potash (potassium), 26, 28, 30, 31, 50, 103–4, 106, 110
potassium-argon dating, 115
potassium feldspar, 168
Pretoria Group, 169, 172
Project Mohole, 75
propagating fracture hypothesis, 124–25
proteinoids, 32
proton magnetometers, 64, 77
pumice, 141, 161
pyroclastic flows
    deposits from, 160–67
    from Mount St. Helens, 138, 140–41

pyroxene, 26, 103, 170–71, 174

radioactive isotopic dating
    of Hawaiian volcanoes, 112, 115
    history of, 21–23, 25–31, 55
    and obsidian dating, 49
    reliability of, 55–62
radioactive nuclides, 26–30
radioactive wastes, storage of, 160, 166
radiocarbon dating, see C14 dating
radon, 43–44, 56
Ralph, Elizabeth K., 55
Rayleigh-Taylor (R-T) instability, 104–5
Rehoboth Seamount, 154, 157–58
Reykjanes Ridge, 146
ridges
    and magnetic anomalies, 69
    mid-ocean, 11, 64–71, 77, 80, 82, 101, 144
    on Moon, 13
    pillow lava erupted at, 144–45
rift zones, 114, 118–19
Rogers, Henry Darwin, 19
Rosenfeld, Charles L., 128
rubidium, 26, 28
Rutherford, Ernest, 21
Ryan Lake, 135

San Andreas Fault, 70, 78, 80–82, 98
San Juan volcanic field, 160
scarps, on Mercury, 8, 13
Scotese, Christopher R., 86
sea-floor spreading, 32, 64–74, 77–78, 80, 153
seamounts, 112–25, 135–59
sediments
    climate revealed by, 29, 80–82, 91, 95–96
    isotopic age determinations using, 30, 120
    on seamounts, 155, 157–58
Sequoia gigantea, 57–59
seismic stations, 64, 68
shear-melting hypothesis, 124
shield volcanoes
    of Emperor Seamounts, 115
    in Hawaii, 114–25
    on Mars, 9–10
    planetary comparisons of, 13
    on Venus, 10
Shoestring glacier, 130, 138
Siberia (paleocontinent), 88–89, 93–96
Sila Fault, 70
silica content, of magma, 12, 103, 115, 160, 162
Silurian period, 93–94
Silver, Eli A., 112
sinuous rills, 12
Smith, Adam, 18
Smith, William, 18
South Africa, Bushveld Complex in, 168–76

South America, 81–82
Spirit Lake, 130, 132, 135–37
steam vents, on Mount St. Helens, 132, 143
strontium, 107
Strutt, R.J., 21
subduction, of plates, 82, 92, 95, 100–106, 109
Suiko Seamount, 120
sun
    energy of, 19
    magnetic field of, 60, 68
    stability of, 92
sunspot activity, 61, 68
supernovae, 60
Swift Reservoir, 138

Taylor, P.T., 153
tectonic activity
    on Mars, 10, 13
    on Mercury, 13
    on Moon, 8, 13
    on Venus, 10, 13
    See also plate tectonics
tektites, 39–40
tephra, 161–62
Tethys Sea, 94
Tharsis region (Mars), 10
thermal energy, 162–63
tholeiitic basalt, 101–2, 107, 115, 121
Thomson, William, 19
thorium
    in isotopic age determinations, 22, 31
    nuclear tracking and, 38, 40, 43–44
    and pyroclastic flows, 160
thrust faulting, 92, 97
tillites, 91, 96
time, geologic
    chart showing, 86
    and evolution of earth, 31–35
    history of awareness of, 18–24, 25–31
    See also dating
Tonga, 101, 102, 107
Tonga trench, 71, 73
toothpaste pillows, 148
Toutle River valley, 132, 135–37, 141
transform faults
    description of, 64, 66–67, 69–71
    as east-west escarpments, 77, 78
    vs. mare ridges, 12
    San Andreas Fault as, 70, 78, 80–82, 98
Transvaal Supergroup, 168
tree ring dating, 49, 57–59, 61
Trembour, Fred W., 47
trenches
    at continental margins, 83
    description of, 65, 71–72
    island-arc volcanoes paralleling, 100
    plates destroyed beneath, 67, 69, 71
tritium, 42
Tuamoto-Line Islands, 122

tuff, 163–66
Turkey, pyroclastic flows in, 164–65

uniformitarianism, 18–19
uranium
    for isotopic age determinations, 21–23, 26, 28, 30
    and nuclear tracking, 39–40, 43–44
    from pyroclastic flows, 160, 166
Urey, Harold C., 23, 30

Valles Marineris (Mars), 10, 13
Vardzia (USSR), 165
varves, 59
vents, of volcanoes, 114, 118, 143
Venus, 10, 12–13
vesiculation, of magma, 161
volatile elements, 31, 161
volcanic ash, 128, 132, 135–38, 140, 143
volcanic conduit, 120
volcanism
    in Hawaii, 112–25
    island-arc, 99–110
    on Mars, 9–10
    on Mercury, 8
    on Moon, 7
    at Mount St. Helens, 128–43
    and obsidian dating, 50–51
    pillow lava from, 144–52
    and Planetary Evolution Index, 12
    and seamounts, 153–59
    on Venus, 10

Waianae volcano, 121
Walcott, C.D., 20
water
    on Earth, 11
    on Mars, 9
    obsidian absorbing, 47–54
    peridotite melted by, 104, 107
    in pillow lava formation, 144
Wegener, Alfred, 76
Werner, Abraham Gottlob, 18
Wolkberg Group, 168
Wood, Charles A., 7

xenoliths, 115, 120

Yellowstone National Park, 51, 160, 162

zeolites, 166
Ziegler, Alfred M., 86
zircon, 29–30